Document Delivery Services:
Issues and Answers
Issues and Answers

Document Delivery Services:
Issues and Answers

By
Eleanor Mitchell
Arizona State University West, Phoenix

and
Sheila A. Walters
Arizona State University, Tempe

Lt

Learned Information, Inc.
Medford, NJ
1995

TABLE OF CONTENTS

PREFACE

O ur original intent in writing this book was to produce a guide for practitioners working in the rapidly changing document delivery environment. We wanted to meet the needs of staff members in libraries where document delivery was evolving from traditional interactions between libraries to a service that also incorporates commercial document suppliers, electronic document retrieval and transmission technology, and end-user activities. When we began our research (late 1992), published literature related to document delivery was limited primarily to journal articles and conference proceedings. We saw a need to collect the scattered information into a single document—anticipating, ingenuously, that we could complete the project within 12 to 18 months.

But the past two years have brought a virtual explosion of information related to document delivery—most of which is reported in professional journals. At the same time, numerous conferences and workshops thematically devoted in whole or part to document delivery are being offered on an almost regular basis (the authors have logged thousands of air miles covering conferences in the United States and Europe). While many readers have surely seen the same articles and attended at least one of the cited conferences, we still believe firmly that the need exists for a book that covers all facets of document delivery. We hope that our effort as presented herein fulfills that mission and finds a receptive audience.

We endeavored to provide a balanced representation of all issues related to document delivery, both practical and theoretical. Although we began with an admitted bias toward the librarians' perspective, our numerous contacts with publishers, subscription agents, commercial document providers, and professionals dealing with copyright issues has led us to a broader appreciation for other points of view. We hope that this work will, in some small way, promote interactions leading to workable solutions for all of those laboring in the information profession.

We would like to acknowledge the many people who have assisted us in compiling this work. The staff of the Interlibrary Loan & Document

Delivery Services at the University Libraries of Arizona State University Main and the Document Delivery staff at Fletcher Library at Arizona State University West proved just how effective and timely interlibrary loan and document delivery can be. We received tremendous support from colleagues in many other departments at the ASU libraries and from the library administration while we researched, wrote, and edited this book.

We are also indebted to many individuals here in the United States and in the United Kingdom, the Netherlands, Australia, and Canada, who provided interviews, tours of library and publishing facilities, or prompt assistance when queried in person, by telephone, mail, or electronically. The book sections related to emerging technologies, networking, copyright, and international trends in document delivery could not have appeared without the help of those whose knowledge and experience in such areas far exceeds our own.

The enthusiastic support of others was exemplified when the authors invited colleagues monitoring ILL-L and ARIE-L listservs to join us for an impromptu discussion of document delivery issues one morning during the American Library Annual Conference in New Orleans in June 1993. Twenty-five people braved the heat and humidity to join us at the crack of dawn before the day's regular conference activities began. We came away from that lively discussion with a strong confirmation that a need definitely exists for a handbook on document delivery. The only suggestion made that day on which we did not follow through was the idea of an appendix listing various commercial document suppliers. Not that the idea wasn't worthy, but we discovered that that would be a work in itself! We did, however, discover numerous guides to commercial information suppliers and document delivery products already existed, so we chose to reference those works rather than merge their contents. Other than that, we have tried to address all other issues that relate to document delivery.

Most importantly, we acknowledge the support of our families, who frequently were left to fend for themselves during this endeavor. Thus, we gratefully dedicate this book to Ben, Oliver, Howard, Scott, Wendy, and Blanche, who gave up a lot during the last two years—often without fully understanding why!

Chapter 1

Document Delivery: Issues and Trends

To be truly effective . . . remote access should have a response time so rapid that requestors are unaware of both the distance from which the materials travel and the vehicle that transports them (White 1992, 12).

Document delivery is coming of age. The convergence of economic pressures and technological capability encourages libraries to seek innovative ways to fulfill, with physical access, the promise of bibliographic access. At the same time, the new scholar, tantalized by the instant visibility of current journal contents and the ability to browse distant library stacks from library, home, and office, has high expectations for local collections. Today's researcher is not content with the lengthy overland route for receipt of remote materials. How to cope with this dilemma, how to bridge the gap between the vision of the "virtual library" and the patron's quite actual and very immediate need has become an interest bordering on obsession among library professionals.

CHANGING LIBRARY ENVIRONMENT: WHEN PUSH COMES TO SHOVE

Four elements are frequently identified in the professional literature as shaping the contemporary library environment: the increase in published information; the expansion of access to that information through technology; the growing demand for physical access that matches bibliographic access; and reduced buying power, forcing libraries to spend more for less.

In reaction to these pressures, a visible commitment is emerging to take an "access" route, whereas in the past, "acquisition" would have been the course. Librarians are strengthening and improving old channels, the traditional networks and reciprocal agreements. Options less used in the past, such as commercial vendors and alternative full-text sources, are being re-evaluated; both circumstances and capabilities have indeed improved suitability for library purposes. Sue Kennedy (1989, 69) wrote: "Demand, economics, and technology: together, these trends have contributed to the growth of networking among libraries and have put increasing pressure on resource sharing as a means of meeting the information needs of library users." Networking, resource sharing, and an increasing dependence upon nonlibrary providers are all hallmarks of the present age of access.

Interlibrary loan as a vital library function and requisite user service is receiving more attention within libraries and larger institutions than ever before. The interlibrary loan department's operations are under greater scrutiny as well: librarians are asked to perform time and cost studies, to study efficiency and acknowledge that of peer and reciprocal institutions, and to make decisions based on increasingly complex information and a growing array of alternatives. Even the location of the interlibrary loan function within the library's environment is changing, as well as its economy. These changes are aptly reflected in the latest draft of the National Interlibrary Loan Code, which includes some radical language. The statement that ILL is an "integral element of collection development for all libraries, not an ancillary operation" (ALA RASD ILL Committee 1994) is in startling contrast to the 1980 code's statement that ILL was "adjunct to, not a substitute for collection development." (Boucher 1984, 139)

NETWORKING AND RESOURCE SHARING

"Before World War 2 interlending was regarded as an optional extra, a grace and favour activity, to be indulged in sparingly; any research library considered it an admission of failure to have to obtain any item from elsewhere." (Line 1989b, 1) Without apologies, without excuses, interlibrary loan and resource sharing are now imperatives for the modern library. Intentional interdependence has become not merely a recourse of last resort, a safety net, but a commitment increasingly central to operations.

Now, interest in resource sharing activity and networking is on the rise; we see this in the statistics and surmise it from the contents of professional meeting agendas and journals. The 1990-91 ARL Statistics included a new graph in which Kendon Stubbs analyzed changes over the past five years in ARL library resources, resource sharing, and numbers of library users: "While the chief audience for ARL libraries, faculty and students, has grown by 10-16%, new information resources available on campus have declined, by 15% for monographs and 2% for serials. At the same time, however, there has been a sharp 47% rise in the number of interlibrary loan transactions." (Pritchard 1992, 3) Other statistics in a recent ARL/RLG Interlibrary Loan Cost Study indicated that, over the last decade, ARL interlibrary activity has increased dramatically: lending by 52 percent, borrowing by 108 percent. Their 1991-92 statistics for ARL libraries showed 4.1 million items loaned and 1.4 million items borrowed by 120 research libraries in North America. The OCLC ILL Subsystem statistics underscore this trend: "The OCLC ILL Subsystem was introduced in 1979, and twenty months later the one millionth loan was transacted. The elapsed time between each one million transactions has decreased steadily over the years. In late 1988, just eighty-nine days elapsed between the eighteen and nineteen-millionth requests." (Kennedy 1989, 68)

Until relatively recently, there was much in the literature to support a contrary view—a firmly entrenched commitment to ownership. Thomas H. Ballard (1991, 22) wrote: "Resource sharing is nothing more than a return to closed stacks with one important difference. Closed stack access was very, very fast." In their summary of six workshops with university directors and library administrators on the future of libraries, held in 1991, Richard M. Dougherty and Carol Hughes (1991, 6) noted that "collection self-sufficiency is still the dominant operational philosophy of most research universities, even in the face of certain and severe financial constraints. Even with the increasing power of technology to improve access to and delivery of information and documents, when push comes to shove, faculty want materials on campus. They don't want to be dependent on other distant libraries for needed materials." They continue: "Cooperation was rarely mentioned by workshop participants as a key strategy for achieving their preferred futures. . . . A general lack of interest in cooperation beyond the most peripheral programs was apparent." (ibid., 16)

David Tyckoson (1991, 39) cited statistics from a 1986 study that supported that perception: "Recent ARL statistics demonstrate the magnitude of the commitment to the paradigm of ownership. Out of an average acquisitions budget of $3,083,287 for ARL libraries in 1986, 97% was allocated for the direct purchase of materials for the library collection. Only 3% of the average budget was spent on other materials, including access to online vendors and bibliographic utilities. It is clear from these figures that ownership still dominates most academic collection development." That analysis, however, does not recognize the relative cheapness of access versus acquisition: the preponderance of budgeted funds for acquisition is much less significant in measuring actual *activity.*

More recent opinions reflect a different perspective in the professional literature. As Joan Blair (1992, 73) wrote, "For the library of the twenty-first century, the stand-alone collection is a starting point rather than a self-contained environment. A library's immediate collection will serve as a core resource, with circles of additional available information grouped around it." The ARL/RLG Interlibrary Loan Cost Study also reported a shift from "dependence on locally-owned collections to an ownership base complemented by resource sharing interdependence among research libraries and an emerging reliance in the 1990's upon fee-based or commercial suppliers." (Roche 1993, 1)

MORE INFORMATION

The "information explosion" is a phrase so shopworn that it has lost its edge; yet, that cliche indeed represents a root cause of our present professional discomfort. Research activity in the U.S. has doubled in the last fifteen years (Cameron 1993, 23); research literature in the field of physics alone has doubled in the past ten years (Leach and Tribble 1993, 360), with similar growth in other scientific and technical areas. Even well-funded libraries struggle to keep up with the increasing fragmentation of the disciplines, the splintering of publications into narrower, more specialized titles, and the sheer volume of scholarly production. There exists "enormous pressure on researchers to identify all applicable information and data. They can no longer limit themselves to staying abreast of the work being done at established, well-known research centers but instead must seek out those small pockets of activity where the application of a little-known method may change the entire focus of their research." (Ditzler, Lefebvre, and Thompson 1990, 377) Librarians have been less than convincing in portraying the "access" philosophy to patrons. "Efforts by librarians to de-emphasize ownership are interpreted as a failure to understand both the political environment and legitimate differences in research methodologies among disciplines." (Dougherty and Hughes 1991, 6)

GREATER ACCESS, INCREASED DEMAND

There is also greater and, perhaps misleadingly, direct electronic access to bibliographic information, which tends to conceal the distinction between what is owned and what is not. Searching tools are easier to use, and cast their nets ever more widely over the information universe; on computer screens, what is here, what is local, what is remote, appear as equivalents. Access to other libraries' catalogs, across town and across the globe, has altered the expectations of users. In a university library, it is not uncommon for a patron to request a title and to inform the staff of where it is available. "This nearly universal access to information, and the new concept of ownership to which it has given birth, has even blurred the distinction between ownership of the information and the very different concept of ownership traditionally applied to the package in which the information resides—the book, journal, videotape, or audio-cassette." (Atkinson 1984, 1103)

Technology has also demystified the interlibrary loan process. It is no longer magic through which libraries make unowned objects appear, but through channels increasingly visible to their users. This may make the established interlibrary loan processes, with reciprocal borrowing agreements in place between specific institutions, more difficult to justify to the user who cannot comprehend why we don't just "get it from Hawaii," or the local public library, or wherever else they have discovered an item. Like advertised merchandise that is "out of stock," desirable items that have been excluded—even intentionally!— from local shelves frustrates users.

DECREASING PURCHASING POWER

Both more information and fewer resources with which to permanently acquire information lead to the current dilemma. "During the 1980s, the inflation rate for library materials exceeded all other services and commodities used by higher education institutions, with the exception of employee fringe benefits." (Leach and Tribble 1993, 361) The "serials crisis" is another critical phrase that has lost its impact through overuse. Library costs have been on the rise since the seventies; indeed, since 1981, the average cost of a book has risen 49 percent, of a journal 105 percent (Baker and Jackson 1993, 3). Subscription prices of journals in particular are leading many libraries to hard choices. Cost-cutting measures, among them massive serial surgery, leave the library user feeling sorely denied: even what was formerly on hand is found no longer, yet the bibliographic possibilities represented on computer screens seem limitless.

INFORMATION IN NEW PACKAGES

One cannot ignore the changes in packaging that have altered the information environment. Both the structure of published information and the time component of information delivery are breaking free of tradition: unbundling, single article delivery, electronic journals and books, and tables of contents indexing with linked instantaneous document ordering exemplify new avenues for publishers. The promise is of immediacy, of personalizing, of collaboration that has the power to change the face of scholarly communication. Information can be purchased in smaller units— articles rather than issues or volumes—and without consideration of permanency in the collection. Different techniques can be called upon in order to identify, locate, and deliver what is requested. Libraries may never physically receive what is ordered and, indeed, as patrons gain experience in the use of delivery systems, may not even do the ordering. Without the role of mediator, librarians will have to identify new ways in which to apply professional expertise to the delivery of information to users: that of pathfinder, for example, or translator, or facilitator.

FOR FREE OR FEE

Another dilemma resulting from the growth in the amount of published information, the increasing complexity of information retrieval, and the financial stresses on our institutions, lies in the management of information delivery costs. Some costs are associated with services rather than directly related to individual documents. There has been a growth in fee-based services within libraries. Populations of users that either do not qualify for free service or whose demands on time or expertise are excessive may purchase services and access to information. Fees sometimes are associated with "value-added" services; other times they are for the same basic level that primary customers receive. Intensive reference and research assistance, online searching, and document delivery are frequently the services provided to "outsiders" for a fee.

Other fees are levied not just upon groups of users but upon types of library activities. This phenomenon may reflect some discomfort with the philosophy of access and the technologies necessary to provide it, causing them to be held at arm's length rather than integrated seamlessly into professional repertoires and institutional budgets. On the other hand, it may simply be due to the ease of measuring the cost of, for example, an online search or a document delivery transaction. Some costs are directly related to document delivery: "During the past fifteen years, ILL service in libraries of all sizes and types has changed from free to overpriced. . . . Now ILL librarians are 'shopping around' for the best prices for their patrons, who are finding it more and more difficult to afford the high fees being charged by some libraries." (Cline 1987, 85) Weighing patron needs against options and costs and making individual determinations that fit institutional parameters are increasingly part of the professional territory.

What was formerly in-house may now need to be obtained from another source; that source, be it a library covering its own increasing costs, or a commercial provider, is likely to charge for its services. Too, with greater bibliographic access, much of what was never intended to be acquired locally is temptingly, immediately visible on OPAC screens. The unreliability of charges associated with obtaining sometimes arcane items brings a whole new dimension to financial planning. While budgeting for access is becoming more primary in the acquisitions process, libraries have not yet comfortably factored in delivery: "There is an expanding trend, especially among academic research libraries, to include access to resources in the same equation as building collections, using acquisitions funds to purchase access to electronic resources and databases. However, a plan for document delivery of electronically identified material has been alarmingly absent. Although the telecommunication path for bibliographic access to the individual is being strengthened, there is little attention being given to the development of a corresponding document delivery path." (Wessling 1992, 2)

Libraries traditionally do not charge individual patrons directly for using what has been purchased for their shelves by assigning a fee for an item consulted or borrowed. Yet many libraries do charge-back costs associated with providing what they have chosen not to acquire. In some instances, an optional fee affects only the speed and method of delivery: a patron may choose to pay for faxing or rush delivery. In other cases, however, fees are associated with the mere provision of a document, serving to favor those who can afford to buy access over those who must make do with a library's shrinking inventory of acquired materials. If access is to be a positive and intentional library strategy, it is inconsistent to discriminate among users according to their ability to pay.

INTERLIBRARY LOAN IN TRANSITION

It's a sign of the times. Interlibrary loan departments around the country are changing the ways in which they work. Even their place within their institutions is changing and they are seeking to encapsulate and advertise these changes through new department names. In a discussion on the interlibrary loan electronic discussion group (ILL-L Listserv) in early 1993, librarians responded to queries about describing this function. One library was changing "Interlibrary Loan" to "Interlibrary Services," reflecting the broader, nonreturnable nature of most transactions; another was renaming the department "Information Delivery Services." Some proposed names were more fanciful: "Library Extended Access Program" (LEAP). Others were more functionally descriptive, such as "Document Access and Retrieval." Virginia Boucher joined the discussion with a practical suggestion: "Perhaps in the best of all possible worlds the new name would be Interlibrary Loan and Document Delivery Services." In a 1993 paper, Mary E. Jackson and Shirley K. Baker commented: "The term 'interlibrary loan' no longer reflects the activities being performed. 'Interlibrary' does not always apply: 'library to library' interactions exist along-side interactions with a variety of non-library suppliers of information and materials. 'Loan' is an inaccurate term for the supply of photocopies and for document delivery services where copies are sent out without expectation of return. Thus the services offered by 'interlibrary loan' departments need to be renamed." (Baker and Jackson 1993, 2)

In their 1983 study, Boss and McQueen identified document delivery as one component of interlibrary loan (Information Systems Consultants 1983,

1). In a posting on the ILL-Listserv in May of 1993, a decade later, W.C. Divens wrote, "Actually, what we have here is a shift of perspective. In the good ol' days, Document Delivery was a subdivision of interlibrary loan. Now I think it is fair to say that the reverse is true." Graham Cornish, too, has noted this change in approach. He writes, "The term 'interlending' is generally used as a shorthand for the supply of a document by one library to another library, although the form in which that supply takes place may be the lending of the original physical text or the provision of a surrogate copy in paper or microform. . . . I think it is time we moved from using 'interlending' as a portmanteau term and began talking about 'document supply' to cover both activities." (Cornish 1991a, 125)

Traditional interlibrary loan has survived, even thrived, on good will among librarians, and on the persistence and patience of patrons. Problems with the system have been cogently identified by Jackson and Baker (1992, 7): incomplete or unspecific holdings on national databases; complications of fees and invoices and the development of parallel mechanisms such as reciprocal agreements to avoid costs; inadequate staffing in size and training; policies that limit cooperation; no interfaces among messaging systems; a system that is library-to-library, rather than supplier-to-end user; a preponderance of reliance upon library suppliers; not enough support for optional lender selection; difficulty of data collection and statistics gathering; unreliability of turnaround. In addition, they cite the "lack of uniform access to and use of the technologies of messaging systems, fax, electronic document transmission systems, and interlibrary loan management software, (which) encourages libraries to choose lenders according to matching technology rather than other, perhaps more pertinent, criteria." (ibid. 8) With document delivery taking a more leading role within libraries today, these impediments must find resolution either through the improvement of the current system or the development of alternative ones.

BEYOND INTERLIBRARY LOAN

In 1983, Boss and McQueen's study defined document delivery as "the transfer of a document or a surrogate from a supplier, whether a library or a document service, to a requesting library." (Information Systems Consultants 1983, 1) A recent definition, from which operational parameters might be drawn, expands upon this: "The provision of documents, published or unpublished, in hard copy or microform, at an established cost upon request, not including the on-campus delivery of documents to patrons' offices." (Jackson and Baker 1993, 2) Document delivery services may, however, encompass more or less than this description allows. Gerry Smith's definition is broader: "A cluster of methods for ordering and delivering information, with a strong emphasis on new technology, flexibility of sourcing and speed of response." (Smith, Gerry n.d., n.p.) (Headland Press) Yet another definition, which places document delivery within a functional context, was posted by Wayne A. Pedersen on the ILL-Listserv in May, 1993: "Document delivery . . . is the transmission of a document by any means between a user and a supplier. All other categories of document delivery—commercial document delivery, fee-based document delivery, interlibrary loan document delivery, etc.—fall under this rubric."

Document delivery systems of today are beginning to address some of the concerns raised about interlibrary loan. First, an emphasis on speed. Librarians are able to identify, locate, order, and notify of availability more rapidly. Speed or cost may be the determining factor in selecting vendors: some on-line document delivery services provide a menu of vendors, with ascending prices based upon delivery time. While turnaround time studies still indicate this as an area needing improvement, overnight or even within-the-hour delivery is no longer uncommon.

Flexibility is another hallmark of document delivery systems. With an array of suppliers from which to select and a variety of technologies to employ every step of the way, users face far more choices and many more options. Alternative requesting and delivery mechanisms are just two visible signs of the effect that technology has had on systems. Ironically, this has led to greater complexity for staff involved with document delivery, even while the systems themselves promise users greater simplicity.

Disintermediation is a natural consequence of the development of quick, easy, and accessible tools and technologies. Increasingly, the user can request an item directly from the search screen, and can receive the article at the location of his choice. The labor, and the quality control, of document ordering are removed from the purview of those who may be footing the bills. The role of the librarian may become one of providing the user with good choices and helping to structure the process so that the user can weigh the benefits and costs of a variety of search and retrieve mechanisms.

Greater dependence upon technology for internal processes and end user access is yet another characteristic of the new document delivery mechanism, and one with far-reaching consequence. Karen Liston Newsome (1990, 636) wrote of the effects of technology on interlibrary loan: "Each new technological advance prompts changes in borrowing and lending patterns, protocols, performance standards, staff training needs, users' perceptions of interlibrary loan, and administrative relationships between (and sometimes within) institutions." Document requesting is better integrated into the research process. New and improved products with full-text and full-images, some mounted locally, others centrally, can make the delivery time short—even instantaneous. Ordering has been simplified by e-mail, uploading, and fax; delivery, by fax, full text online, and locally installed, free-standing workstations.

Even traditional interlibrary loan services are enhanced by the use of electronic means to communicate and deliver. "Technology is making geography and time non-issues for document delivery." (Finnigan 1992, 106) The omnipresence of fax machines, due to their improved speed and quality, contrasts sharply with what Boss and McQueen found in 1983: only 2 percent of libraries and 10 percent of document services had ever used telefacsimile for document delivery and only on an experimental basis (Information Systems Consultants 1983, 34). While improving delivery time and making life on some levels easier for interlibrary loan staff, problems in terms of incompatibility, costs, and staff and user training attend the introduction of new technologies into the process.

Increased networking, possibly with very different partners, is another element of document delivery. We are seeing new alliances, sometimes between strange bedfellows, as librarians from different types of institutions, and publishers, vendors, and brokers cooperate and collaborate to provide enhanced access and service. Within the library, a blurring of departmental and functional lines occurs where document delivery is concerned, with both public and technical services units, as well as information technology staffs, participating in the construction of new paths for these services. "In the past, the typical interlibrary loan department handled requests for borrowing or photocopying material from other institutions, and acquisitions and serials departments were responsible primarily for ordering, receiving, and processing library materials being added to the collection. In the revised paradigm, we may find librarians from a variety of backgrounds and with a variety of interests working together—researching and ordering document delivery services, placing subscriptions for CD-ROM products, managing deposit accounts for article delivery, tracking copyright compliance, and handling accounts receivable for material supplied to external sites. No longer will this task be the sole responsibility of one unit." (Bluh 1993, 51)

Joan Blair (1992, 71) pointed out that the managers of today's libraries are increasingly called upon to consider forging alliances that might have been impossible just a few years ago. Perhaps in response to these relationships and to the growing complexity of the processes, library professionals involved in document delivery are fine-tuning their own approaches: some authors have predicted that "entrepreneurial attitudes, values and production practices will be adopted" increasingly by librarians involved in the document delivery process (Ditzler, Lefebvre, and Thompson 1990, 384).

An increasing reliance upon nonlibrary and commercial providers marks today's document delivery service. "Rapid technological advances are changing the methods used for document delivery services, and, although once almost exclusively the domain of libraries, document delivery is also becoming 'big business' in the private sector." (Love 1990, 5) Vendors are linking delivery mechanisms to indexing and abstracting services, making ordering a few keystrokes (and a deposit account or credit card) away for librarians or end users. They are introducing tailored products, such as SDI and current awareness services, and are adopting the technologies and conduits most suited to libraries. Vendors also provide copyright clearance, eliminating related delay and uncertainty.

As reliance upon outside sources becomes commonplace, document delivery, as Jackson and Baker describe it, becomes less a nicety and more a necessity. The commercial services represent a movement away from the imposition of lending and borrowing, and a change from requesting to ordering, with all that that implies about the primacy of your need over all other activities of the provider: "The relationship between a commercial vendor and a customer is itself an advantage. It is a business relationship. The transaction is an 'order' rather than a 'request.' When we request material from other libraries, we are dependent on cooperation, good will, and the restrictions and pressures under which that library is operating at that moment. Of course, our need is secondary to that library's mission to support its own institution. In contrast, commercial vendors are dependent on their customers. When we place an order with a vendor, successfully filling that order is very important to them." (Williams 1992, 221)

The increasing dependence upon nonlibrary suppliers for information and documents is not without risk: "Be they in books, films, electronic impulses, or other types of media, these symbolic representations of information cannot be supplied unless they are preserved. For a library to rely on private business interests to preserve humanities legacies is to ignore the realities of economic forces and the profit motive. . . . A library with an established need in a given area of research can never assume that 'the other guy,' particularly a for-profit vendor, will continue to fill those patron needs in perpetuity." (Hacken 1988, 489) The scenario envisioned by Sheila Intner (1991, 109) that "failing to acquire something through purchase does not mean being unable to acquire it rapidly and easily" may be unworkable unless libraries recognize a continuing role for interlibrary supply, for coordinated collection development, and for the planned resource sharing that will ensure this future.

REDEFINING THE LIBRARY

The library's role will be redefined by the access philosophy. Dougherty and Hughes (1991, 6) state that "the period of forthcoming change could be characterized as the transition from the physical library to the logical library." What makes an "excellent" library, and excellent service, will also be reevaluated: "Although access and ownership are not mutually incompatible, they do produce different sets of criteria for measuring the success and failure of the library." (Tyckoson 1991, 37) Traditional measures of accomplishments, the fodder for annual reports and fiscal plans, no longer tell the complete story. "The library's quality of service will be measured by entirely new types of service units, while circulation and library holdings statistics will play a less important role." (Epstein 1989, 65) Even library professionals not immediately and directly responsible for document delivery will feel the effects of the change that rock the foundations of librarian training, values, and daily routines. "With a shift to the paradigm of access, collection development librarians will need to allocate significantly more resources for materials that may not become part of the library collection . . . [and] will need to select among several different levels of access to information. . . . Ownership thus becomes a subset of access." (Tyckoson 1991, 41)

Others foresee a shift in focus from maintenance of large collections to obtaining specific information as needed. "Libraries will move from a function of collecting materials in anticipation of user needs to one of acquisition upon presentation of need. . . . a major preoccupation of librarians has been with ways to anticipate what these needs might be and with processes of evaluation and selection of these sources. . . . Libraries will shift from a process of identification of information sources to the delivery of information." (Summers 1989, 27-28)

Thomas W. Shaughnessy (1991, 4) makes a distinction between the supply-oriented and the demand-driven library:

> The supply-oriented library operates on the philosophy that the user is best served by assembling large collections of materials across a broad range of disciplines. . . . According to this calculus, potential use is just as important as actual use. . . . The demand driven library recognizes that either it can no longer afford to meet the standards of the previous model, and/or decides to adopt a new approach to meeting the information needs of its constituents. . . . The library tries to acquire materials which it has reason to believe are needed fairly immediately by specific clientele groups. Even ten years ago, Jeffrey Saldinger (1984, 640) wrote that "information technology in general and bibliographic databases in particular have made reactive access to information more effective than proactive ownership of information."

Despite the conclusions drawn by Dougherty and Hughes about the deeply rooted commitment to traditional notions of library self-sufficiency on the part of the academic powers-that-be, there are certainly indications to the contrary. In a 1992 survey of ARL library directors, 85 percent of those responding reported that they were using or developing electronic document delivery services; 49 percent are subscribing or planning to subscribe to electronic journals, with 60 percent providing access to electronic full texts (Schiller 1992, 3-4). While not yet the fully realized "virtual library" as envisioned by the most optimistic observers, at the very least those results indicate recognition that not all resources must be in-house, in print-and-paper format.

In their 1983 study for the Council on Library Resources, Boss and McQueen (Information Systems Consultants 1983, 60-63) drew some pessimistic conclusions about the state of document delivery:

> While there are not standards for what constitutes adequate document delivery service, the consultants have formed the opinion that document delivery is presently weak. It should not require an average of 10 to 16 days for a patron to get an item held by another library, nor for the transit time to be an average of 6 days or more. Nevertheless, document delivery does not appear to be a high priority concern of the library profession as a whole, but rather of a limited number of concerned librarians who are actively assessing the future of libraries. . . . The actual performance of the system has done an excellent job of convincing users to wait. . . . What has generally been characterized as satisfaction is actually acceptance of a pattern which has existed for decades.

Almost a decade later, in their study of document delivery and interlibrary loan for the ARL Committee on Access to Information Resources, Shirley K. Baker and Mary E. Jackson (1993, 4) identified a number of problems with interlibrary loan services. They found that limitations are unsatisfactory to patrons; access is not assured; timeliness isn't adequate; there is unpredictability; and costs are arbitrary. They concluded with a cautionary note:

> It is our opinion that the current system is reaching the breaking point and needs re-examination and redesign. It is also our opinion that an updated system could carry more traffic with existing staffing. It can do this with a rethinking of local processes, substitution of user-initiated activity for some current staff- and paper-intensive functions, integration of alternative suppliers into the interlibrary loan stream, and innovative uses of technology.

Richard M. Dougherty (1991, 59) sees a major transformation in libraries of the future. "At the heart of this transformation will be a fundamental change in what is expected of libraries. Researchers will attach more importance to locating and obtaining information and less importance to where the information was obtained. . . . Library users will be less concerned about the size of the local library collection and more concerned about the timeliness of document delivery." The realization of this vision depends upon the development of systems to make both the distance traveled by a document and the vehicle of transport invisible to the user.

Chapter 2

Networks for Document Delivery

The largest libraries, both academic and public, were afraid that the great unwashed (the small libraries) would decimate their collections and deprive their patrons with frivolous requests (a small library is any library smaller than the one in which you are employed, while a peer library is one you wish would accept your library as an equal). (Morris 1991, 83)

"Network" has acquired several meanings in a library context: it has been applied to agencies, to projects, to loose affiliations and formal arrangements, to hardware and software, and, even, to good intentions. In a discussion of document delivery, "network" often refers to the agreements, structures, and systems that support and make possible library resource sharing or other cooperative ventures. Increasingly, the term has come to encompass more of the gloss of the computing world as well, referring to the "connectivity" itself.

"The word network in the library field carries the connotation of cooperating membership, resource sharing, linkage, and use of new technologies." (DeJohn 1989, 81) The National Commission on Libraries and Information Science has defined networks even more functionally (Rouse and Rouse 1980, 4):

> Two or more libraries and/or other organizations engaged in a common pattern of information exchange, through communications, for some functional purpose. A network usually consists of a formal arrangement whereby materi-

15

als, information and services provided by a variety of
libraries and/or other organizations are made available to
all potential users.

And, what is not a network? Graham Cornish (1991b, 273) wrote:
"Networking . . . is a relationship between more than two organisations by
which they provide some mutual benefit. A purely centralised model
where the centre provides everything and the other libraries take from it is
not a network." (Cornish 1991b, 273)

The formal library networks, whose number in the United States alone
is conservatively represented by more than 500 listed in the 1993 *Bowker
Annual* (Barr 1993, 577-606), crystallize a longtime tradition of cooperation
in the library profession. In the 1960s, the term "networking" became popular in the professional literature, coming to signify more structured
arrangements than those formerly represented as "sharing" and "cooperation" (Angle 1982, 235). Networking activities were supported by increasingly sophisticated bibliographic retrieval, cataloging, and locational tools,
and by rapidly evolving communications systems. Other factors also were
key in the development of the library networking movement: in the 1960's,
Title III LSCA funding to promote multi-type library cooperation; in the
1970's, the international machine-readable cataloging standard MARC format; the creation of the National Advisory Commission on Libraries and
Information Science; and the increase in quantity and cost of information,
and its processing, control, and access. The capabilities of computerized
environments spurred the development of networks, promising enhanced
internal processes and reduced costs, eliminating the duplication of staff,
effort, and materials, and improving access to automation.

TYPES OF LIBRARY NETWORKS

Library networks have been broadly categorized by purpose, location,
and membership. Networks originating around a single purpose, for example OCLC at its inception, have grown to have broader missions and functions made possible by the fulfillment of that original purpose. Regional,
single-type networks, such as the public library systems in the post-war
years, broadened to include more varied partners as LSCA support and
recognition of mutual benefits promoted the acceptance of special and
other libraries into the networks. Statewide networks are flourishing, with
more than half of the fifty states reporting library networks in place or in
progress (Rogers 1993, 46).

In 1985, Carolyn Bucknall (1985, 68) wrote:

Issues of access, particularly to the traditional formats,
books and journals, have been enormously conditioned

by resource sharing, especially as enhanced by networking. Despite the wonders of the electronic world, resource
sharing still ideally takes place among institutions in geographical proximity, so that document delivery time or
user travel time is minimized.

Regional and local networks relying upon a backbone of surface delivery still thrive, even in the face of the immediacy of delivery offered by the
expansion of fax and electronics. Resource sharing networks have histori-
cally been based upon proximity. Following a hierarchical ILL model,
agreements establishing provisions for document delivery and the mechanisms to support these systems have tended toward the local or regional.
System financial arrangements (sometimes municipal or state-based),
administrative requirements, and the logistics of transportation supported
this pattern. Considerations that affect such systems consist of the following: size of geographic area; roads and terrain; weather; population in
delivery area and traffic; type of library and hours of operation; type,
shape, and weight of materials and their packaging; and patterns of lending. (Smith, C. 1991) These factors are particularly important in developing
nations where the dependence upon "technology" for document delivery
may involve little more than a bicycle and, possibly, a telephone. These
factors may seem mundane in contrast to the more "high-tech" factors
involved in electronic delivery, yet they are essential to reliable, efficient
delivery, particularly of book and nonprint format materials

Increasingly, networks evolve around method of delivery and compatible time-related priorities. We are seeing fax networks, such as
Pennsylvania State University's FAXNET of twenty-two sites, and Ariel networks, linking libraries committed to use of a given technology for delivery of information. A network committed to rapid document delivery might
be formed around this methodology. In some instances, the use of particular software or other technology is a prerequisite for network participation. Adherence to agreed-upon procedures, priorities, turnaround time,
etc., may be components as well.

Edward M. Walters (1987, 21) developed a taxonomy of networking
activity, through which he identified distinct characteristics of five types of
networks. These categories, helpful in identifying the structure and priorities of a network, consist of: (1) unit cost networks—the national networks
and utilities, such as OCLC, whose existence reduces the costs to individual libraries of certain processing functions; (2) multi-state regional auxil-
iary enterprise networks—regional brokers, the middle men, for the
national utilities; (3) authority-sanctioned networks, usually created and
supported by the state; (4) discipline and type-of-library networks; and (5)
local consortia or proximity networks, sustained by "favorable distance
relationships among the members." Document delivery, even when not the

driving reason behind the formation of a network, is an important function in many networks of all types.

Growth and Change in a National Network

OCLC, perhaps the best known of the library networks, links over 15,000 libraries in 47 countries and territories. It was established in 1967 as the 48-member Ohio College Library Center, with the goal of providing a shared cataloging program having a central "bank" of standardized biblio-graphic records. The Interlibrary Loan Subsystem was introduced in 1979, and 4,643 libraries were using the system by 1991 to request items while 3,958 were supplying items (Nevins and Lang 1993, 38). Although most ILL activity through OCLC (not taking into account other local and regional interlibrary loan network activities) flows between libraries of the same type, substantial activity involves borrowing and lending between different types of libraries (ibid., 39).

A name change in 1977 to Online Computer Library Center (Repp 1990, 263) signaled a recognition of a broadening of organizational mission and perspective. Since the early seventies, regional networks had contracted with OCLC for cataloging access, some abandoning automation projects already in the works (ibid., 267). In 1977, membership and participation in governance was extended to these out-of-state networks. Through these regional affiliates, member libraries could now "pool their efforts and resources in setting policy, solving problems and dealing with the OCLC system. . . . These regional networks provide local training, implementa-tion, documentation and liaison support between member libraries and major utilities." (Ferrall n.d., 4)

Perhaps due to the product-driven orientation of OCLC in the 1980s, relationships with the regional systems have at times been frustrating: "The interests of OCLC appeared to be focused on creating a stream of products and product extensions that would ensure the growth of the organization. Regionals . . . believed that much of OCLC's product development was wrong or irrelevant to the trends they were observing." (Keys 1992, 13) And, as Keys points out, alternatives had emerged by the late 1980s to the formerly exclusive services that OCLC offered to member libraries, such as cataloging and ILL capabilities.

During the past two decades, OCLC has changed from a cataloging con-sortium to an information conglomerate. It has become an "information research center; an innovative marketing agency; a major information and database holder, supplier, distributor, and creator; a national network; and a huge membership organization." (Berry 1993, 28) It has introduced tai-lored products and services that enhance and improve library operations and innovations directed toward the end user: e-journals, article citation and delivery services, and access to commercial vendors. Perhaps mindful of past unsuccessful marketing strategies that had contributed to the shift of libraries away from OCLC and other utilities to more local systems—what Keys (1992, 14) calls "the new feudalism"—plans for future networking are more sensitive to the concerns of libraries: CEO K. Wayne Smith has said that OCLC will "keep libraries paramount as we approach Internet and NREN decisions. We are committed to broadening access to OCLC services while leveraging the investments libraries have made in terminals, local area networks, and local systems." (Rogers 1992a, 24)

Recent trends in library cooperation have revealed a movement to networking on quite another scale: "Today, the concept that 'OCLC is my consortium' no longer works for most libraries that are seeking closer, smaller, special arrangements. Libraries may well belong to very large systems like OCLC, RLG, and many state networks, but many are looking to meet a large part of their resource sharing needs as members of much smaller groups, groups that sometimes consist of only two or three other libraries." (Higginbotham and Bowdoin 1993, 49)

Networking Case Study: OhioLINK

In 1986, an Ohio Board of Regents Library Study Committee was charged with exploring space issues for the 13 "state-assisted" college and university libraries. The Committee suggested a cooperative solution, and one based upon technological innovation, with two recommendations: regional high-density storage sites and a shared online library catalog. These recommendations led to the development of OhioLINK, the Ohio Library and Information Network, which is becoming "a single paperless catalog and a physical and electronic delivery service." (Dunn 1993, 14)

In 1987, the libraries involved were at different stages in the automation process. "Each institution had moved independently in terms of timetables, systems, and functions. . . . By 1987, four of the thirteen were still without significant automation other than OCLC, two were in serious contract negotiations with commercial vendors, seven had fully automated circulation systems, acquisitions systems, and public access catalogs, or were well underway. . . . In all, seven different commercial vendors . . ." were represented. (Repp 1990, 268) In 1988, committees formed in response to the Board of Regents' charge examined the needs of the various library functions; task forces represented users', librarians' and systems managers' perspectives. After formally soliciting vendor information, a 3,000 specification RFP was released to fifty vendors in 1989. The major goals of the project emerged: creation of a shared electronic online catalog, with circulation and document delivery capabilities; provision of a gateway to informational databases; and development of a scholar's workstation to facilitate access (ibid., 269). Innovative Interfaces was selected from the eight responding vendors; the system implementation began in 1991 (Hawks 1992, 62).

The evolving system is described by Kohl (1993, 44) as "basically a hybrid system that includes individual campus systems linked to a common central facility." Each library has system software to support its online catalog, circulation, serials control, and acquisitions functions. The OhioLINK database provides access to these resources: a union catalog of participating libraries' holdings; commercial bibliographic databases; and a GOPHER gateway to the Internet. "Patrons using the local databases can move easily to the central resource to search for a book throughout the system and request its delivery, to investigate a commercial database, or to surf Internet." (ibid.) Patrons can search for items and determine availability from their home computers, can request materials and pick them up at any member library, or have them delivered to another member library, or can have articles sent by fax (Dunn 1993, 14).

The OhioLINK network now consists of eighteen two- and four-year colleges, with others planning to join in 1995. It provides 24-hour-a-day access to more than thirteen million volumes as well as other resources. David F. Kohl (1993, 44), Dean and University librarian at the University of Cincinnati, concluded from the OhioLINK experience that "cooperation is no longer a marginal nicety but a central necessity. . . . we are not so much cooperating libraries in the traditional sense as something much more extreme: mutual branch libraries. Creating a statewide virtual collection for any patron to access replaces the concept of local ownership with something closer to a sense of stewardship for the larger library community."

Networking Case Study: Development of a Statewide Multitype Network—Pennsylvania's IDS Model

Pennsylvania's document delivery network is a good example of one that, beginning regionally and dependent upon courier delivery, demonstrates flexibility and responsiveness to changing needs and capabilities. IDS developed from a limited, regional delivery service among 14 primarily academic member libraries and the Lancaster County Library in 1969, to a flexible, statewide network of over 190 "full" member libraries of various types in 1993 (with an additional 601 affiliate, "receive-only" sites). Intended originally as an alternative to the U.S. mail system for interlibrary loan, IDS now includes an extensive fax network as well as wide-ranging courier services. From an average of 34,800 items annually in its first two years of service, the volume of materials provided grew to 77,600 in 1972, and to 317,689 pieces in 1992 (Phillips 1993b).

From its beginnings with one, then two, library-operated vehicles making three round trips weekly among the thirty-five member libraries in south and central Pennsylvania, the delivery service expanded in 1976 with statewide membership. In each of three service regions, two routes operated, with thrice-weekly service. In 1979, financial and operational concerns led to the abandonment of library-operated courier vehicles in favor of contracted courier service with Purolator Courier; rate increases led to a change again in 1986, to UPS, and an increase to five-day-a-week delivery.

A variety of bibliographic and union tools have supported the development of the IDS in Pennsylvania. A Union Library Catalogue begun in the 1930's (and now superseded by PALINET's interlibrary loan component), a Pennsylvania Union List of Serials, and film and audio-visual holdings catalogs have all, by promoting bibliographic access, reenforced the demand for IDS services. Access Pennsylvania, the state's CD-ROM union catalog of the records of over 750 public, school, academic and special libraries, has served to promote and support the automation of Pennsylvania's libraries, to define their resource sharing commitment, and to structure its implementation.

LSCA funding provided initial support for the service and, with state funding, has provided ongoing operating funds. The 1993 budget for IDS was $861,000, with $500,000 subsidized by the State library. Member libraries' annual fees, initially identical, now are based upon the level of usage. The base member fee is $935, with the highest-volume borrowing library paying $5,610. The average cost per shipment in the first quarter of 1993 was $2.51, an 11.5 percent increase over the 1992 cost (Phillips 1993a).

"Throughout its twenty years of existence, IDS has maintained its viability amid changing telecommunications and resource delivery technologies." (Deekle 1990, 33) With this in mind, a recent survey of IDS full members was undertaken, the intent of which was "to formulate a strategy and guidelines for the fuller use of telefacsimile or other electronic delivery means." (Deekle 1993) A snapshot of borrowing activity for the week of May 10-14, 1993 revealed that member libraries' first choice for photocopy delivery was the U.S. mail; only 7 percent of photocopies are sent through IDS (and these usually accompany books being delivered). Of "other" delivery methods noted for photocopies—fax, Ariel, air mail, federal express, and "pick-up"— fax was used in 34 percent of the transactions, and Ariel in 45 percent. With an IDS goal of complete fax availability by 1995, 600 facsimile machines purchased with state funds are already in place in Pennsylvania libraries. What remains to develop is a commitment that fax will be the primary means of article delivery (Jackson 1991, 14).

The ACLCP (Associated College Libraries of Central Pennsylvania) consortium, now 15 academic library members, has continued the resource sharing efforts that provided the impetus for the IDS. A Title III grant in 1987 provided funding for telefacsimile machines for each member in support of rapid document delivery and reference services among member libraries and other Commonwealth and regional library telefacsimile services. In addition, the state provided for fax equipment at the four major research libraries and took upon itself the commitment, included in the

ACLCP proposal, of producing a statewide fax directory. ACLCP also proposed the development of telefacsimile protocols, which were the basis of the fax component of the Pennsylvania Interlibrary Loan Code (Wilson 1988, 16).

Fax traffic among ACLCP libraries and reciprocal-arrangement consortia is not presently monitored, but is felt to be limited to filling rush requests at this time (Soo Lee, 1993). While there is interest in liberalizing fax use, a draft of recommended guidelines specify use of fax for communications functions, such as submitting requests, responding to rush requests, advising of unfillable requests, and providing status on ILL requests. According to draft ACLCP guidelines, "Only articles needed on a 'RUSH' basis will be transmitted by telefacsimile. All other materials will be sent via IDS or regular mail."

Case Study: Document Delivery in Illinois—from LCS to ILLINET Online and Dial-Access Circulation

> The extension of interlibrary loan services across library-type categories, as well as beyond regional boundaries, has been the basis on which access to the comprehensive collection of Illinois' libraries by every citizen in the State has been built so successfully (Newsome 1990, 636).

ILLINET, the Illinois Library and Information Network, is a "true multi-layered, multipurpose, multitype library network" (LaCroix 1987, 20) of school, public, special, and academic libraries. Eighteen regional systems provide coordinated services and programs for the member libraries. Research and Reference Centers, including the University of Illinois at Urbana-Champaign, Southern Illinois University at Carbondale, Chicago Public Library, and the Illinois State Library, provide the systems another tier of interlibrary loan and reference services. Yet another level of resources, those of private or special collections within selected libraries, can be tapped through these Research and Reference Centers.

The Illinois automated library union catalog, ILLINET Online (IO), now serving as an interlibrary loan backbone for over 2,600 Illinois libraries, grew from an automated circulation system, called LCS, put into place at the university library at Urbana-Champaign in 1978. Fourteen additional academic libraries joined the University of Illinois libraries to form a network in 1979-80. The LCS provided these libraries with both local and interlibrary circulation capabilities, permitting brief citation searching and access to library locations, holdings and circulation status information from the records of all member libraries. A second component, the online catalog at the University of Illinois, is now a shared, full MARC record union catalog for the OCLC participating libraries in the state. The ILLINET Online system can be searched globally, regionally, or locally for libraries' holdings; the identified materials can be requested directly, and are delivered through the statewide delivery system (Starratt, Varner and Cline 1992, 229).

Three categories of network participation exist: direct, for the libraries that maintain a current, ILLINET Online circulation database and whose patrons can borrow directly; reciprocal, for libraries with different automated circulation systems that can link to IO; and indirect, for all other ILLINET libraries which can dial in. Direct membership in LCS (now Illinois Library Computer Systems Organization, or ILCSO) has now grown to forty, including public and private college and university libraries, community college libraries, a special science and mathematics high school library, and the Illinois State Library (Sloan 1992, 26). In addition to these libraries, which use the system in support of local operations as OPACs and for circulation, 800 libraries participate by their contribution of catalog records to OCLC, which are loaded weekly into the ILLINET online database.

About 2,600 Illinois school, public, special, and academic libraries (members of the Illinois Library and Information Network) have dial-in access and can initiate document delivery requests from the ILCSO member libraries' collections directly. Each of these network libraries has been identified with a patron record that enables it to initiate a charge transaction online. The charge transaction is printed in the holding library and the item is retrieved and delivered through the document delivery system. (Two interconnecting surface delivery systems serve Illinois libraries. One, dating back to the early days of the Illinois regional systems in the 1960s, consists of system-operated couriers that travel between libraries and headquarters in each region. The second, the Intersystems Library Delivery Services connects the system's headquarters and high volume libraries through its daily routes.) From 104 such transactions in March 1990, the first month of this service, usage grew to 3,227 transactions in the ninth month (Sloan 1991b, 101). In the first full year of expanded dial access capabilities, 400 libraries initiated 43,192 such ILL transactions (Sloan 1992, 27).

ILLINET members use a variety of means—telephone, OCLC, LCS, and dial access—to transmit requests to other libraries, the regional headquarters, or the Research and Reference Centers. A hierarchical approach to interlibrary loan is giving way to more direct transactions (LaCroix 1987, 21). In addition, ILLINET libraries are increasing their use of facsimile for the document delivery process. By mid-1990, over 1,000 fax machines, funded through LSCA matching grants, were in operation in Illinois libraries. Fax and dial access have reduced the dependence of local libraries upon the regional systems as intermediaries. "With this increased capability for rapid information transmission between libraries, many are choosing to go directly to a lending library, thus flattening out the borrowing patterns in the state." (Smith , C. 1991, 24)

Since 1979, 1.1 million transactions have been initiated through ILLINET Online; the ILCSO libraries have initiated 4.1 million direct interlibrary circulation requests against each others' holdings (Sloan 1993a). The University of Illinois, a net lender in 1978-9, had become a net borrower by 1988-9 (Newsome 1990, 638). As Bernard Sloan (1991, 107) points out, "These figures serve to further underscore the fact that resource sharing activity does not necessarily involve wholesale raiding of the collections of larger libraries."

The 1992 fiscal year budget for Illinois imposed severe cuts in the operations of the network, but the ILDS and ILLINET Online were preserved (Sloan 1992, 28). Decreases in regional library system funding are expected to promote direct borrowing by member libraries, and may involve changes in the traditionally free-of-charge interlibrary loan and faxing services for state residents.

The implementation of the Linked Systems for Resource Sharing project is the next step in networking in Illinois. Underway since 1990, this project is intended to interconnect ILLINET Online with other automated library systems in the state (Sloan 1993b, 62). Connectivity will be through each system's linkage to the Internet; after mutual searching capability is established in this way, a messaging system for interlibrary loan will be incorporated into ILLINET Online. The second phase of the project would allow users to search other systems using their own "local" language and strategies, and to initiate ILL requests in the local, familiar format.

Case Study: From Regional to International Networking—LASER

In the United Kingdom, most interlibrary loan requests are filled by the BLDSC, with regional library systems filling less than 20 percent of the requests that originate within their regions (Bonk 1990, 231). As Sharon Bonk (ibid.) has written, "Interlending in the United Kingdom, on the whole, is a one-way system. . . . Until recently, there was little point-to-point borrowing." This is a largely centralized model that does not truly fit a definition of networking as being "mutual."

The Regional Library Systems in Great Britain were first established in the 1920s and 1930s to enable libraries within the same geographical area to borrow material—normally books—from each other. Regional headquarters maintained the union catalogs; requests were sent there, then forwarded to holding libraries. "In addition to providing the key to interlending by developing the union catalogue, many Regional Library Systems also become involved in the transportation of stock—true document delivery services!" (Edmonds 1991, 251)

LASER, the London and South Eastern Library Region, was formed in 1969 as an amalgamation of LUC (London Union Catalogue) and SERLS

(South Eastern Regional Library System). Although primarily composed of public libraries, LASER membership is open to nonpublic libraries as well. LASER's commitment to automation to serve interlibrary loan was evidenced in early projects. In conjunction with the British National Bibliography, LASER converted the BNB union catalogues into machine readable form. Its minicomputer system, developed in the mid 1970s, was one of the first online union catalogs in the world (LASER 1992). Three other library regions sought access to this system, joining LASER in the development of Viscount, a "national co-ordinated interlending and bibliographical network." (LASER n.d.)

Seven United Kingdom Library Regions and the BLDSC are now linked to the Viscount system, which provides online access for item location identification, ILL messaging within and between regions, and access to the BLDSC's ARTTel service (an interlibrary loan and document delivery messaging service). Materials requested, both books and photocopies, are delivered in one work day between libraries within the LASER area by seventeen vans; material comes by rail from BLDSC to London where the vans take over. Nonmember libraries can purchase vouchers for use in this LASER Transport Scheme.

As the UK participant in Project Ion (see below), LASER is again providing impetus for resource sharing, though on an international rather than regional level, and in a less homogeneous system. "Viscount reflects a commonality of system and function, whose origins lie in the proprietary computing and networking development of LASER. Ion reflects the need to network between different computer networks in order to provide the same climate for resource sharing which can be provided by a centralised database." (Smith 1993a, 94)

Case Study: International Networking through Project Ion

Project Ion is a pilot project begun in 1990 to link the interlibrary loan systems of the United Kingdom (LASER's Viscount and ARTTel), France (SDB/SUNIST's PEB service), and the Netherlands (PICA). Financed by the three participants and the Commission of the European Communities, its objectives are to 1) achieve interconnection between three computerized library networks to support and develop international interlending and messaging services; 2) improve the efficiency of international interlending services; and 3) demonstrate the capabilities of Open Systems Interconnection (OSI) communication protocols in a message-oriented environment for interlending services by the interconnection of computerized networks with different technical characteristics. OSI is a set of standards developed by the International Standards Organization (ISO).

The OSI networking model, adopted in the late 1970s (Arms 1989, 40), permits communication between computers of different types, from different manufacturers, based upon internationally accepted protocols. In Project Ion, an international interlending service, based on OSI library standards, was integrated with existing national services, resulting in a mixture of proprietary and open systems protocols. The three networks were linked and provide international loan messaging and identification services, which will permit the searching of computerized union catalogs with different search mechanisms (OSI 1990, 491). A gateway known as the National Focal Point permits the interconnections between the national systems (Plaister 1991b, 298) and routes the international traffic, but it is transparent to the user.

The third phase of Project Ion, the user evaluation stage, is expected to be completed in 1994. Project Ion will also survey other European library networks and organizations to determine their interest in membership. Other market and product developments will also be evaluated; for example, electronic document delivery is one area of expected future experimentation (Pastine 1992, 123).

NETWORKS BASED ON COMMON GOALS, COMMON INTERESTS

Networks where proximity is unimportant are increasingly prevalent: "Partnerships are now forming around common goals or common notions, rather than geographic locations." (Lenzini and Shaw 1991, 39) Libraries of similar type, or with common subject interests or other concerns, are using technology to overcome geographic boundaries to resource sharing. For example, university libraries, however remote from one another, may enter into reciprocal borrowing agreements based upon usage and collection compatibility; similarly, medical, agricultural or other specialized libraries also find that networking on a more global scale meets their subject interests and fits their mutual time frame more than a reliance upon collections and processes locally available. The increased bibliographic access that encourages this remote sharing is accompanied by greater dilemmas of delivery.

PICA and RAPDOC: A Dutch Network for Rapid Article Delivery

PICA was founded in 1969, with the aim of promoting library services and efficiency through automation and cooperation between the Royal Library of the Netherlands and several university libraries. Its components include an online shared catalog, access to various online databases, local library system support, and an interlibrary loan system, which serves as "a central facility to handle the interlibrary loan of books and periodicals between more than 300 connected organizations." (Costers 1992, 2)

The PICA RAPDOC project began in 1991 when the Ministry of Education and Science agreed to subsidize "Fast Delivery of Documents," a project with the following service goals: improving accessibility of journal articles; guaranteed delivery of 85 percent of journal photocopy requests within twenty-four hours; delivery to institutions and end-users in printed or electronic form; and possible direct submission of requests by end-users. Administrative goals also included minimizing procedures, maximizing applications of existing standards and solutions, and maximizing the possibility of international connectivity (van Marle 1993, 54-70).

An automated union catalog for periodicals, designed by PICA, has been in operation since 1984; the RAPDOC plan was based upon an analysis of interlibrary loan periodical requests received through this automated system. A core collection of 7,000 journal titles, from which 85 percent of the requests could be filled, was identified. This collection is dispersed among the 19 participating libraries, which have agreed to fill a quota of requests—rapidly—from identified titles held. PICA software directs the incoming requests to holding libraries based upon the libraries' previous level of participation in fulfilling article requests. PICA has purchased commercially and made centrally available table of contents data. The data will be linked to local holdings information and can be loaded and searched on members' OPACs. Requests for titles not locally held are handled through the national automated interlibrary loan network operated by PICA.

Initially, remote requests will be delivered through the postal service. A technical infrastructure for electronic document delivery, both between participating libraries and to end-users, is in the works. For delivery of items held locally and requested through the local OPAC, this involves scanning of documents, storage in a "mailbox" on the document server, notification of patron, and removal of document from the server upon the completion of the transaction. For items not locally held, the requests will also be generated through the OPACs which are linked nationally, and the scanned documents transferred via the Dutch national electronic research network. The realization of this phase of the RAPDOC project depends upon the development of local capabilities. The success of library networking endeavors will hinge upon the circumstances of each of the participants: "There is no general solution for this aspect, but it is library and university dependent." (ibid., 67)

Case Study: Subject-type Library Network—The National Network of Libraries of Medicine

The information climate surrounding the biomedical community in the 1960's is now mirrored in the global information climate. There is a proliferation of published information coupled both with great urgency to get access to that information and varying resources on the part of information seekers.

The Medical Library Assistance Act of 1965 enabled the National Library of Medicine to respond to this climate with the development of a Regional

Medical Library program, a hierarchical network of medical libraries supported by a variety of highly computerized reference, referral, bibliographic access, and document delivery services. The recently renamed National Network of Libraries of Medicine now includes over 3,600 hospital and other health science libraries, the "basic units"; 130 major medical "resource" libraries, most often in academic institutions; eight Regional Medical Libraries, institutions contracted with NNLM to provide services to the libraries in their regions; and the National Library of Medicine. Beyond institutional resource sharing, the current emphasis is on outreach to medical community members who, due to location, lack of information, or lack of institutional affiliation, do not have access to the network's services (Mehnert 1992, 152-155).

The network's document delivery component rests upon an established standard and protocol in which requests are systematically routed within the local area, then to the local resource libraries within the region, to the Regional Medical Library, and only then to the NLM. Requests not fillable at the NLM level are then routed to other regional libraries or beyond. When direct funding for the document delivery system (which reimbursed only appropriately-referred requests) was eliminated in 1981, all member libraries could take advantage of direct access to all libraries within their regions, and to NLM directly when their regional resources failed to fill their requests. Despite this option for somewhat decentralized access, allegiance to the hierarchical standards for referral prevails (Hill 1989, 51).

When the OCLC ILL Subsystem became available in 1979, many biomedical libraries that could afford to use it, did; one advantage that it offered over the Regional Medical Library Network was access to the related but nonmedical literature excluded from many of the medical collections. Due to the costs of OCLC participation, the incompatibility with their established technology, their volume, and the nature of the health information work flow, other libraries found an ILL alternative in the NLM's introduction of DOCLINE.

DOCLINE is an automated interlibrary loan system that routes journal article requests automatically to holding libraries. Developed over nearly a decade, DOCLINE was implemented in 1985, with unit libraries joining in 1986-7, and most biomedical libraries in the country able to participate by 1987. (ibid., 56) It links bibliographic information, from MEDLINE and HEALTH databases; serials holdings information from SERHOLD, the periodical holdings of RMLN libraries; and institutional information, including a routing list of libraries to which each institution wants its ILL requests sent. These linkages support the accurate and rapid flow of requests: for example, the requesting library keys in its identification number and a "unique identifier" number and the system responds with the complete information for the institution and for the citation. (Full citations may be entered for items not in MEDLINE or HEALTH.) Other information necessary to completing the transaction, such as patron's name, copyright information, etc., can be attached to the request or can be bypassed and a default accepted. Requests for monographs, serials not in SERHOLD, and audio-visual materials are accepted, although, without holdings information, they will either be routed to a named library or to NLM (Colaianni 1989, 35).

The introduction of user-friendly "Grateful Med" software encouraged end-user searching of National Library of Medicine databases; the development of "Loansome Doc" in 1991 gave individual users the ability to generate online orders for Medline articles directly. This is signaling a trend toward bypassing the library as intermediary in the document delivery process in favor of direct requester access.

NETWORKING ON THE NETWORKS: ELECTRONIC DOCUMENT DELIVERY

In his 1991 Research Networks and Libraries, Gary Cleveland (1991b, 1-2) wrote, "Research networks are distinct from what is commonly known as networks among librarians." Research networks, according to Cleveland, "provide the infrastructure for a variety of networking tasks . . .", while library networks "support the nuts and bolts operations of libraries."

Library networks rely increasingly upon research and computing networks for access, communication, and even delivery functions relating to resource sharing. The vehicle, and the service it transports are not always clearly differentiated. Caroline R. Arms (1990, 15) elaborated upon those blurred distinctions. Networks like OCLC, RLG, and library consortia "provide access to a shared resource . . . [and] link terminals to a host computer that provided a particular service. . . . To the librarian, the service and the network were almost indistinguishable." Arms contrasts this with computing networks, which she likens to highway systems carrying traffic, "independent of the services that can be offered by third parties because it is there."

Internet

The Internet is an international grid of computer networks. Almost a million host mainframe computers, in over fifty-two countries (Raish 1993, n.p.) provide access to five to ten million individual users (McClure, et al. 1993, 26) through their terminals or personal computers. Gateways connect the host computers to over 7,000 separately run networks located all over the world. The Internet provides "a common protocol suite for communications, physical gateways between networks, and common naming and addressing conventions." (Engle 1991, 7)

"Internet is an existing, operational network of networks. NREN is a program, a concept, and a vision of an interconnected future." (McClure, et al. 1993, 26) Although Vice-President Gore has referred to the National Research and Education Network (NREN) as a "superhighway," John Perry Barlow of the Electronic Frontier Foundation disagrees with this metaphor: "The superhighways already exist, because we already built the system; only we have forgotten to build any on-ramps, county roads, or two-lane highways." (NREN 1993, 52) The High-Performance Computing and Communications Act was signed into law in 1991 by President Bush, providing authorization to fund a high-speed computer network of computers, data and users, in order to promote communications among scholars and researchers; one component of this is the NREN. Through more recent legislation now in Congress, the National Information Infrastructure Act of 1993, some of the "on-ramps" can be created to extend access to this network to a broader audience. Over $800 million in federal funding has been authorized to develop networked applications for health care, education, and libraries; additional funds will support the local connections such as to libraries and schools (ibid., 51).

Over the past few years, many librarians, particularly in academic, government, and research settings, have become familiar with the Internet, and have begun to seek ways to integrate use of Internet functions into their procedures and routines. Internet has not made similar inroads into the public libraries. A survey of public librarians found many barriers to the development of national networking in public libraries: cited were "limited knowledge about the Internet, inadequate equipment, and limited staff knowledge in the use of computers and telecommunications; confusing and contradictory information about how to connect to the network; no 'systems' people to implement the network in their libraries; and no time to commit to such activities. In addition, some public librarians are unconvinced that there is (and will be) public library 'stuff' useful to them via the Internet." (McClure, Ryan, and Moen 1993, 23)

We are seeing, however, the growth of "Free-Nets," through which libraries may provide a variety of community and public service information to their publics. Some public libraries, such as the Seattle Public, are providing, through their OPACs, Internet connections to a variety of databases, from other libraries' catalogs to National Weather Service information to UnCover.

As academic libraries increase their use of the national networks for resource sharing and communication, there is the danger that the public libraries will find themselves out of the loop. As Marshall Keys (1992, 14) points out, the Internet, like cable television, has "created whole new classes of the disadvantaged, whose economic situation restricts their access."

Librarians use the Internet in many ways. The e-mail function enables users to communicate directly with other libraries, users, vendors, publishers, and others, quickly and at low cost, and to receive and share information through discussion groups and bulletin boards. Users may search other published and unpublished remote databases and commercial online bibliographic systems such as OCLC's EPIC, DIALOG and BRS, often at lower cost and faster (Tenopir 1992b, 103). Electronic journals and books and a host of other reference sources are also available on the Internet.

The availability of library catalogs from around the world on the Internet is tantalizing; a recent estimate found as many as 450 just within North America (Saunders 1993, 45). As with many electronic bibliographic tools, the illusion of "instant" access raises user expectations and increases demands for delivery. Lack of standardization among online catalogs, from command structure to methods of access to various databases, and inconsistent information about what is held by a library and whether holdings listed are complete, complicates the process of searching remotely. Cautions have been expressed about use of the OPACs on the Internet: "Internet allows one to look at more library catalogs than one could manage in any other way, but the sum of the catalogs available on Internet does not equal a national bibliography." (Keys 1992, 10)

Each of these capabilities may factor in the realm of document delivery. The e-mail (communications) function may be used to send or receive document requests to other libraries as well as to vendors, and to communicate information on the status of requests; some libraries have implemented electronic systems for receipt of patron requests as well. Libraries also use the e-mail to expedite requests to publishers for permission to photocopy. In addition, e-mail through the Internet provides access to the ILL-L listserv discussion group, an electronic forum for interlibrary loan librarians, where both theoretical musings and very practical information are shared. Similar discussion groups focus on concerns of fee-based services (FISC-L), Ariel users (ARIE-L), and those interested in copyright issues (CNI-COPYR).

Document delivery librarians need to be aware of the proliferation of electronic publications available on the Internet (in some instances, only in this format). While electronic book projects such as Project Gutenberg from the University of Illinois offer electronic versions of books available otherwise in paper, electronic journals may not necessarily have print counterparts. The e-journal *Postmodern Culture*, indexed in *MLA Bibliography* is available in electronic format only on the Internet: delivery of an article from this publication requires a departure from the "normal" ILL/document delivery processes. Questions of how to indicate "access" to specific electronic journal titles are now being raised among serials librarians: some libraries have begun to include the titles of e-journals to which they provide access in their serials lists.

Gary Cleveland (1991b, 51) wrote: "The quantum leap in bibliographic access to be provided by distributed information retrieval over research networks will make the disparity between information access and document supply even more severe. The balance can potentially be redressed by using research networks to send documents in electronic form directly to the user's workstation where it can then be read on screen or printed off-line." For interlibrary loan functions, the Internet has been more widely used for communication of requests than for the delivery of articles to date: "Some users share electronic documents, but much of the ILL or document delivery via Internet is for document requests only—the actual documents are then FAXed or mailed." (Tenopir 1992b, 102) There are, however, several document delivery projects in existence now that rely upon the Internet for transmission of text.

Research Libraries Group and Ariel

The Research Libraries Group began in 1975 as a regional consortium of three major academic libraries (Harvard, Columbia, and Yale), and the New York Public Library. It has grown into a network of over 120 libraries, with a changing focus from cooperative automation and technical processing to information provision. Resource sharing, with expedited delivery and priority service for members, has been a hallmark of RLG since its beginnings (Jackson 1993b, 2).

RLIN is the international computer network that serves as a technical infrastructure for the programs of RLG. Informational and bibliographic databases are available on RLIN, the largest being the union catalog of holdings of member and other contributing libraries, with citations for over twenty million titles. [David Richards' essay, "The Research Libraries Group," in *Campus Strategies for Libraries and Electronic Information*, provides an introduction to this network (Richards 1990, 57-75)]. RLIN databases have been available through Internet since 1989. Like OCLC, RLG has begun to offer citation and document delivery services, called CitaDel, with a new end-user component; the recently released patron-oriented search system, Eureka, permits a host of search approaches and functions: browsing, limiting, viewing location and call number information, searching for print and nonprint materials simultaneously, downloading in a choice of bibliographic and display formats, sending search results to an e-mail account, and ordering from CitaDel directly. Fixed annual fees as well as per-search fee options and blocks of searches are available.

Ariel is a software package, independent of the RLIN interlibrary loan system, that permits rapid transmission of documents and images over the Internet; the images are higher quality, with greater resolution than telefacsimile copies, and the delivery is less expensive than through telecommunications systems. Ariel was developed by the Research Libraries Group as a component of its "Shared Resources Program" (now called ShaRes). After a beta test in 1990-1991 at six sites, Ariel was released for sale, and is now being used worldwide for document delivery. As part of a "document transmission workstation," which includes an IBM-compatible microcomputer, scanner, and a laser printer, Ariel permits the user to scan, store, transmit, and print any material. With the Ariel software, an article in a paper journal may be scanned into a PC, and the image transmitted over the Internet to the requesting library's workstation. While the benefits of using Ariel (e.g., time and cost savings, high-quality copies, and flexibility of installation) encourage more libraries to install the system, there are some disadvantages as well. Ariel software must be in place at both the supplying and at the receiving site, which limits participation of libraries and reduces options for delivery to the end user; and end user delivery is primarily in print rather than electronic format. "Emphasis is on giving libraries a production delivery system for printed materials. . . . The emphasis is on printing: receiving and printing the document." (Ulmschneider 1992, 79)

Libraries need not be members of RLG to use Ariel; without formal membership, the "network" is fluid. An Ariel directory is available on the Internet through RLG, which lists registered libraries (those that have purchased the software). An electronic discussion group dedicated to Ariel (ARIE-L Listserv—electronic addresses are ARIEL@IDBSU or Ariel@idbsu.idbsu.edu) allows new users of Ariel generally to announce that they are up and running, and provides a forum where problems and solutions can be shared. New Ariel users have been known to post requests on the ILL-L Listserv, such as, "Hi, I'm looking for Michigan Ariel users." Interlibrary loan departments will note use of Ariel in the libraries on their routing lists; when requesting from these sites, a note can be added to the note field requesting Ariel delivery and providing the requester's Ariel IP address. One Ariel library has added a note to all photocopy requests to "Please fax or Ariel if cost does not exceed maximum cost listed."

Recent discussion on the ILL-L Listserv has taken issue with one library's announced surcharge for Ariel service. Questions of convenience and cost to the supplying library are being considered. A recent study of Ariel undertaken by Deakin University in Australia found that Ariel was the cheapest way to supply documents: "For transmission of documents by Ariel the only direct cost is the staff time involved. . . . Costs are incurred at the receiving workstation in terms of consumables. For net suppliers of documents there appears to be a significant cost benefit in supplying documents by Ariel." (Kosa and Tucker 1993, 2)

The NAL-NCSU Project on the Transmission of Digitized Text

In 1989, the National Agricultural Library and the Libraries and the Computing Center of North Carolina State University launched a demonstration project to investigate the "technical feasibility and the administrative structures involved in an NSFnet/Internet-based document delivery system for library materials." (Casorso 1991, 17) The pilot involved the transmission of requested documents over the Internet: the borrowing library transmitted a document request via OCLC to the National Agricultural Library, with a note requesting scanning and Internet delivery. NAL staff retrieved, scanned and electronically filed the document; the file was compressed and transmitted to a file server at the requesting library, where it was retrieved and printed, downloaded, or transmitted to the requester's own computer. The pilot study established that it was possible to "successfully capture, transmit, receive and output digitized image files via the Internet." (Casorso 1992) Significantly, the project also demonstrated that it was possible to hurdle one major obstacle to all automated document sharing; i.e., "overcome dissimilar computing environments." (Casorso 1991, 21)

The Digitized Document Transmission Project (DDTP), an expanded two-year project, began in 1990. Fifteen libraries from different regional networks installed the project platform in their interlibrary loan offices, including "off-the-shelf, graphics-capable, networked desktop computers, scanners, and laser printers," according to an NCSU Libraries information release, and began transmitting requests in July of 1991.

The second phase of the DDTP project, Electronic Document Delivery Service, the end-user delivery system, was recently tested at North Carolina State University. Users could submit their requests via campus e-mail; requests were filled by the DDTP sites which then transmitted the scanned or electronic documents over the Internet. The received file was automatically placed in the account of the individual user on the document server, and an e-mail notification sent, informing the patron that the document was in. The requester could then retrieve the article electronically on the campus network, and transfer the file to his or her own computer. "Through EDDS, the end-user is able to receive a machine-readable copy of the journal article, a feature not yet found on other electronic document transmission systems." (Jackson 1993a, 18)

One issue that arose from the NCSU project was that of speed. John Ulmschneider of the NCSU Libraries reported that an eight-page article took the sending library about twelve minutes to process, which would allow for perhaps four requests per hour; it took the receiving library nineteen minutes, however, to "retrieve a document, uncompress it, and print it, NOT including the time it takes the printer to actually produce a printed page." (Ulmschneider 1992, 82) This factor provided greater incentive to deliver documents electronically to the end-user's computer screen, avoiding the printing bottleneck. Tracy Casorso (1992) reported broader concerns: "Preliminary findings of the NCSU Libraries initiative indicate that the research library community is not as well prepared to investigate networking issues as one might expect and that Internet reliability is questionable. The Internet, as it is presently constructed, may not be suitable for production work."

With this demonstration project completed, the universities no longer use the EDDS software for interlibrary transmission between institutions; one library, NCSU's Natural Resources library, is still using it to send articles to the main library at NCSU, however. The software developed for the project is expected to be released for use by other libraries. NCSU is now developing another document delivery system that will work with its online catalog products, a "barrier-free" model automated document delivery system with which patrons will interact directly. The system, now in early planning stages, will accommodate check-out and delivery of books from any of the state universities' library catalogs, as well as journal articles from a variety of indexing and abstracting services accessible through the online catalog.

JANET: Electronic Networking in Great Britain

Library use of electronic networks is worldwide. JANET, the Joint Academic Network, like the Internet, "supports electronic mail, library OPACs, conferences, bulletin boards, and gateways to other networks and services." (Kesselman 1993, 134) JANET provides access to the European academic network of 18 different countries, to the British Library's BLAISE-Line (the online bibliographic database service), OCLC, and other sources; all major bibliographic utilities in Britain have linked their databases to JANET (Cleveland 1991b, 73).

Most interlibrary loan requests in Britain are handled by the British Library's Document Supply Centre; the online service for transmitting requests, ARTTel, has been linked to JANET since 1987 (ibid., 78). With the availability of library catalogs on JANET, and the increasing use of e-mail for interlibrary loan messaging, the pattern of centralized lending may shift. "In the United Kingdom, the JANET network will allow users to send full-text articles to any other JANET user. And with Super-Janet, which permits a much wider range of high-speed data transmission, one will be able to send an article with good quality graphic images not only to other JANET users, but also to the thousands of US users on the open network system INTERNET." (Barden 1992, 435)

The Future of Networks

As other electronic document transmission systems are developed, questions of communication between networks will arise; for example, Ariel users can only send to another Ariel workstation, thus forming a technology-defined, and limited, network for document delivery. Mary E. Jackson (1993a, 21) wrote: "Standards must be set if compatibility . . . [is] to be achieved. The introduction of these document delivery systems can be compared to the late 1970s when interlibrary loan messaging systems were introduced by OCLC, RLG, and WLN. Interlibrary loan librarians are still waiting for compatibility among these messaging systems, and are hoping for better compatibility among the new delivery systems."

CONCLUSION

In a *Wilson Library Bulletin* article, Harold Billings (1993, 34) describes the new alliances that the "renaissance librarian" will form. Urging a recognition of information as a commodity and a more businesslike approach to its management, Billings said that "the national and regional networks at their grass roots . . . continue to apply more of the same old solutions to problems that may have mutated beyond cure with the common elixirs." Alliances with other libraries, as well as with other information and computing professionals, with the world of scholarly communication, with publishers, and with commercial information providers are components of this new technological age. As Billings (ibid., 33) puts it: "Home alone is just not good for libraries."

Chapter 3

Integrating Document Delivery Systems

Timely document procurement and delivery should be viewed as a viable and better alternative to ownership of necessary but infrequently used information sources. . . . Information which is available more quickly is of greater value to the user (Allen and Alexander 1986, 6-8).

With these two tenets as inspiration, this chapter begins by examining the phases of the delivery process and determining ways to improve turnaround. The focus then shifts to integrating delivery services into library services overall.

Technology is making access to large bibliographic databases progressively easier. Even remote access is being simplified, making possible the proverbial library without walls. An increasing number of "do-it-yourself" document delivery opportunities exist. Because electronic document delivery complements bibliographic citation retrieval, many libraries now provide document delivery systems with end-user access. That is the fastest way to acquire items not in a library's collections, especially if the documents are delivered electronically directly to requesters.

Are such services popular with end-users? The UnCover Company's UnCover database available through the CARL Corporation Network and the Internet was one of the first integrated online document delivery services available. According to Martha Whittaker, more than half of the articles delivered through UnCover are charged to credit cards (James 1993, 95). Not every library user carries a credit card, however, nor does everyone have ready access to a fax machine. DIY activity, therefore, siphons off only a small percentage of requests, and document delivery is still very much the domain of the library, with ILL continuing to serve a large clientele.

Waldhart reported an annual 3.1 percent growth rate for photocopies "loaned" and an annual 1.5 percent increase in originals loaned among ARL libraries from 1974-75 through 1982-83 (Waldhart 1984, 206), a reflection of the early years of the "Information Age." A more recent review of ARL statistics covering 1985-86 through 1992-93 showed a 60 percent increase in borrowing by research libraries (Stubbs 1994, 6). In 1992-93, ARL libraries reported 1,531,149 borrowing transactions and 4,226,134 lending transactions (Daval and Brennan 1994, 39).

Nevins and Lang also studied interlending, this time among multitype libraries in the OCLC system. They found in the five-year period from 1985-86 to 1990-91 the number of borrowing requests *doubled* from 2.7 million to 5.4 million, and the number of items loaned increased 98 percent, from 2.3 million to 4.5 million. The good news was that during this era of tremendously escalating volume, the fill rate remained consistent at about 86 percent (Nevins and Lang 1993, 38). Another factor that has remained constant in libraries is the staffing levels (Stubbs 1994, 7). Faced with decreasing budgets or buying power to increase collections, stable staffing levels, and increasing ILL/DDS traffic, libraries welcome anything to speed the document delivery process.

Luckily, service does not have to go from a fast track to a snail's pace when the library is the mediator. Improved document procurement has been credited to several major technological advancements (Kennedy 1989, 77-78):

- Automated ILL subsystems from major bibliographic utilities that speed transmission of requests and communication between borrowers and lenders
- Commercial online and CD-ROM databases, particularly those with document ordering or full-text retrieval
- Improved and affordable telefacsimile equipment
- Optical disk storage and electronic transmission of digitized images
- Expansion of national research networks and international linking of the networks
- Software packages designed exclusively for improved efficiency in ILL
- Automated information delivery systems that can integrate online/ondisc searching, ordering and delivery with a library's OPAC, serials holdings, and even ILL/DDS.

THE DOCUMENT DELIVERY PROCESS

The choice of supplier and the mode of shipment are often seen as the crux to improved turnaround, but the entire document delivery process must be reassessed if meaningful improvement is to occur. Boss and McQueen (Information System Consultants 1983, 60) determined that while those two aspects take up approximately half of the total time for satisfying a document request, even reducing supplier turnaround time by 50 percent might only change overall satisfaction time by 25 percent. John Budd (1986, 76-80) and the ILL Department at Southeastern Louisiana University (SLU) conducted a year-long study in 1986, which tracked the segments of the ILL process. Budd's study provides suggestions for improving turnaround at various processing phases. Using Budd's findings as a benchmark for traditional interlibrary loan processes, we will examine the following steps in the document delivery process to determine ways to improve service:

1. Submitting request
2. Initial processing—logging in, verification, and identifying potential suppliers
3. Transmitting request to potential suppliers
4. Processing by suppliers, including locating item in collections, retrieval and preparation for shipment
5. Shipment of material
6. Receiving processes
7. Delivery to patron.

Submitting Request

For patrons desiring library-mediated document supply, the process begins with submitting a request. In most libraries this involves filling out appropriate paper work, such as an ILL or document delivery request form—a tedious and laborious process. Patrons with several requests are likely to abbreviate citations or leave off relevant information. Patrons may even be required to turn in the request in person, so that it can be checked for completeness. Occasionally, in sparsely staffed ILL/DDS offices, the signature of a professor, teacher, boss, or parent must be obtained before ILL requests are processed! Except in special libraries, telephone requests are rarely accepted unless the patron is disabled. The few libraries that do accept phone requests usually limit the number of requests that can be called in; some may also impose language restrictions on phone requests. Most libraries do accept faxed requests. A fax still requires filling out a proper request form (plus a cover sheet), but alleviates the need to make a trip to the library just to turn in requests.

In the ideal "in-person ILL interview," staff members advise of alternate sources or recognize uncataloged titles or items cataloged in such a way that only a librarian can locate the entry! This cuts down on the volume of ILL requests and may encourage willing patrons to do further verification on incomplete citations. Too often, instead of being offered alternative suggestions, patrons unable to locate items are routinely shuffled to ILL/DDS. And, if the docdel process involves a lot of paperwork and inflexible regulations, library users are discouraged.

In small libraries where reference and ILL are closely linked, the reference interview is useful; but if both reference and ILL staff are equally busy, and shuffling patrons rather than providing service becomes a pattern, the ILL interview serves little purpose. Assistance should always be available to those who desire it, but no reason exists to turn away patrons who may be unskilled or too busy (or even too lazy) to verify citations. Patrons should be informed, however, that unverified citations have a lower fill rate and are likely to take longer.

Many libraries accept document delivery requests submitted electronically. This practice prevails particularly in academic institutions where library users often have campus e-mail accounts. Electronic requests have also been incorporated into statewide library networks, such as Access Pennsylvania and Illinois' ILLINET. Patrons at one library who see that an unowned title is available elsewhere in, for example, the ILLINET system, can automatically generate a request for the item. The request is sent to both the ILL department of the patron's library and the potential lending library (Higginbotham and Bowdoin 1993, 41-42).

Colorado State University (CSU) is one of several ILL/DDS units to develop a patron-directed electronic ILL request system. When a library user locates an article cited in one of the online databases featured on the library's OPAC, the patron can initiate a request using "GRAB-IT" software, which captures the citation on the screen onto an ILL request form. Personal data is added and with the stroke of a key, the request is sent to CSU's ILL Service. Users do not have to have e-mail accounts since CSU's system does not use passwords (Wessling 1993, 24). If CSU owns the title requested, the program prints the citation on screen with the CSU call number; such requests are not sent to ILL (Dearie and Steel 1992, 123-128). OCLC's ILL Link also allows the end-user to direct an electronic request to the ILL department.

Electronic ILL request systems ease the disgruntlement of patrons when a desired work is not available in the library. Requests get to the ILL department faster—and more legibly. Systems that do not automatically capture the citation from another database are subject to keying errors, but even typos are easier to contend with than trying to decipher illegible penmanship. Requests that are automatically captured from a standard bibliographic database are already verified; potential suppliers may even be listed. This cuts processing time and improves turnaround.

Requests turned in at branch libraries can be faxed each day to a central ILL/DDS unit. If fax equipment is not available in the branches, "runners" should pick up the requests daily. Couriers between branch facilities may also speed the delivery of requests, but internal mail systems and standard postal services are usually avoided unless they are known for exceptional service. Pickup once or twice daily from all locations where requests are regularly submitted should be standard operating procedure—not dependent on catching someone who happens to be going in the right direction. A daily pickup schedule may be impractical for branches where requests are infrequent; a phone call or e-mail message can alert ILL/DDS staff when requests are on hand.

Initial Processing

The borrowing library has complete control over the initial processing stage (from logging in through basic search and verification to determine possible suppliers). Not every request can be filled following "standard" procedures, but each library can determine what is "standard" and how much time to devote to requests that are difficult to verify or not readily available. Occasional rush requests may require immediate attention, but batch processing as much as possible is more efficient than trying to process individual requests as they are submitted. However, libraries that begin the process only once daily add one to three days (counting weekends) if requests are collected all day for batch processing, which begins at the start of the next work day. The log-in process should be on the day of submittal for accurate turnaround accounting.

Statistical record keeping should be as simple as possible. Automated ILL tracking software, such as SAVEIT (Interlibrary Software and Services, Inc.), ILL Patron Request System (Brigham Young University Library), or AVISO (ISM Library Information Systems) reduce filing and record keeping. Automated tracking systems may prolong initial log-in processes, but reduce overall processing. Management reports provided by such systems provide clues to other means of reducing turnaround.

Unless it is done automatically by an integrated library system, avoid checking each request against a master file of qualified patrons. Requests from an occasional blocked patron should not delay processing valid requests. This step may be necessary if there is regularly a high percentage of unqualified requests (submitted by people who owe fines or who are not registered library users).

Also, unless the percentage of requests found in-house warrants it, every request need not be checked against one's own holdings as the first searching step; 10 percent is about average. If regulations do not require that patrons check library holdings before submitting an ILL request, then the library catalog is the place to start. This check may also be worthwhile if ILL is subsidized, because some patrons will try to get a free copy via ILL rather than make a copy at their own expense from the library's collection—an ulterior meaning of a library "user"!

Copyright Monitoring

At this point in the initial processing, copyright monitoring should be limited to those cases where requests from a single patron exceed copyright guidelines. Such requests can only be processed if permission to copy is obtained from the copyright holder and royalties are paid for each copy acquired.

The fastest way to clear copyrights is registration with the Copyright Clearance Center (CCC) or a similar reproduction rights organization (RRO) in other countries. Every title or publisher listed in CCC's *Catalog of Publisher Information* (COPI) is copyright cleared for registered, reporting, royalty-paying organizations. Unless COPI indicates otherwise, prior permission is not needed.

An alternative is to send such requests to a copyright cleared commercial document supplier. It does not matter whether one uses a library or commercial supplier. Just remember that the "rule of five" only applies to the first copy when a single patron's requests exceed copyright guidelines.

Royalties are owed on *all* other copies acquired for the same patron from the same journal, including future requests from that patron for additional copies from the same journal. If requested material is not copyright cleared via CCC registration or available from a commercial document supplier with copyright clearance, no alternative exists but to return such requests or get prior permission from the copyright holder. (Monitoring "same patron" compliance is especially difficult when requests are not submitted together. Even if some slip by initially, they are fairly easy to catch later when monitoring annual copyright records.)

Some ILL/DDS departments (particularly in special libraries) handle the "prior permission process" for patrons; some even absorb royalty payments. Most units are too understaffed, however, and require that the requester handle the process, especially since educating the public about copyright is part of a library's responsibilities. Getting prior permission from anyone other than an RRO is time consuming, labor intensive, and frequently fruitless. That is why many libraries just send such requests to a commercial document supplier.

For all the ballyhoo publishers make over copyright abuse, those not registered with an RRO or PUBNET rarely provide a rapid means for libraries and independent document providers to acquire permission to copy. In libraries that emphasize "rapid document delivery," the patron's "need before date" often expires before permission to copy is granted. Besides failing the needs of the requester, publishers, authors, and document suppliers all suffer economic losses.

Most ILL/DDS units try diligently to comply, but copyright guidelines are vague and vary from country to country. There is no guarantee that a document request will be filled or whether the item supplied will be an original or a copy, so it is worth delaying processing by trying to obtain *prior* permission on requests that *might* exceed the "rule of five." Be prepared, however, to report and pay royalties on *all* copies acquired that do exceed copyright guidelines. Most publishers prefer receiving the royalties rather than a barrage of requests for permission to copy. [At least, that was the advice of several publishers queried as to which is more important—prior permission or payment of royalties; the responses unanimously favored payments!]

Of course, one must budget for "after the fact" royalty payments but the annual royalties owed are usually less than the internal costs involved with monitoring compliance on every request prior to processing it. Libraries that have no budget for paying royalties are forced to monitor copyright compliance before procuring copies. If publishers want to have single article document delivery become "big business," they will have to lighten up on prior permission requirements and foster royalty payments after the fact, or be more willing to allow commercial document suppliers the rights to sell copies.

Search and Verification

Requests should be sorted into batches of monographs and serials for preliminary searching. Although some monographs will end up being photocopy requests and some serials will become loan requests, dividing the two works best for initial batch searching to verify titles and to locate holdings.

Monograph searching can begin immediately using standard union catalogs where the title can be verified at the same time that locations are found. In large online union lists, such as OCLC, RLIN, or Pica, monographic searching is a breeze. The search and order process for serials on national bibliographic ILL utilities is more time consuming and costly unless the title is unique or the ISSN number known. Searching is thwarted by title changes; multiple entries depending on cataloging practices for microform, paper, and electronic editions; and the necessity of checking multiple regional union lists. Serial requests are more likely to require decisions or detective work by higher level staff. Initial verification in a comprehensive print, microfiche, or CD-ROM union list can be faster and less costly than searching online without a unique identifier.

Searching the catalogs of commercial suppliers can be cost effective, especially when a citation can be verified and the document ordered at the same time. Commercial catalogs come in all formats—from print to online. OCLC and RLIN each have databases (FirstSearch and CitaDel, respectively) that allow the concurrent search of holdings of several commercial suppliers.

Database vendors, such as DIALOG, provide lists of titles available full-text. Full-text retrieval is best left to staff trained to search and retrieve files from the database vendor, but at this stage of processing one can at least determine which titles are available, providing the online searcher with the vendor, database name, and file number(s). *BiblioData Fulltext Sources Online* (Orenstein 1994), *DIALOG Full-text Sources Alpha List* (Dialog Information Services 1992), and the *Directory of Electronic Journals, Newsletters and Academic Discussion Lists* (Strangelove and Kovacs 1993) are excellent for this purpose. A source found in most ILL/DDS units, *Ulrich's International Periodicals Directory* (Ulrich's 1992, 5031, 5045) includes sections on "Serials Available on CD-ROM" and "Serials Available Online."

Even if a citation proves too expensive to order from a commercial supplier, having verified the title—and, one hopes, discovering an ISSN number—should speed additional location searching on national ILL subsystems. The volume of requests and the skill level of the searching staff determine whether sending requests to potential suppliers as they are identified or simply recording or printing locations for batch sending is more cost efficient.

Student assistants, library aides, and library assistants can efficiently perform all processing up to this point. The tasks include logging in, searching standard bibliographic resources, and sending requests to regular suppliers. A supervisor may make routing decisions selecting the suppliers to try first. Requests for titles available in full-text databases should be turned over to trained searchers. The selection of suppliers should be based upon known track records and actual potential for filling the request, rather than strictly on delivery costs. Requests should be sent to likely suppliers using the fastest means available. Within 48 hours of receiving a batch of requests, at least 60 to 80 percent should be on their way to potential suppliers.

Paraprofessional staff members should be trained to verify and locate potential suppliers on most of the remaining requests. These include requests with complex cataloging entries, publications that are too old to be included in some of the electronic databases, esoteric publications held by few libraries, document types that are often uncataloged, and bad citations. Virginia Boucher's *Interlibrary Loan Practices Handbook* [1984 (2d ed. forthcoming)] provides resources for verifying difficult requests.

Any requests still unverified after two to four days or those requests with no suppliers identified should be referred to a professional librarian for additional searching or to an information broker. Failing all else, they should be returned to the patron. Requests unfilled because they were "not found as cited" or unfilled by all potential lenders may also occur in this processing phase. Some problem requests may have to be sent "blind" to libraries with good collections in the subject area, especially if the item is likely to be uncataloged. This often occurs where specific libraries are known to have good collections such as government documents, patents, standards, and microfiche collections. Blind requests, however, are not encouraged.

Unverified requests are often sent to the publishing agency's library, the publisher, or even the author. Publishers tend to take the wording on ILL/DDS forms literally and to interpret them as a request to "borrow" or "photocopy"—even when an adequate MAXCOST is listed to cover purchase costs. They often respond with a letter quoting the price rather than sending the document and an invoice. Perhaps a future ILL/DDS work form will include the category "purchase"; but, until then, if an ALA or IFLA ILL form is being mailed to a publisher, it is wise to add a note specifying that it is a purchase order. The national library in the country of origin is always a possible supplier when no other holdings have been determined. Even if the national library can not fill the request from its own collection, it can often provide other locations.

Libraries with poor bibliographic resources may opt to send unverified requests to a full-service information broker that verifies citations and locates copies for a fee. This process may delay turnaround and raise delivery costs, but probably not any more than if library staff were to expend extensive time and effort trying to verify the title and locate suppliers.

There is no "average" time for processing problem requests as each one is unique. Success depends upon available bibliographic resources and the skill, persistence, and luck of the hunter, but no more than five to seven work days should pass without sending a request somewhere. "Problem requests" rarely account for more than ten percent of the ILL/DDS volume, but such requests invariably skew turnaround averages. At SLU the elapsed time between receiving a request and its submission to potential suppliers varied from none to 46 days, but the mean was 6.46 days and the median 5 days (Budd 1986, 76).

Online search procedures. Access to online or CD-ROM databases of commercial vendors or nonprofit suppliers provides more possibilities for verifying citations and ordering documents online—possibly even retrieving a full-text version. Identifying the most likely database is not always easy. Requesters often fail to provide the source of a citation, even though it may have come from a database with online document retrieval or ordering capabilities. Many users are unaware which file was used, especially if a search was mediated or if access to multiple databases is "seamless" on the library's OPAC. The source is often identified as "DIALOG," "CARL," or the "online catalog," naming the system or vendor instead of the database. Being able to cite a database accession number often provides a discount on delivery costs, but it would be "penny-wise and pound foolish" to attempt to verify every request in online commercial databases! The following is an embellished list of candidates for online verification originally identified by Lorin M. Hawley (1993, 70-73; 1992, 47):

1. Papers cited as "in press" or "forthcoming." Online databases have less "lag time" between publication and indexing than most of their print counterparts.
2. Citations for conference presentations without page numbers—a clue to possibly unpublished documents.
3. Technical reports—NTIS online is cheap and easy to use, while print/microfiche counterparts inspire procrastination among searching staff!
4. Dissertations and theses.
5. Education-related reports and conference papers.
6. Obviously incomplete citations.
7. Miscellaneous citations that require staff to go to other libraries to verify. Online verification is less expensive than personnel costs, and faster.

In most cases online verification and retrieval in multiple commercial databases is a task for expert searchers. Every database, vendor, and online catalog has unique features. Screens vary; coverage varies; so do search commands and log on/off procedures. Most important, costs for search and retrieval vary.

Large databases can be divided into smaller segments by date of publication. When the date is known, select the smallest file to speed searching. Currie and Olsen (1985, 6) noted that the years available may vary depending on the language of the article. They found Chemical Abstracts Service, for example, covered documents in English for 10 years but Russian language documents for 15 years. Timely online searching may not be achieved unless searchers participate daily in regular ILL procedures. Backlogs are likely if only one person is responsible for the process; using subject specialists in other departments may work unless staff outside ILL/DDS give requests a lower priority compared to regular duties. At Arizona State University, ILL/DDS paraprofessionals are being trained to search DIALOG databases. It remains to be seen whether a suficient volume of documents will be available online for staff to develop and maintain searching proficiency.

Bjørner (1990, 111) cites three rules to follow when searching online. "Be quick. Be creative. Be flexible." If that fails, "Be quick to get off the system." Here are a few tricks of the trade for effective online search and verification:

1. If unfamiliar with a file, check the database search manual to determine search commands and procedures. Even experienced searchers take time to glance through search guides before going online and write out a search strategy before the clock starts ticking and charges begin. Because citation searching seems simpler than subject searching, a written strategy may seem a waste of time; but if one is batch searching several requests or incomplete citations, it pays to prepare. Without a plan, searching in full-text databases is particularly tricky.

2. Batch requests for the same databases, developing a search strategy that encompasses as many requests as possible in one logon. Do this for each batch of requests sorted by subject. Many vendors allow searching across several databases at once, which may be cost effective when several requests are to be searched or when the exact database is not known in advance.

3. When searching unfamiliar subjects, use the vendor's "index" databases (i.e., Dialog's DIALINDEX or BRS' CROSS) to determine the most likely databases.

4. The fastest way to access a specific work is with the database accession number, if known in advance. Barring that, use the Boolean operator "and" to search unique key words in the title field. A title that has no "unique" words (e.g., "History of Science in America"), requires more creativity. "And" key words in the title field with the ISSN and/or year of publication to narrow the search quickly. When batch searching several requests on the same subject, it may be faster to search using key descriptors, if the same key words are not common to all titles. Author, source, or date are useful in familiar databases, but search commands for these fields vary considerably, even in databases from the same vendor, so they may be less useful for inexperienced searchers.

5. Use the power of microcomputer terminals and software to input complex searches before going online (Tenopir 1992a, 66). For example, the "type ahead" feature available on Dialog's DIALOGLINK software, allows one to input and check search strategy before logging on. Pre-input searches are uploaded online much faster than average typists can key. If searching on a "dumb terminal," not a microcomputer, develop a search strategy in low-cost databases, "save" it and then "execute" the saved strategy in the desired database. Many vendors offer "introductory rates" on different files each month. This feature provides a chance to practice in new files and to use those files to set up a strategy for executing in more expensive databases.

6. Use a search strategy that provides "set numbers" for each term; then, any original terms that must be combined with new ones can be done without repeating parts of the search.

7. When searching mixed databases where only selected articles are avail-

able in full text; do a quick search in the least expensive of likely databases. If it is necessary to print more than the "title only" format to complete the verification process, select the least expensive format available that will both verify which citation is correct and describe whether it is available in full text. The search manual should advise which formats provide full-text information.

8. Determining the length of an article may be more difficult than finding its full-text availability. Even from the same vendor, databases show little consistency as to where (or whether) the length of the document is given; often it is included only in higher priced formats. No one can read a screen at 9600 or even 2400 baud, so a searcher must download interim search results to review off line. Once the length of the file is known, a decision can be made whether to download immediately, order a file to be sent via the vendor's e-mail system to download the next day, or use other delivery options. Each option affects turnaround and cost. Print options vary in cost in each database. In some, it costs less to download online; in others, more—another reason for checking vendor's documentation and writing out a search strategy based on the least expensive search and retrieval options. To download, turn off the printer; download to disk at as high a baud rate as possible. When off line, print the file.

9. Response time is fastest during nonpeak hours for database vendors— sometimes a reason to select a foreign-based vendor in a different time zone if files and search protocols are similar and telecommunication costs are reasonable. The time spent online also varies with the equipment in use and the stability of telecommunication networks. The minimum recommendation for regular full-text downloading is 2400 baud; 9600 baud is preferable, but is not compatible with every database vendor or telephone system. Check with local online user groups and experiment to find the best combination.

Using the Internet to search and retrieve. The possibilities for verifying difficult citations online has improved tremendously through access to library catalogs via the Internet. Verifying foreign citations is more likely to be successful when one can access major research collections or the national library in the country in which a work was published. Specialized and esoteric materials may also be found in catalogs not available in printed form, but available through remote access online. The number of OPACs available over research networks is mushrooming.

Locating online catalogs is not as difficult as it once was, especially if one can access a "gopher." A gopher is analogous to an electronic directory or classification scheme. It provides menus that guide users through various categories—including library catalogs. Selecting that category brings a new menu of geographical choices. Select a region and at the next menu, select from online catalogs in libraries from that area of the globe. Each catalog available is likely to have unique log-in procedures and search protocols but there is usually a "document" that provides the basic information. Read or print the instructions for the catalog(s) of choice. Then select the telnet function to access the catalog and follow the previously given instructions to use the catalog.

Gophers employ a search function called "Veronica." Use it to pull up a variety of guides to online catalogs. Enter the search string "Internet accessible library catalogs" [without quotation marks]. Several catalog guides can be retrieved. One comprehensive guide is *Internet Accessible Library Catalogs and Databases*, by Art St. George and Ron Larson (1992, copyright 1991). It contains detailed instructions on accessing numerous catalogs. It separates U.S. entries into catalogs and databases accessible with or without charge. It provides an international section, a section on library catalogs with dial-up access, and other non-OPAC resources. Another popular guide is *Accessing On-Line Bibliographic Databases*, by Billy Barron and Marie-Christine Mahe. Without a gopher that file can be retrieved by anonymous FTP from : ftp.utdallas.edu:/pub/staff/billy/libguide. While these guides are helpful, accessing library catalogs is easiest using a gopher client rather than by following the instructions from any of the printed guides. A gopher client appropriate for your workstation can be retrieved by anonymous FTP from the University of Minnesota at: boombox.micor.umn.edu:/pub/gopher (Barron and Mahe 1993, lines 48-61).

The most seamless access of all to other library catalogs occurs when a library is part of a system that provides a gateway service. The OPAC then accesses not just the catalog of the home library, but also other bibliographic indexes and abstracts. Some of these resources are strictly useful for verification purposes, but many offer locations or provide delivery options. A sampling of the Arizona State University Libraries Online Catalog includes direct access to an encyclopedia, some H. W. Wilson Indexes, ERIC, IAC's *Expanded Academic Index*, *ABI Inform Index*, and UnCover. Selecting UnCover provides a gateway into even more databases available through CARL Corporation, including several library catalogs, one for U.S. government documents, two serial union lists, *Choice Book Reviews*, *Magazine Index*, *Business Index*, the serials catalog from the British Library Document Supply Center, and *Journal Graphics Online*. By selecting MELVYL, the University of California's OPAC, for example, one could then gateway into even more databases. There is considerable overlap among the databases offered at various libraries. Most commercial files block users not registered with the leasing library, but the variety of publicly accessible files is enormous.

Libraries without an Internet or e-mail account can gateway to other libraries via remote dial-up access. For those who need dial-up access to various OPACs, *OPAC Directory*, formerly *Dial IN* (Schuyler 1992), provides information on collection strengths, network memberships, Internet addresses, loan policies and restrictions on online services, along with the modem-in phone numbers for OPACs worldwide. A similar guide is Search Sheets for *OPACs on the Internet: A Selective Guide to U.S. OPACs Utilizing VT100 Emulation* (Henry, Keenan, and Regan 1991).

Blind, unverified requests will seldom need to be sent to suppliers with so many resources available. This should improve turnaround. That's good news—the bad news is that searching OPACs on a catalog-by-catalog basis is time consuming unless familiar catalogs are being accessed. Also, OPACs do not always represent the total holdings of their respective libraries. Most are fairly complete for materials acquired since the mid-1970s when the surge towards online catalogs began, but older holdings will not be included unless the library has retroconverted prior holdings to machine-readable format. ILL/DDS staff will be delving into old, musty national bibliographies and indices for years to come, but the online catalogs do provide current updates that can save hours of manual searching and expand the resources available within one's own library.

Routing Requests/Selecting Potential Suppliers

Until the material is received or the request returned unfilled, the process is out of the control of the requesting library. The only action that can control this phase is initially selecting the suppliers most likely to fill the request. At SLU, John Budd found that 83.4 percent of the items requested were filled by the first or second supplier selected; only 2.6 percent needed to be routed to more than five suppliers (Budd 1986, 76-77).

In selecting the most likely suppliers, libraries bound by cost restrictions may be locked into lending protocols established by regional networks. The "Model Interlibrary Loan Code for Regional, State, Local, or Other Special Groups of Libraries" (Boucher 1984, 137) states: "Requests should be routed through channels established by libraries participating in this agreement. These channels are . . . ." From there each group develops protocols to protect net lenders from being overwhelmed by net borrowers.

In regional networks, ILL requests are often routed first to libraries within the same system, then to the smallest library nearest the requesting library, on up to the larger libraries in the group. Requests are sent out of the network only if no library within is able to fill a request. Ideally, requests from libraries within the network get priority over those from outside. Many regional resource sharing networks are able to improve turnaround between member libraries by providing a shared catalog or union list, possibly an electronic message system to member libraries, and perhaps mechanisms for speeding delivery, either by fax, electronic transmission, or with network couriers.

Resource sharing agreements are designed to distribute lending as fairly and as quickly as possible among participating libraries. Most achieve that goal and have a high fill rate; however, protocols favor load leveling more than speed. Members should be free to decide when to use the network or go outside. Large libraries, likely to be net lenders in the group, often will not participate without a hierarchical lending protocol, unless net lenders are otherwise compensated. Those libraries that recognize the actual value of their service are more likely to participate in resource sharing agreements when they see a chance to recover costs—either through net lending subsidies or through balanced reciprocity. Libraries that absolutely must get ILL materials without paying lending fees may be strapped into network lending protocols. Libraries that are willing to pay have much more versatility in selecting a lending string based on potential fulfillment.

Case Study

A senior at a local high school working on an honors project for a statewide science competition needs a book not owned by his school library. The librarian checks that state's union list and chooses a lending string: first, another library in the school district; second, the local public library; third, the largest public library in the county; fourth, the state library; and fifth, the state university library. The book is in use, lost, or noncirculating in the first four libraries (not unusual, since every school is knee-deep in such projects at the same time). By the time the request reaches the university library, a whole week has gone by (even longer, if the request was mailed between libraries).

Based on average ILL costs, the original requesting library spent at least $16 locating potential lenders and processing the request. Four libraries have handled the request. At the least, each library checked holdings and availability status before determining that the book was not available to loan. The fifth library has pulled the book off the shelves, circulated it and shipped it to the requesting library.

The average cost to each library was around $10 to $12. If the school librarian had gone directly to the university library, paying the average lending charge of $10 to $15 a non-network request, the student would have had the book within a few days at a much lower cost to the local taxpayers and the six libraries involved. This situation exists because the average taxpayer might consider a "document delivery" budget frivolous in a school library, but would support a "resource sharing network."

Once a library has a fair idea of how much it costs to process an interlibrary loan request, it may discover that a free versus fee lending string may no longer look like a bargain. Perhaps then, emphasis can be redirected towards rapid turnaround rather than misdirected towards false economies.

Oddly enough, even patrons who complain about the slowness of ILL are rarely willing to pay for faster service. The library, therefore, must take the initiative. The average American citizen thinks nothing of ordering a pizza for ten or fifteen dollars to be delivered (even tipping the driver), but balks at paying for the delivery of information to be consumed. Maybe libraries should take a cue from pizza businesses and charge fifteen dollars for every ILL request and offer a refund if the material isn't received within one week! Libraries that rely strictly on the patron to pay all lending charges will almost always have a lower fill rate and higher internal processing costs than if ILL costs are subsidized and requests sent first to the supplier most likely to fill the request. Jan Fullerton (1991, 105) describes this as "direct-to-best-location interlending."

In considering the value of access versus acquisitions, a library's document supply function provides savings on the total expenditures required to acquire, process, store, and maintain information sources. Instead of acquiring every item desired, each library needs only those items for which the number of users or the frequency of use make outside access impractical. The community as a whole benefits when only a few institutions acquire little-used, but particularly expensive, items.

While an effective interlending system achieves a saving to the community, allocating the savings to individual institutions is difficult because it requires a collaborative agreement among institutions not to duplicate resources. Each duplication of a resource, however, reduces the community-wide savings. Jan Fullerton (ibid., 107) summarizes: "Charging fees throughout the system recognizes that borrowing institutions are also achieving a private saving, and compensates those who make this possible, while setting the fee at the marginal cost of supply ensures that a community-wide net saving is still achieved." Fullerton also indicated that subsidizing such a service "is justifiable so long as a margin of net benefits to the community remains after the cost of the subsidy is added."

A distinct advantage of a national union catalog database, such as the OCLC network, is the ability to quickly determine which libraries own a title. With one search many potential suppliers can be located. Online serial union lists add even more precision to a holdings search, allowing the requesting library to know in advance when the exact volume desired is owned. By confirming holdings ahead of time, the chances of having a request filled by the first potential supplier improve, cutting turnaround time and lowering processing costs.

Turnaround improves even more if OPACs are linked to circulation records. This is usually only possible within small local area networks or by moving from one online catalog system to another electronically.

When the day arrives that national cataloging databases, such as OCLC, are routinely integrated with local circulation records, turnaround will indeed improve. When a request has just been checked electronically and the OPAC indicates it is on the shelf and not in a noncirculating collection, fill rates from the first lender in a lending string should soar.

We see the difference at Arizona State University Libraries. The fill rate for ILL borrowing requests was 84 percent in 1992/93, while the fill rate for intra- and inter-campus document delivery requests was 90 percent. ASU's lending fill rate was 63 percent for requests received from within Arizona (where at least some libraries check ASU's OPAC directly) as compared to only 45 percent for requests received from out-of-state (ASU ILL/DDS [1993]). Knowing in advance whether a title is actually available does indeed improve fill rates and reduce turnaround.

Transmitting Requests

The method of transmitting the request to potential suppliers plays an important part in reducing turnaround. Online, fax, e-mail, and telephone are the fastest methods. Requests sent by mail average 3 to 5 days to arrive; even air mail can take from 3 to 14 days if foreign suppliers are involved.

Nothing is more tedious in the document delivery process than typing out requests and maintaining address files. Anyone who can type and proofread a mail request can be trained to send requests electronically. Often the same equipment can be used to receive documents electronically.

While office supplies are easier to acquire in comparison to a computer or a fax machine, almost any library can do a cost study to justify an investment in some type of electronic equipment for transmitting requests and receiving materials. Richard P. McKenzie compared fax costs using standard residential and commercial rates and WATS lines at day, evening, night, and weekend rates against the cost of mailing the same document by first-class mail within the United States or to the United Kingdom, Japan, Canada, or Mexico. He also compared Express Mail rates. The cost of supplies and equipment were considered in determining what volume of faxing would be needed before the equipment would pay for itself (McKenzie 1993, 18-22). McKenzie's data provides strong justification for the immediate installation of good fax equipment.

Regardless of how the request is sent, be sure to include as much of the following data as possible so that the supplier can process the request as quickly as possible:

1. Complete **bibliographic data**, including numbered series, accession numbers of documents in microfiche sets, and report or classification numbers of government documents and standards.
2. The **source of the original citation** or the citation verification source—including an accession number, if known. The original citation

source can be critical in speeding delivery. It may be the only clue to locating a document in a large microfiche collection (e.g., Early American Drama), or set of printed volumes (e.g., *The British Parliamentary Papers*), or a collection of patents or standards.

3. A **need-before date**.

4. **Shipping instructions** (if other than Library Rate for book loans or first-class mail for photocopies). Include your fax number, e-mail or IP address if documents are to be faxed or transmitted electronically.

5. The **maximum authorized cost**. Do not send a request to a nonreciprocal library or a commercial supplier with the maxcost blank or "zero" indicated. Clearly indicate if a request is simply for a price quote. Lending rates for most libraries are available online, such as in the *Name-Address Directory* in the OCLC ILL subsystem, or in the *Interlibrary Loan Policies Directory* (Morris 1995 in press). Do not delay a request or waste a supplier's time by failing to use such resources prior to sending the request. Some suppliers—both commercial and library—charge for price quotes, especially if they must pull a document to confirm the number of pages to be copied, or if they have to verify a request before being able to respond.

6. Any **numbers needed for billing or tracking** (either for your purposes or the suppliers').

7. The **lending library's call number or commercial supplier's accession number**, if known. Be sure that a call number supplied is the actual number of the lending library. If a lending string lists more than one library, a call number may be more confusing than helpful.

Regular suppliers should be given pre-addressed mailing labels if documents are not to be sent electronically. This saves the provider's time and assures that packages are addressed accurately and not lost en route.

One advantage in using information brokers is their willingness to accept requests in any format. Libraries prefer single requests on standard ALA or IFLA ILL forms, but an information broker will also accept lists of requests from computer print-outs or even hand-written original citations. Sending unverified requests in nonstandard formats saves the requesting library processing time, but slows the delivery process at the supplier's end. Lorin M. Hawley (1992, 47) has identified some steps to improve turnaround from information brokers:

1. **Send the original citation in full**, particularly if it was verified online. Full database citations include useful data, such as authors' affiliations, conference locations, ISSN and ISBN.

2. **Identify the database** so the broker can contact its producer directly, if necessary. If the original reference is not from an

online database, a copy of the original is better than a retyped or paraphrased citation, with less potential for error.

3. **Use care when retyping citations**. Train staff to recognize those parts of a citation that are most important to a broker. Unverified citations should be keyed in as they originally appeared without alteration. Journal title abbreviations should be left intact. Author names should include middle initials when known. Article titles may be truncated, but should include at least one searchable key word in case online verification is necessary (Hawley 1992, 47).

Automation has greatly speeded up the transmission and tracking of ILL/DDS requests through various message systems operated by national bibliographic utilities. The most widely used is OCLC's ILL subsystem which handled over seven million requests in fiscal year 1993/94 (OCLC 1994, n.p.). The OCLC subsystem allows a borrowing library to locate a document online and then pull up an ILL work form, on which the library can designate up to five potential lenders. From then on all communications regarding the status of the request or potential problems in filling it occur electronically over the network. This system allows a "borrowing" library to track the progress of each request at every step. Mail requests on the other hand, are totally untraceable except by phone, fax, or additional mail follow-up, which usually doesn't occur for at least three weeks.

Commands vary slightly in each electronic ILL system, but most operate along the same principles as OCLC. Each system does have unique features, however. DOCLINE, for example, allows borrowing libraries to set up a lending profile for the preferred order in the lending string of libraries frequently used. That feature is particularly convenient when the majority of ILL requests are filled within a network or regional consortium; the profile can be overridden when a different lending string is desirable.

Unfortunately, because most bibliographic utilities use proprietary protocols and data structures for their ILL message systems, communication is usually not possible between different ILL networks. Some require dedicated equipment. An RLIN library, for example, can not submit an ILL request electronically to an OCLC library. Requests have to be mailed or faxed unless the sender subscribes to both systems—expensive duplication.

On the other hand, communications barriers are being eliminated as more systems become accessible over national research networks such as the Internet. Electronic mail networks with message handling systems based on the X.400 standard can send requests to libraries and document supply centers, as well as deliver documents to requesters (Cleveland 1991, 55). Future ILL electronic mail message systems should use the ILL protocol, based on the OSI reference model developed by the National Library of Canada, in cooperation with the Library of Congress and the International Organization for Standardization. It defines a standard that makes it possible for different ILL systems to communicate regardless of the ILL software or hardware used (Turner 1990, 110).

The ILL protocol, based on ISO 10160, "Interlibrary Loan Application Protocol Definition" and ISO 10161, "Interlibrary Loan Application Protocol Specification," standardizes four aspects of ILL communication (Turner 1990, 114):

- The number and type of messages to be exchanged in an ILL transaction (shipped, received, returned, renewal requests and responses, overdue notices and recall messages, etc.)
- The data elements in the messages (date, time, ISBN or ISSN, etc.)
- The sequence of messages (i.e., a renewal response could not be sent before a renewal request was received)
- The transfer syntax which allows the receiving library's computer to identify all of the elements of an ILL message.

Well-established ILL utilities find it expensive to change to a new standard when the current system works well, but the prospects for linking ILL subsystems with commercial document delivery provide an economic incentive for standardization. The National Library of Canada, in particular, has championed the standard and has encouraged a number of vendors to produce products that implement it (Cleveland 1991b, 101). Many current international resource sharing projects are using and testing the standard ILL protocol. Canadian products that have incorporated the standard include ENVOY 100, an electronic message service; AVISO, a PC-based ILL management system; and ROMULUS 2, a CD-ROM system for locating serials and ordering documents from Canadian libraries. All are gaining wide use in Canada (Smale 1994, n.p.). In the United States, the Research Libraries Group has developed a Z39.50 service called "Zephyr," which lets users of other online systems search RLIN and CitaDel files using the same commands as their local system. Search results are also displayed in the same format as the local OPAC (RLG 1994). RLG has also teamed with ISM Library Information Services to test AVISO's compatibility with RLN libraries.

Processing by Suppliers

In his turnaround time study of libraries using the OCLC ILL subsystem, Budd found 85.4 percent of the supplying libraries updated the record to "shipped" in less than one week. The mean time was 4.58 days; the median, 3 days. Photocopy requests had a mean of 5.39 days; books, a mean of 3.52 days. Photocopies take longer due to the time required for the actual copying prior to shipment (Budd 1986, 77-78). The *Interlibrary Loan Practices Handbook* [Boucher 1984 (2d ed., forthcoming)] and the protocols for various ILL networks provide standard operational procedures, but most systems refine internal processes to improve turnaround.

Receiving Requests

Requests being received electronically should be printed at night when equipment (computer terminals, fax, etc.) is not otherwise in high use. High volume lending units, however, may need to download requests more than once daily. Some document suppliers prioritize requests according to how they are received. Telephone, fax, and electronic requests are deemed more urgent than mail requests, for example. This punishes the "have nots," poorer libraries that must rely on mailed ILL/DDS request forms. At least in the initial processing (checking call numbers and pulling materials, etc., there is no reason for such prioritizing.

Preliminary Processing/Holdings Check

Code requests for faster processing along the way. Letter or color codes may be used to tag requests that are rush; to be faxed; to be sent over Ariel (or other electronic document transmittal systems); to be shipped by courier; in a special location; to be invoiced; etc. Coding usually requires input from staff at a supervisory level. While coding, a supervisor should also be able to spot requests which might be problematic for lesser trained staff and can set those aside for later attention. (These include requests for items in uncataloged collections; requests from nonreciprocal libraries that listed a cost below the library's standard lending fee; requests impossible to fill by the need-before date; and loan requests for materials in noncirculating collections.)

If the requester supplied a call number, trust its accuracy. More libraries have access to specific OPACs and can provide the lending library's call number. This was done to save the supplier time and speed the document to the requester. Don't negate that effort by "verifying" the call number. The few that turn out to be "not on shelf" can be rechecked. Set aside requests for items that are not available or require follow-up (e.g., not owned, noncirculating, checked out) to notify the requesting library as soon as possible.

Retrieving Materials

Sort requests by location and disperse staff to retrieve materials. As items are pulled, they should be sorted for the next process—photocopy, scan, fax, microcopy, circulate, etc. Electronic retrieval of documents is becoming a part of this pattern, and will certainly play a bigger role in the future of document delivery as well as affect staffing levels in document supply units. Materials which must be retrieved from off-site locations (branch libraries, branch campuses, remote storage facilities, etc.) may not be available on a "rush" basis.

Circulation/Copying

ILL/DDS staff members should handle as much as possible of the work—circulation, copying, faxing, electronic transmission, invoicing, and packaging for shipment—allowing the unit to control turnaround and monitor quality. High volume lending/document supply units should have a circulation workstation in the ILL/DDS unit. The unit should also have state-of-the-art equipment for photocopying, microcopying, telefaxing, and electronic scanning, storing and transmission of documents. Some equipment may need to be duplicated in high-volume branch libraries. It is more efficient to make copies in the branch to avoid transporting materials to the central ILL/DDS office for copying and then having them returned.

Suppliers using scanners for fax and other electronic shipping modes improve their turnaround because they do not have to package copies for mailing; however, time is lost if documents are photocopied prior to faxing or electronic transmission. Scanning from an original document is slower than photocopying, but staff familiar with the equipment and trained in proper techniques can be as adept at scanning from originals as they are at photocopying. Operators must pay close attention in aligning documents so that edges are not cut off or tightly bound borders blurred. Unless a test copy is printed or viewed on-screen, the sender is not likely to detect scanning errors.

Some libraries photocopy prior to scanning because some fax machines and computer scanning workstations are faster when using an automatic sheet feeder rather than operating in the book mode. This is particularly true when using computer interfaces to expand the memory of fax machines or for other electronic storage of documents. [This area of technology still has considerable room for improvement].

If the requester has asked for fax or electronic transmission, one can reasonably expect that equipment at the receiving end is turned on and in working condition. Suppliers are not obligated to repeatedly resend due to problems at the receiver's end. The baseball rule "three strikes and you're out" usually applies; after three tries (other than automatic programmed resends) a supplier is justified in resorting to another mode of shipping.

Many commercial suppliers use scanners and electronic storage to develop document inventories ready for almost instant delivery. Libraries, on the other hand, are more likely to retrieve a document each time it is requested and then scan and send the requested pages. Bit-mapped page images, particularly those with graphics or half-tone images, even when compressed, require huge amounts of storage space. Digital data required to reproduce a facsimile of a page with graphic images can be twenty times more than that required for the same page in machine-readable text. Computer storage capacity is a premium commodity in most libraries. Since standard file retrieval techniques (key-word) do not work, most ILL/DDS units have little interest in storing scanned files any longer than necessary.

It is too expensive and labor intensive to convert complete volumes to electronic format just for interlending. A few "virtual libraries" are beginning to convert and store works in an electronic format immediately upon acquisition, but most libraries are only converting noncopyrighted, retrospective materials which require immediate preservation. Some libraries subscribe to electronic journals and other electronically stored document collections, but licensing agreements may restrict lending to other libraries.

What the future holds regarding interlending of electronically stored documents still protected by copyright is yet to be determined. Librarians argue that the same rules for copying and lending should apply to electronic copies as for any other copy; only the format and transmission medium is different. Publishers disagree on the grounds that electronic documents are too easily revised, and that a networked environment provides the potential to send one scanned document to millions of users.

Updating Status Reports

It is more cost efficient to batch all updating and do it once daily after all available materials have been retrieved. Suppliers dedicated to rapid document delivery know that even a few hours makes a difference to the requester, and may update more frequently if staffing and equipment allow. Those suppliers that charge for service certainly should notify requesters as soon as possible if they cannot supply an item.

With the possible exception of retrieving materials from off-site locations, most of the batch processing up to this point should have occurred within forty-eight hours after receipt of a request. The processes related to copying, faxing, and electronic transmission may require additional time, depending on the volume of requests, staffing, and equipment. Ideally, either the material or a status report should be on its way for at least 85 percent of the requests within four days.

Document suppliers, whether commercial or library, should employ effective quality control standards. Whether pages are photocopied and mailed, faxed, or transmitted electronically, it is disheartening to receive an illegible copy or one with pages missing or edges cut off. The entire process for both the requester and the supplier has been a complete waste and must be repeated, adding to costs and delaying fulfillment.

Shipment of Material

Boss and McQueen (Information Systems Consultants 1983, 53) lamented the "steady increase in online searching and requesting of materials, but . . . no corresponding shift to faster document delivery techniques." The biggest hurdle in document delivery is often the actual process of getting an item from point to point (Allen & Alexander 1986, 18). The study conducted by John Budd (1986, 77) indicated that the period from shipment to receipt had a mean of 8.21 days, a median of 7 days and a range of 0 to 29 days; 89.2 percent arrived within two weeks of shipment.

Budd's study at SLU was done before electronic transmission of documents—even by telefacsimile—was widely used. Today libraries and commercial suppliers that regularly transmit electronically or by fax notice great improvement in shipping time over items shipped regularly by mail or even by couriers when long distances are involved. The shipping mechanisms most often employed include:

- Postal services
- Courier services—either library sponsored or commercial
- Telefacsimile
- Electronic transmission
- Full-text online.

Comparative turnaround studies show delivery by mail perpetually the slowest. The postal service, however, remains the method most widely used around the world for delivering items from one point to another. Sometimes there is no other choice, but often staff members are not aware of cost-effective alternatives. Even when the postal service must be used there are ways to speed shipment. In countries where automated sorting equipment is used, mailed requests and packages should always be typed, not hand addressed. High speed sorting equipment can process 30,000 typed addresses per hour at a cost of $3 per thousand. Handwritten addresses must be manually sorted at an average rate of only 800 per hour and at a cost of $35 per thousand (Winick 1991, 26).

Addressing guidelines that speed mail delivery include:

- Print the address in block uppercase letters
- Keep a uniform left margin
- Eliminate all punctuation except the hyphen in the postal mail (ZIP) code
- Put one space between words, and between letters and numbers
- If mail is addressed to the someone's attention, the name must be the first line of the address
- Use standard abbreviations—N, S, E, W, AVE, ST, LN, RD, PO Box, APT, BLDG, FLR, STE, RM, DEPT—with no punctuation.
- Spell out the entire city name, but use two-letter abbreviations for states and (Canada) provinces. For any mail going in or out of the United States, spell out the country in full; but the country is not required on mail between European nations as long as the postal code is correct.

- Place the last line of the address at least one inch above the envelope's bottom edge with nothing printed below
- In using a computer to address envelopes, use a program that prints the bar code. This saves a step at the post office and reduces the chance of error.
- If possible, bring mail directly to the nearest mail processing center, rather than having it picked up (and delivered) at the library (Associated Press 1994, A7).

For delivery between libraries in the same metropolitan area, courier services are reliable and definitely speed delivery. Many library systems run library-sponsored couriers, but this can be expensive, requiring additional staff and incurring costs for procuring, maintaining, licensing, and insuring a vehicle. Using commercial couriers may be more cost efficient.

Among commercial couriers, UPS (United Parcel Service), Emery, Purolator, and Federal Express are the services most frequently considered—at least according to professional literature. Libraries, however, would do well to check the "yellow pages" and compare rates and services for all couriers listed. Small local companies may underbid the larger national carriers for deliveries within a relatively small region.

The three state universities in Arizona, for example, wished to provide daily courier service between the three libraries. Commercial directories list many couriers for the three metropolitan areas (Flagstaff, Phoenix, and Tucson), but few were common to all three cities. Rates varied widely but services were surprisingly similar. The company selected provides daily service between each university at a rate that would have only paid for twice-weekly deliveries from some of the other companies.

Not only are the rates lower but the company required less paperwork and wrapping. Books and journal articles are simply put in lightweight, but sturdy, covered bins. Two bins (one for each library) are picked up daily and transported to the two other libraries. Each bin holds up to thirty pounds and is shipped for a flat fee of $4.25/day (whether the bin is full or almost empty). The three libraries reduced turnaround by an average of five days compared to former shipments by mail while reducing shipping and packaging costs.

Some libraries use unique transport systems. The library at the Armidale College of Advanced Education (N.S.W., Australia) served for many years as the base library for the New England Regional Medical Library Network. For a quite reasonable cost, the system employed a linen service and a blood pathology service that made regular stops at each institution in the network to deliver library materials as well. Even over mountainous terrain and wide distances delivery time averaged 48 hours. [Newling Library at ACAE has since merged its collections with Dixon Library at the University of New England, Armidale, and no longer provides this service (Williamson 1994, e-mail note.)]

Probably many local options exist besides the more obvious commercial carriers. Large national services are likely to be the only option, however, when couriers are used for transporting books beyond state lines.

As early as the 1960s, telefacsimile was being touted as the delivery medium of the future for photocopies, but the existing technology forced most suppliers to reserve fax for rush only. Transmissions were often interrupted and always agonizingly slow, making fax a nightmare for the sender—and not always popular on the receiving end either due to use of thermal paper. Now, plain paper fax machines with batch processing, delayed transmission capabilities, and computer interfaces to expand memory capacity have made fax an acceptable alternative to shipment by mail. For libraries within the same telephone zone, fax transmission is an inexpensive way to speed document transmissions. With the right equipment, even long-distance faxing is faster and more cost effective than packaging and mailing photocopies.

An even higher copy quality is achieved when articles are scanned and digitally transmitted. European nations led the way in early experiments with this technology. The first trial was ARTEMIS (Automatic Retrieval of Text from Europe's Multinational Information Service), a project of the Commission of European Communities. Although ARTEMIS was technically feasible, it was not economically marketable (Information Systems Consultants 1983, 47).

Next came the ADONIS project, sponsored by a group of major periodical publishers who wished to protect the copyrighted distribution of their publications. In 1984, requests to BLDSC were analyzed. A high percentage were from the ADONIS group's journals and a large proportion were for articles published within the latest five years. The text and graphics of articles were digitized onto optical disks. Upon request, BLDSC retrieved the stored images and distributed copies printed on a high quality copier. Copyright accounting was automatically recorded by the system (Compier 1993—interview). From this trial project, ADONIS now enjoys a worldwide market in single article document delivery via a leased CD-ROM database. It is used primarily by medical and pharmaceutical libraries and by other commercial document suppliers (Ashton 1994—telephone conversation). A similar image print server was developed about the same time in the United States by University Microfilms International (UMI). Now marketed as ProQuest Multiaccess system, it provides local article retrieval with remote printing or fax delivery of previously scanned images.

More recent developments allow libraries to scan their own holdings using software such as RLG's Ariel. The digitally scanned image is then transmitted over the Internet. Ariel provides rapid electronic resource sharing of library collections. Scanned documents are stored only until the receiving library confirms that is has an acceptable copy. This system avoids using vast amounts of disk space for storing huge files, and copyright compliance remains the responsibility of the receiving library, not the supplier.

Full-text Retrieval

Libraries can also reduce turnaround by eliminating dependence on other suppliers. This can be done through electronically stored, full-text databases containing documents that can be retrieved directly by the requester. Full-text, electronically retrievable documents are becoming more widely available. This method is certainly fast, but most libraries do not have staff trained to know what is available and how to access the variety of electronic resources. Chapter Two provides an introduction to electronic networks and describes selected electronically stored collections available for full-text retrieval. Another guide to such resources is the *Directory to Fulltext Online Resources*, which provides instructions for retrieving a variety of sources and downloading the files to a microcomputer (Kessler 1992, 21-50).

Training staff for full-text electronic retrieval is difficult because each document may require a different retrieval protocol. Access to full-text files on electronic networks is usually via a gopher or the TCP/IP log-in command "TELNET" with the remote system's address. An address has two forms—a domain name address and/or an IP address. An example follows: TELNET Melvyl.UCOP.EDU to connect to the University of California's online catalog. From that point on procedures vary considerably. Database files of commercial suppliers may also be accessed, but use is usually restricted to licensed users and document retrieval often requires a credit card or a deposit account.

Efficient online retrieval is a developed skill acquired with practice and easily forgotten with disuse. Bjørner describes full-text search and retrieval procedures as "search" and "output" processes: "Full-text *output* for document delivery . . . involves a brief bibliographic search for a known item and printing the full text of that item in an expeditious manner." She admits to being "occasionally surprised at how much money the output of a specific item can cost if the article is unusually long, or if technical difficulties manifest themselves."—problems novice searchers are likely to encounter in unfamiliar databases (Bjørner 1990, 109). Total turnaround and costs for documents retrieved full-text include the time and costs involved in both search and retrieval. The database vendor is the supplier. Supplier records can be used to compile a list of titles that are consistently cost effective for online retrieval. Such a list is even more useful if the database(s) to use are included.

Receiving Processes

If documents are to be received electronically, receiving equipment should be set up whenever it is not in use for other purposes. Equipment should not be turned off even when the library is closed. Paper supplies and toner should be adequate for receiving when equipment is unattended. Most suppliers fax at night to take advantage of lower telephone rates. Computer workstations used to receive documents by e-mail are often in use for other purposes throughout the work day, so much of the receiving of electronic documents occurs during nonwork hours.

Whenever a document is received electronically—either by fax or e-mail—check it immediately for legibility and to be sure no pages are missing or cut off. Notify the sender immediately if there is a problem. Since electronic shipping methods are so rapid, the supplier may maintain access to the original document and can quickly resend any bad pages. Delaying even a few hours may mean the volume has been sent to be reshelved, making it difficult to intercept and delaying the re-send. If the problem was a scanning error, alert the sender to re-scan, not simply re-transmit or yet another follow-up will be required. Scanning errors include missing pages, top or bottom margins cut off, inner margin blurred, and poor legibility of half-tone images.

If couriers deliver materials each day, don't rely on mail room staff for delivery after the regular mail run. Have ILL/DDS staff retrieve deliveries. Even if processing can't be completed on the day of receipt, at least the materials are on hand should any patron arrive to check on the status of a request.

Copyright records should be updated at this stage. Sometimes copyright information is published on the title page of each article. ILL/DDS staff can record the published royalty fee on those articles as the material is checked in, just in case CONTU guidelines are eventually exceeded. However, publishers often establish a sliding fee with royalties highest during the first year of publication, and decreasing each year thereafter. Paying based on what was printed at the time of publication may mean paying too much! An automated ILL tracking system will record copyright data when records are updated to "received." Libraries that track copyright manually simply keep an alphabetical title file of every photocopy acquired for documents published within the last five years.

Delivery to Patron

Document supply is expedited when documents acquired electronically can be sent directly to the end-user in electronic form. While this method is the ideal, most libraries today receive the document and then notify the patron of its arrival and availability for pick up. The speed with which the document was acquired means nothing if the patron does not pick up the material promptly.

More ILL/DDS units are taking responsibility for direct delivery. Academic and special libraries are the most likely to provide in-house or on-campus delivery. While electronic transmission and telefax play a role in some internal delivery programs, the requester may not have direct access to fax or Ethernet connection. Thus, delivery through internal mail systems or by ILL/DDS staff is likely. If confidentiality is an issue, requiring wrapping and addressing for delivery, it is probably just as fast to continue notifying each patron to pick up materials. In such cases, notification is usually by telephone or e-mail; regular mail, because it is so slow, should be used only as a last resort.

Fee or Free

Many libraries face the issue of whether to charge for document delivery. When a library provides service strictly for the convenience of the user, then fees sufficient at least to recoup costs for copying and staff are not unreasonable. On the other hand, if document delivery services are an alternative to acquisitions and patrons have no other choice, the library should assume the costs—most likely with funding from the materials budget. Even under those circumstances, it is not unreasonable to expect patrons to pay print charges. In most cases, had the library owned the journals, patrons would have paid for copying articles themselves.

Library mission statements often promote the belief that information should be freely available—they view interlending and document supply as a public "service," not a business transaction. Nonetheless, it is an expensive public service and, to achieve rapid turnaround, funding of one sort or another must be available. A few libraries institute high lending fees to discourage the use of their collections. Other libraries view materials not in use by their own clientele as a marketable asset and are likely to set a fee based on cost-recovery, gladly providing fast, high quality service. Those libraries should be able to expect equally efficient service from other fee-based suppliers—whether library or commercial.

INTEGRATING DOCUMENT DELIVERY INTO LIBRARY SERVICES

ILL/DDS processing has always been entwined with other departments—reference, circulation, stacks, copy service, mail room, instructional services, special collections, and collection development. Alternative access leads to additional interaction. Stronger links are forged with collection development and subject specialists regarding the selection of electronic databases, resource sharing agreements, and serial cancellations. Interaction with information technology staff increases as ILL/DDS becomes more automated and uses research networks more, and as public access to electronic document delivery databases expands. The move towards alternative access involves close cooperation with acquisitions and accounting units as materials budgets are diverted from permanent acquisitions and used for document delivery.

Cataloging expertise is necessary for providing access to the contents of electronic resources. When the contents of large sets are not analyzed or indexed, usage is usually low. The medium may be different, but the problem is the same. When a library acquires an electronic database to provide access to works not otherwise owned, the titles available electronically must be identifiable by the public (just as if the volumes were in the stacks waiting for use), even if this entails future problems with withdrawals should the library cease to subscribe to a database.

Record Keeping

Record keeping in ILL/DDS falls into three categories: (1) financial accounting; (2) copyright reporting; and (3) management reports for statistical and collection development purposes.

Financial Records

Financial accounting for document delivery is nothing new to ILL/DDS units. Revenues generated by lending materials often subsidize costs for items borrowed or they recover operational and personnel costs. Lending fees are often collected from patrons to reimburse suppliers; the lending unit invoices nonreciprocal libraries and may generate substantial annual revenues. Fee-based document delivery units also have monies coming in daily from clients. Payments between nonreciprocal libraries may include library-produced coupons, prepaid vouchers, international reply coupons, credit cards, deposit accounts, and the issuing of checks.

In general, ILL/DDS staff members dislike the "business" side of the delivery process. Net borrowers try to establish reciprocal agreements, claiming to be understaffed or having a volume too low to warrant bookkeeping and accounting procedures. Even large libraries tend to give low priority to following up on unpaid invoices unless large amounts are involved or the payer involved is a repeat offender. Few ILL departments have been subjected to an audit of "operational expenses," but an audit is more likely when ILL/DDS uses funds from the materials budget. More detailed accounting procedures may be required when payments include deposit accounts and pre-payments for materials not to be kept by the institution. Traditionally, ILL/DDS units are service oriented, rather than business oriented, but it is time to recognize that it is a service *business* and must be managed cost-effectively!

Copyright Records

Record keeping for copyright purposes is a high priority in an ILL/DDS unit. Many publishers believe that libraries shirk on paying royalties; actually, libraries are just as likely to overpay because of misunderstanding regarding copyright. When in doubt regarding copyright, most ILL/DDS units either refuse to process a request, or they pay royalties on every copy acquired in excess of the "rule of five" without analyzing records case-by-case. ILL/DDS units bemoaned the daily record keeping made necessary when CONTU guidelines were developed, but those records provide a wealth of data for collections management far beyond their purposes for copyright compliance. With good records, a library can actually economize on delivery costs related to copyright.

First and foremost, a good record keeping system is required, whether manual or automated. The file should be maintained regularly. Records should be reported and royalties paid annually. (Libraries with high volumes or costs may prefer a monthly or quarterly schedule). The following procedures will assure copyright compliance for U.S. libraries.

1. Identify all requests in the copyright file that exceed copyright guidelines. In U.S. libraries, this usually includes:

   a. Titles from which more than five copies have been acquired for *various patrons* within a 12-month period from issues published within the latest five years of a specific journal or monograph.
   b. All copies except the first acquired from single works when requests from a *single individual* exceeded copyright guidelines. (The first copy counts in 1.a.) This problem had been relatively minor, but now occurs more frequently as individuals increasingly use electronic tables of contents services.
   c. All copies from titles for which the library regularly exceeds copyright guidelines, but has elected not to subscribe.

   Automated copyright software programs may not record the two latter categories, so a manual file may need to be kept for these two groups.

2. Check the first group of titles against library holdings. Royalties are not owed if the library has a current subscription to a journal or owns the monographs involved. Requests for replacement pages, articles acquired when a journal was at the bindery, or articles acquired due to errors made when holdings were checked can be eliminated. This may represent substantial savings on fees that many libraries pay without checking.

3. Remove records for documents acquired from commercial suppliers where royalties were paid upon delivery. This is another area that some ILL software packages fail to identify. Without checking the supplier, one could pay royalties twice on the same request.

4. Check the remaining titles against the Copyright Clearance Center's *Catalog of Publisher Information* listing of titles *and* publishers covered by the CCC (or its equivalent in other countries). Fill out the CCC log (or provide other acceptable documentation) and submit payment to CCC. (Some libraries prefer to just send the log to CCC and have them calculate the amount owed and bill the library.)

5. Send a form letter to all remaining publishers, providing the complete bibliographic citations for all copies acquired, and requesting a statement for royalties. If and when such statements arrive (often in the form of a letter rather than an invoice), submit payment to the publisher.
6. Submit a list of titles exceeding copyright guidelines to consider for purchase. Providing documentation on the costs to procure the articles (including royalties) and the status of the requesters allows for more definitive decisions.
7. Compare the annual list of titles exceeding copyright with lists from the previous five years to see if titles are repeating. If titles do repeat regularly, and a decision is made *not* to purchase a subscription, a record of such titles should be made in the copyright file to ensure that in the future royalties will be paid on *all* copies acquired.

When an ILL/DDS unit serves a multi-site library system, record keeping is more complex, especially if operations are centralized but budgets are separated. When libraries share services, the combined activity for all libraries served may be used to determine whether copyright was exceeded on any title. If budgets are separate and all of the requests exceeding CONTU guidelines were from one library, then the royalties are paid from that library's budget. If the combined libraries exceeded the guidelines, royalties may be assessed evenly across the board; or, to whichever library received copies of articles after the original five (regardless of which library acquired copies 1-5); or, assessed according to an agreed-upon formula.

U.S. libraries often question whether to register with CCC and pay royalties through this central agency or just deal directly with publishers. Not having to get prior permission to copy from the publishers registered with CCC is reason enough for most libraries to register. Any library that acquires articles mainly from popular or trade journals will almost certainly find it cost-effective to consolidate copyright processing through CCC (or an equivalent RRO in other countries). Libraries in which the titles exceeding copyright are more esoteric will usually need to contact many publishers directly. The various RROs are attracting more publishers each year, however, so fewer publishers must be tracked down individually—a laborious, time-consuming operation for ILL/DDS staff.

One benefit from direct contact is that many publishers waive royalty fees if copying was not excessive. Some never respond at all; of those that do, it is not surprising to have them ask the library for advice on a "fair royalty fee"! CCC staff advised that the average royalties from the publishers registered with them ranges from $3.50 to $5.00 per document copied. (Gadbois 1994—telephone conversation)

A few publishers send scathing letters demanding to know why prior permission wasn't obtained, and some publishers will demand royalties that match subscription rates, charging as high as the market will bear. Libraries that opt to forego prior permission, therefore, face the possibility of some after-the-fact royalty surprises.

Publishers sometimes advise that the royalties will be "forgiven" when a subscription is entered into immediately. If the use pattern and costs for alternative access (including royalties) indicate that a subscription is the better choice, then this is certainly a good time to subscribe. If alternative access is still the preferred choice, just pay the royalty. It is still the library's decision as to what will be acquired and what will be accessed. As long as royalties are paid, the publisher has no legitimate gripe.

Publishers would promote increased compliance if royalty fees were standardized at an established flat fee. Then libraries could logically budget for this annual expense. The copyright royalty situation is an area where two mutually dependent parties really need to meet and develop ground rules for cooperation.

Collection Management Records

ILL records of titles or subject classifications requested provide excellent documentation for collection development. ILL records point to those areas of the collection where significant gaps in satisfying current user needs may exist. Access is a temporary alternative to acquisitions. It would not make sense to borrow books continually or acquire photocopies if the same titles were regularly requested. On the other hand, there is little reason to acquire expensive works when records show that those titles are seldom requested. Monitoring ILL/DDS records and making acquisitions decisions based on documented user need is a reliable method for selecting library materials.

Adjustments for Staff

Using CD-ROM, online databases, and other electronic resources on research networks for document retrieval and ordering requires increased computer knowledge and skills among ILL/DDS staff. Even personnel with a solid foundation in using online bibliographic databases and national ILL subsystems must broaden skills to include the innovative use of research networks.

Helping staff adjust to new equipment and procedures is the most important part of upgrading one's document delivery system, but this is not always easy. Occasionally difficulties are encountered due to resistance to change. More often, frustrations relate to the fact that ILL/DDS may be the only unit in an organization trying to use "standard" office equipment in a nonstandard way. Information technology staff may provide ready assistance when troubleshooting word processing and spreadsheet programs, but probably few have the expertise to solve technical problems with equipment designed specifically for ILL/DDS applications, such as that used to store, retrieve, and transmit large volumes of data. The ILL-L and ARIE-L listservs are indispensable in trouble-shooting hardware and software problems, especially when internal or vendor support turn out to be less than ideal.

Vendor/Supplier Relations

The familiar caveat, "let the buyer beware," holds nowhere more true than in acquiring equipment for rapid document delivery. One should question closely, for example, this common advertising claim: "A dedicated workstation is not required except in high volume operations." Someone with technical computer knowledge, who knows the requirements for all software programs to be run on the same terminal, should assess whether a computer workstation can indeed be multi-tasking in your work environment. Other software programs in use on the same terminal may have equal or higher storage requirements, and one program can interfere with another. The solution may require reconfiguring the various programs to allow them to work on the same terminal, but determining this may need the skills of a computer expert. Such problems were among the major complaints posted in messages on the Ariel-L Listserv from libraries trying to install early versions of RLG's Ariel software. (Many of the problems were resolved in later versions.)

Unless an operation is quite limited, ignore "minimum" requirements, and go for the "recommended" requirements. Scanned documents take up vast amounts of memory, so buy the largest capacity hard disk and RAM that fit your budget. The same goes for anything that speeds up routine operations, such as scanning, faxing, and printing; otherwise, what is saved on equipment costs could be lost many times over in staff costs and increased turnaround. When purchasing an equipment "package," be sure to order everything (including cables, interfaces, peripherals, etc.) needed to install and operate the system.

Libraries can often acquire hardware less expensively from local suppliers, but it is important to specify exactly what is required. When dealing with equipment suppliers from the business sector rather than library suppliers, be sure that they thoroughly understand your operation. For example, a salesperson for a telefacsimile machine will praise the "extra capacity memory" which will store x pages to be sent in a batch transmission to x destinations. Be forewarned, the pages cited refer to business letters with lots of white space, not journal articles that use up memory three times faster. Also, being able to send the same document to several locations simultaneously is not the same as a batch transmission of several different articles to one location.

Likewise, computer salespersons will laud a product's capability "to scan a page or a photo and incorporate it into another document using your favorite word processing software." They will be dumbfounded to discover you don't incorporate the scanned pages into anything else, and your concern isn't with how well the product can scan "a" page, but rather multiple pages. Try explaining that you want to scan 8 to 15 pages at a time, *without* photocopying first (with both text and photographs coming out perfectly on the same page), store and transmit the pages as one file (not 8 to 15 page files), delete the file soon after sending it, and then repeat the process for hours on end every day! Just to make it interesting, add that you want to batch transmit documents scanned at intervals all day to several designated recipients at the same time every day— preferably at night from an unattended machine when the equipment is not needed for anything else, *except* that it should be able to receive documents at the same time that others are being scanned or transmitted!

When equipment is demonstrated, test it on documents typical for your ILL/DDS unit. Test it on a tightly bound, brittle, over-sized journal, and on pages with scientific notations and illustrations. Test scanning, transmitting, and receiving several articles. The average "state-of-the-art" fax machine will not handle more than 60 to 70 such pages before its "extra capacity" memory is filled. Then, the machine is almost useless (except for receiving), until the stored documents are transmitted. This means that you may not be able to delay transmissions until phone rates are lowest.

A good computer interface may solve the memory problem and also store received documents if the paper supply runs out when the machine is unattended. Determine before purchasing whether one page from an article can be retrieved from a file and re-sent, or whether it will be necessary to re-scan and transmit the whole article if there is a problem. If a paper jam occurs or toner runs out when receiving, can affected pages be retrieved from memory or must the sender re-transmit? Negotiate for a trial period, or lease equipment until you know it works in your situation.

Getting and installing equipment is only half the battle. Learning to use it effectively is the other half. Customer support from the supplier is crucial, and it should include help in installing equipment, staff training, well-written user manuals, tutorials, site visits from customer reps, and a toll-free help line. Maintenance contracts should be available, with repairs guaranteed within a designated timeframe. Equipment loans should be available if the system is likely to be down long enough to have a detrimental affect on service or work flow.

It is unfortunate that the following scenario is familiar to many ILL/DDS staff struggling in a new electronic environment. Software manuals are written with the assumption that the user is a computer expert; technical terms are tossed around with no explanation. Staff may read a basic computer text, but even recent publications often describe outdated hardware, software, and techniques. Time is wasted learning about yesterday's technology rather than today's applications.

Approach the local computer "geek" for instructions on how to "Telnet" to a file on the Internet, which must be retrieved by "FTP." You will likely be informed that you can FTP if you want, but there is this new "gopher" on the system that has telnet and FTP protocols already built in, so it will be easier to retrieve the file that way. Of course, your instructor's computer has the gopher already installed; yours may not.

Then you will probably be introduced to Veronica (and possibly Archie if no gopher is available) who, in spite of having a cartoon character name, can help you locate a file without going through a lot of funny stuff. If you are really lucky, you may come away with a copy of the desired file, or instructions on how to repeat the process and download the file in your own office computer. In the meantime, you will have been shown how to access the weather in the Amazon rain forest, introduced to numerous exotic, erotic and hobby-related bulletin boards, and maybe even how to do a few things relevant to your job. Returning to the ILL/DDS office you ponder whether to relay your new-found knowledge to the staff or keep it to yourself. You don't want people tying up terminals checking area ski conditions or the going rate on McDonald's collectibles instead of retrieving requested documents!

The computer expert probably also warned you to be sure you have enough memory available before transferring a big file, but probably didn't tell you how to determine the size of the file before beginning the transfer! You may also be advised to use a gopher in a time zone where most people are sleeping, as the file transfer is faster when there are fewer users on the system. So what does it all mean?

Navigating the Electronic Byways

Obviously, gaining connectivity to the Internet expands the bibliographic resources and document retrieval opportunities in an ILL/DDS unit. Online databases are proliferating on networks—commercial bibliographic files, library catalogs, full-text reference sources, documents in archival files, and electronic bulletin boards and listservs. How are these used to locate holdings; verify esoteric, incomplete citations; or retrieve an electronic document? The first step is getting connected. If the library is already "wired" and the department already has a computer workstation, access may be as simple as acquiring a computer account and password. Depending on the hardware in place and the connection, some type of circuit board may be needed. If the library and/or department has never been wired for access, consult computing services at your institution for local requirements.

Seamless access and retrieval is still in the pioneering stage for many libraries. Staff must deal with diverse resources, wading through an alphabet soup of computer terminology, trying to determine what is relevant. That is why one should become familiar with "gophers," and "Veronica," which allow one to search for a specific topic or file on research networks. Gophers let one burrow through multiple menus of library catalogs and other file categories stored on computers connected to the Internet. "There are more than 750 Gopher servers in use around the world." (Polterock 1993, 11) Veronica pulls the item off its electronic shelf and puts it on the screen before one's very eyes.

Veronica employs a client/server system, which mnemonically means that Veronica searches gopher server menus and retrieves electronic documents—"serving" them to the "client" (the searcher's computer screen). The client software allows different machines to access the server and then use local resources to manipulate the information sent in response to the query. (Brandt 1992, 18-19) Veronica allows cross-searching for addresses (by personal name or institutional affiliation) and subject searching from the menu text of file titles stored on research networks. It is up to gopher administrators to use descriptive titles in menus (Polterock 1993, 11).

Four heavily used information retrieval, client/server systems consist of the following: Gopher, developed at the University of Minnesota; TechInfo; WorldWideWeb (WWW); and Wide Area Information System (WAIS). WAIS translates queries into the Z39 protocol and is an information retrieval system that allows natural language keyword searching. Gopher, TechInfo, and WWW are larger campus-wide information systems. (Brandt 1992, 18-19) WWW provides hypertext access to Internet files, and has been expanded to include a hypermedia format called "mosaic" that is becoming quite popular. It is available for UNIX, Macintosh, and IBM PC platforms (Polterock 1993, 12).

File transfer protocol (FTP) lets the user move files from one computer to another, usually from a remote host computer to the user's workstation (De Forrest 1993, 3). File transfer refers to the ability to retrieve a file (an electronically stored document) from a remote host computer into your computer. The types of files transferred by FTP include shareware, freeware, and public domain software. Once retrieved, files can be printed as is, stored as a file on a computer workstation, or downloaded onto floppy disks. Retrieved files can be manipulated (changed), which worries publishers who fear illegal copying and revisions. Online editing of electronic documents also has librarians worried, but for different reasons. How does one know whether the document retrieved is the original or a revised edition? Once a document is revised, are earlier versions lost forever?

Christinger Tomer (1992, 8) defines an anonymous FTP server as "a machine at which a system's administrator has enabled remote users to copy files and archived programs that have been posted in order to be shared." Are many people actually transferring full-text electronic files, programs or software from FTP server sites? Tomer reported that in 1991 at least 900 such sites provided a significant degree of service. One site alone, Washington University's Network Services recorded 7.5 million transferred files, averaging 20,503 files/day in 1991 (up from an average 5,995 files/day in 1990). Users from 71,289 systems were identified by their e-mail addresses (ibid.).

According to Gary Cleveland (1991b, 31), "file transfer" via e-mail is unsatisfactory for anything but simple files. He states:

> Difficulties arise because: 1) the file will usually need to be manipulated to place it into a message; 2) many protocols will allow only text files to be included in a message; and 3) only short files are appropriate for this kind of transfer.

Cleveland (ibid., 31-32) describes other possibilities for direct transfer of files of almost any length and structure in his book Research Networks and Libraries. Other basic guides for staff learning to navigate the Internet include:

- Dern, Daniel. 1994. *The Internet Guide for New Users*. New York: McGraw-Hill
- Hahn, Harley, and Rick Stout. 1994. *The Internet Complete Reference*. 2d ed. Berkeley, CA: Osborne McGraw-Hill
- Krol, Ed. 1994. *The Whole Internet User's Guide and Catalog*. Sebastopol, CA: O'Reilly & Associates.

Staff Skill Requirements

In some cases, position descriptions may need to be rewritten to accommodate changes in procedures and staff capabilities. Staff members retrieving and sending documents over the Internet or other research networks must receive proper training. Someone trained to use a photocopier may not adapt instantly to facsimile machines or computers with scanners and memory capacity. When vacancies occur, what skills should one pursue in new staff hirings? J. K. Olsen (1990, 231) identified several information literacy skills required for coping in a computerized environment, including the ability to:

- search bibliographic, full-text, numeric, and directory databases
- use information services on CD-ROM and online systems
- use telecommunications software and systems

- search computerized databases from remote sites, plus instruct library users how to do so
- download information to floppy disk; and organize, manipulate, and analyze the information on a microcomputer
- use citation-management, spreadsheet, word-processing and digital scanning software
- use online current awareness services
- access library holdings using online catalogs.

Gary Cleveland (1991b, 121) noted additional skill requirements related to using networks, including the ability to:

- navigate in a research network environment including remote log-in procedures such as TELNET
- log into (and exit) specific systems after making the terminal connection
- use network directories.

Staffing Levels

The volume of ILL/DDS requests will surely continue to increase as library users gain access to ever-widening bibliographical resources. Libraries should anticipate the need for extra staffing for affected units. If end-user access and document ordering is significant, extra staffing may also be needed in instructional and reference services for end-user training and support. On the technical side, personnel will be required to install and maintain CD-ROM workstations, maintain tape drives, disc drives and other devices on minicomputer- and mainframe-based networks used for bibliographic access and document retrieval or ordering (Lippert 1992, 135).

The service unit that handles government documents likewise is greatly impacted by electronic document retrieval technology. Over 200 government publications have been issued in machine-readable formats. (York and Haight 1992, 14) Software ranges from user friendly to programs requiring sophisticated user skills for retrieval and manipulation. Assisting patrons using electronically-stored government documents— many with no print counterpart—may tax staff skills and require longer patron/staff interactions, possibly requiring double-staffing. Some libraries may decide that rapid document delivery from commercial suppliers specializing in government documents may be preferable to in-house access (Adkins 1993, 29-32).

Integrating Document Delivery with Collection Development and Resource Sharing

The economic restraints of the past few years, among other negative effects, have meant that monographs are not being purchased in as large numbers as they were in the '60s and early '70s. Monographs—even popular titles—are more likely to be purchased as single copies. It is more difficult to find owning libraries for recently published monographs and, when located, those titles are more likely to be already in use. While the quantity of new books being published is on the rise, many publishers have reduced the number of copies printed. It isn't uncommon for books to be out of print within one or two years of publication. This encourages resource sharing agreements where libraries agree to collect in specific areas and share materials.

Serial cancellations by libraries also make it more difficult to locate some journal titles. Even if a library owns a title, it may not own complete holdings. Serial union lists must be kept up to date so that ILL/DDS requests are not sent needlessly to suppliers who can not supply a requested item. Such information is also critical in purchase and cancellation decisions. Commercial document suppliers help to fill the gap when serials must be canceled; however, they tend to stock only the most popular or most current journals—those most likely to be so heavily used in major public or academic research libraries that they are the least likely to be canceled. Journals with limited circulation and readership are often more expensive and among the first to be canceled—even though they may be the most difficult to acquire via alternative access.

An analysis of library serial cancellations undertaken at five mid-western libraries between fiscal years 1988 through 1990 confirmed that journals most likely to be canceled are English language science titles costing over $200 that fall outside the scope of high-use, core serial titles. Those high-use, core serial titles are being maintained while low-use, perhaps unique titles are susceptible to cancellation (Chrzastowski and Schmidt 1992, 2). A study by Ann O'Neill (Alexander 1993, 90—summarizing a presentation by O'Neill) at the University of North Carolina at Chapel Hill indicated that use, rather than cost, is more likely to be the deciding factor in determining whether to keep or cancel a subscription. While publishers tend to lambaste resource sharing agreements as leading to cancellations of serials and limiting book purchases, O'Neill's study indicates that such agreements may actually protect titles that otherwise might be canceled if renewals were based strictly on use by one specific library rather than group needs.

It is easy then to see that document delivery and alternative access affect more than just the interlibrary loan processes. Hardly one library department remains unaffected in some way by that library's document delivery policies and procedures. Vendors of library management systems are beginning to recognize the role of document delivery functions in a totally integrated library. Many systems now being marketed include ILL/DDS modules. Appended to this work you can find a list of suggested requirements that might be included in a standard request for proposals (RFP) for an automated document delivery module of a library's online system.

Chapter 4

Alternative Access:
Commercial Document
Suppliers and Products

*The fact that libraries desire access to infor-
mation as opposed to ownership does not signal
the end of an industry, rather the transformation of one (Collins n.d., 15).*

In considering potential document suppliers, one often distinguishes
between "commercial" and "library" suppliers. But library users may best
be served by a combination of document delivery processes employing
both library and commercial suppliers—and statistics indicate that this is
happening. When Waldhart compared ILL activity from 1974/75 through
1982/83, 63.5 percent of the lending transactions were for photocopies
(Waldhart 1984, 206). When Nevins and Lang compared ILL activity from
1985/86 through 1990/91, only 44 percent of the total lending transactions
were photocopies (Nevins and Lang 1993, 38).

This shift in ILL patterns reflects collection development trends over the
last twenty years. Monies for library materials were diverted to serial
acquisitions throughout the '70s and '80s. It is no surprise, therefore, that
ILL book loans are increasing among libraries, nor that users can find
more of the serials needed from their own libraries, even though
increased access to online indexes and abstracts has led to higher
demand. The shift in ILL patterns also reflects the increasing role of commercial document supply and end-user document ordering; some photo-

copy requests that might have once gone through ILL are now being acquired directly by the end-user.

The pendulum may soon swing the other way as serial retrenchments begin to cut into core journal collections. In a June 1992 survey, Mounir Khalil reported that the two reasons given most often by librarians for using commercial document delivery services were the lack of journal holdings due to budget cuts, and the growing use of online and CD-ROM databases leading to increased demand for materials not held locally (Khalil 1993, 43-44). At least library users now have options other than ILL for acquiring materials not owned by the library. This activity affirms that together ILL and commercial document supply form a complete document delivery package.

A full-service ILL/DDS department's performance may compare well with service from commercial suppliers. The average lending request will be filled and en route to the requester within one to four days, assuming that the item is available. Most books are shipped "Library Rate" and photocopies by first class mail. Many libraries also utilize couriers, especially for in-state deliveries of books, and copy requests are often delivered routinely using fax or electronic document transmission over the Internet. Items may even be delivered directly to the requester.

Not every library, however, has a full-service ILL/DDS department. Many ILL units are decentralized. Requests are received and verified in reference, where staff members may also locate potential suppliers. Photocopying may be done by the library's copy service, packaging by the mail room staff, and distribution of materials handled in the circulation area. ILL staff (often classified and/or part-time) may only be involved with sending requests and receiving materials. Fax may be reserved for "rush" requests, although rush in such circumstances is almost impossible.

In some libraries (most often school and smaller public libraries), ILL might not be offered at all. This lack is usually due to the labor-intensiveness of ILL or the dearth of bibliographic resources to verify citations or locate potential lenders.

Full-text databases (online or CD-ROM) and tables of content services with delivery options directly to end-users can improve access while lessening the need for an interlibrary loan department. Ten percent of the libraries surveyed by Khalil were using commercial suppliers due to staff cuts in ILL (Khalil 1993, 44). Fee-based document delivery can relieve the economic burden of ILL. Commercial services cut internal costs by decreasing processing related to copyright clearance, search and verification, and the need to produce ILL requests in ALA format. Information brokers, for example, accept requests in various formats (a list or computer print-out, or phone requests).

Commercial document supply is not a new phenomena. University

Microfilm's Article Clearinghouse (UMI), the Institute for Scientific Information's (ISI) "The Genuine Article" (TGA), the American Chemical Society's CAS Document Delivery Service (CAS), Information on Demand (IOD), The Information Store (TIS), ERIC, NTIS, and many other document suppliers have been around for years.With such potential for fast, efficient, value-added services, one would expect ILL/DDS units and end-users to be beating a path to the doors of commercial document suppliers. Figuratively speaking, the path is well trod, but libraries often advance to knock on the door and then retreat because service fails to meet expectations.

These and other commercial suppliers technologically equipped and staffed to receive and process document delivery requests should have the edge over low-staffed or decentralized ILL units. They should also be able to offer a wider range of shipping options than most libraries. Vendors working from private, noncirculating inventories or from electronically stored collections available for immediate transmission can be expected to have a higher fill rate than libraries. Document suppliers, with inventories based on a current awareness database, are often used when currency is vital to product development or research because little lag time occurs between publication and availability.

Why then, has it taken so long for commercial document delivery to be accepted as a viable alternative to interlibrary loan. Why do commercial suppliers so often come out on the short end of the stick when compared with ILL? And, why have the comparisons created so little concern?

In 1985, Wiggins [(1985) 1990, 197] concluded that "no really good comparative data exist on the delivery times of interlibrary loan services versus those of commercial document delivery services." That observation prompted a flurry of research to fill the void. The majority of the ensuing studies compared academic ILL units or large regional ILL clearinghouses, both of which are likely to be better staffed and equipped than ILL units in other types of libraries. In most of those comparisons, the ILL unit scored higher or equal to the commercial suppliers. Commercial suppliers fare better when compared with ILL in special libraries.

Most of the studies feature libraries that had fairly high fill rates and low delivery costs, but slow turnaround. Substantial improvement in turnaround for a reasonable fee were key concerns. Unfortunately, "standard" turnaround from commercial suppliers is about the same as that from a good library supplier. It isn't that commercial suppliers aren't staffed or equipped for fast service, but most commercial services expect customers to pay more for rapid turnaround or special handling. Many suppliers could undoubtedly have most requested documents ready for same day or next day shipment, but unless the requester has indicated "rush," a commercial supplier is likely to delay shipment until the company's "rush time frame" has passed. Otherwise, there is no incentive for the client to request and pay for special handling!

Published statements from various document suppliers imply that only a small percentage of requests received are for "rush delivery," and that they get few complaints from regular library clients about the speed of services. Motivation for improvement was nil. Traditionally, the "regular" library clients of commercial document suppliers have been special libraries, which often have comparatively few document delivery requests, and they are willing to pay whatever it takes to acquire needed items quickly.

Few studies feature small college, public, or school libraries, and few feature libraries outside the United States that may have a regular need to acquire materials from abroad. All such libraries would make an ideal market for commercial suppliers or document delivery products—if the price were right. In the United Kingdom in 1978, J. A. Pickup (1978, 28) challenged commercial suppliers "to develop this market for the benefit of the user, the supplier, and the nation as a whole." The challenge was bypassed by most commercial suppliers despite early publisher-based studies that suggested 82 percent of users were willing to pay for rapid document delivery (Gurnsey and Henderson 1984, 70). What remained unknown was 1) how much were users willing to pay, 2) for what were they willing to pay, and 3) who comprised the 82 percent—corporate users, the general public, or libraries.

Libraries are not unwilling to pay for document delivery; nonreciprocal libraries charge each other every day. One advantage to dealing with libraries, even those that do charge, is that library lending and copy fees are usually known in advance. Price structures from commercial firms, on the other hand, can be difficult to decipher, with various "add on" fees beyond the "basic" fee. Libraries (or their patrons) on tight budgets must often restrict use of commercial suppliers to those requests that meet the criterion: "unavailable elsewhere," "needed rush," or "exceeds copyright restrictions." Suppliers that have made one hour to forty-eight hours turn-around "standard" without added fees for special handling have discovered a market really does exist.

Price is not the only reason why libraries look to other libraries first. Most commercial suppliers do not have a "product line" comparable to the collections of large libraries. Inventories of available titles are often small, overlapping among vendors, and likely to be readily available in most large libraries. Retrospective journal collections beyond five years are uncommon among commercial suppliers, and even when advertised as available, often require extra days for fulfillment.

In analyzing the results of the initial trial of the ADONIS CD-ROM database, Campbell and Stern (1990, 20) stated: "Usage was fairly evenly spread over the 50,000 items recorded, with very few articles receiving more than five requests; the 50,000 requests only utilized twenty-five percent of the 199,440 articles on the discs." The FOUDRE Project in France had only two percent recall for the same article (Ménil 1993, 34).

Kent, Merry, and Russon (1987, 15) studied the ILL requests received by five international document supply centers—the British Library Document Supply Centre (BLDSC); Centre de Documentation Scientifique et Technique (CDST) in France; and from the United States, the National Library of Medicine (NLM), CAS, and OCLC member libraries. Sixty percent of the requests received by these services were for serials not more than three to four years old. They developed a core list of the 500 most frequently requested titles (ibid., 39-42). Most commercial suppliers use core lists, lists of frequently cited journals, and lists of journals included in major indices and abstracts to determine which journals they will stock. Libraries use the same core lists for collection development purposes, so the percentage of a commercial supplier's inventory being unique is fairly low. Kent, Merry, and Russon (ibid., 25) also confirmed that even of the 500 most frequently requested journals, any single article was likely to be requested less than once per year, a fact that has been validated repeatedly by records kept by United States libraries for copyright compliance purposes. Since commercial suppliers can not afford to invest in a "just-in-case" inventory, and no ILL/DDS department is likely to settle for a 60 percent fill rate, libraries will always have to rely on resource sharing to acquire that elusive 40 percent beyond "core commercial collections."

Trade document suppliers have long grumbled that resource sharing among libraries is unfair to the publishing trade and that fee-based document delivery services within libraries are unfair competition with private enterprise. The commercial trade, however, did not seem interested in pursuing a library market by developing moderately priced products or services to provide timely document delivery. Oddly enough, even with study after study showing ILL as effective as most of the "old timers" among commercial document suppliers, no one other than ILL librarians ever seemed to take such studies seriously. Until recently, vendors did little to improve their standings, and regardless of how it ranked in comparative studies, ILL was still perceived as slow and cumbersome by non-ILL librarians and library users.

The first glimmer of interest in developing a larger document delivery market came from database producers, who realized the benefits of being able to deliver articles indexed in their own databases—particularly, if the end-user could bypass ILL and order directly. This development began with online services, such as Dialog, BRS and SDC, offering online ordering capabilities. Eventually, CD-ROM databases became the preferred medium for database searching, and vendors that were just beginning to realize the profits of online ordering probably saw a drop in business as mediated searching decreased. Library users enjoyed the less constrained searching available on CD-ROM databases (no clock ticking with charges building). Patrons were likely to end a search with reams of citations—relevance to be determined after the search session; and interlibrary loan was usually the next step to get everything desired, but not owned, by the library.

Database producers made a quick comeback in identifying a market. Information Access Company (IAC) linked InfoTrac, a CD-ROM biblio-graphic database, to a journal collection (sold separately) stored on micro-form cassettes. Others developed full-text electronic editions of standard reference works (i.e., *Grolier's Academic American Encyclopedia*) or data-bases with full-text article retrieval capabilities. The early products were often limited to a single subject and the number of users at any one time was limited by the number of workstations available.

The first jukebox CD-ROM carousels arrived in 1989. NEXT Technology in Cambridge, England, developed a jukebox for the ADONIS project, with an average disc retrieval time of three seconds (Barden 1990, 89). UMI also developed a jukebox arrangement in 1989 for its first periodical image database, Business Periodicals Ondisc; it allowed 240 CD-ROMS to be accessed at a single workstation. It had an "Image Print Server" that could support approximately 100 article print requests per hour—a capacity greatly expanded by the second generation in 1992 (Barnes 1992, 19-20). Jukeboxes improved housekeeping problems related to CD-ROMs, allow-ing discs to be quickly added to the database by junior staff, while improv-ing retrieval time as well (Barden 1990, 89). Databases loaded onto a library's online public access catalog (OPAC) or local area network (LAN) were the next step in reaching a larger audience of end-users.

CARL (Colorado Alliance of Research Libraries) Systems (now CARL Corporation) linked document delivery to an integrated library system database of journals received regularly by libraries on the system. As a journal issue is received by CARL, the table of contents is scanned into UnCover, a current awareness database available to all libraries on the sys-tem. Scanning the whole issue and storing the bit-mapped images elec-tronically was the next step towards rapid document delivery.

Electronic storage of documents for regular document delivery can only be done with a publisher's permission; royalty fees established by the publishers are included as part of the total cost for the delivered doc-ument. The standard fee and the copyright fee appear on screen so users know the total cost in advance.

Document delivery through UnCover was introduced in October 1992 and quickly gained in popularity with end-users who order directly, and with ILL/DDS departments. At first, ILL/DDS units were more likely to use UnCover as a backup when libraries owning a title had high lending fees. However, with UnCover's costs averaging under $10 per request and turnaround averaging one to four days, more libraries began to use the service regularly.

Like the veteran actor who achieves "overnight stardom" with one successful role, UnCover's success brought commercial document delivery into the limelight. Every company with a link to the serials industry—publishers, subscription agents, article clearinghouses—seemed to recognize a potential new market, and commercial document delivery services began to proliferate. "Access versus acquisitions" were no longer fighting words between publishers and libraries, but the basis for a whole new business relationship—the single article market—with everyone wanting in on the action!

Libraries seem ready to jump onto the commercial document delivery bandwagon. After years of rising subscription costs devouring materials budgets, most libraries have faced serial cancellations and collections that are becoming noticeably short on books. There is no easy way to tell users that currently received journals will be cancelled; nor to explain why a library has only one copy of the latest best seller. Library patrons do, however, generally accept the concept of rapid document access. Instead of negative publicity over cancellations, positive public relations is based on speedy access and stretching tax dollars. Consumers tend to like trying new products, and the general public often accepts alternative access more readily when individuals control the order process rather than being shuffled to ILL! Library staff need not worry. ILL/DDS will always be there for the patron who is not into DIY (do-it-yourself), and for library users without a credit card, access to a fax machine, or who can not (or will not) pay for service. The volume of mediated requests is unlikely to slow down anytime soon, even with the rise in nonmediated docdel.

A mind-boggling array of new commercial products and services is being developed. Purchase decisions are often made outside ILL/DDS, especially when end-user document delivery options are a consideration. Reference, collection development and information technology department members may determine which online databases to acquire and how to "network" them. ILL/DDS staff should be aware of developments, know what is available, and maintain a key role in the total delivery process.

A current trend is reorganizing the ILL unit of yesterday into the document delivery services of today, with ILL as one function of the new service. This hybrid unit may generate requests for patrons; serve as a receiving area for items ordered directly by end-users who lack access to a fax machine; serve as a relay station, forwarding electronically received documents to requesters; or provide document delivery within the institution.

Commercial suppliers would like to become the primary supplier in the DDS process, eliminating ILL as much as possible. To do this, they must provide products that combine current awareness services with more timely delivery than ILL. Commercial document suppliers are more likely to reach this goal when costs are about the same as traditional ILL, advertised turnaround times are met, the fill rate is high, and value-added services reduce internal ILL procedures.

To decide whether commercial suppliers should supplement or replace currently used suppliers, consider the following:

- Current costs to acquire, maintain, and preserve permanent collections as compared to acquiring information as needed—"just-in-case" acquisitions vs. "just-in-time" access
- Determine desired level of document delivery services
- Review the variety of commercial services and products available.

ACQUISITION OR ACCESS

Ownership of serial titles involves more than simply paying an annual subscription price. Additional costs associated with binding, circulation, record-keeping and housing are often overlooked when determining total expenditures for serials. The desire to add new titles . . . also has an impact because the budget must also increase to accommodate these new titles (Ardis and Croneis 1987, 624).

Because most budgets are shrinking in the midst of an information explosion, many libraries are shifting to a philosophy of just-in-time collection development. The situation did not develop overnight, but the urge for a fast solution sometimes leads to cancellations based on cost factors alone. Content and relevance to the collection are primary selection criteria; cost is secondary in the development of most library collections. The same order should prevail in deselection processes. Expensive titles should not be dropped first, just to get through the pain as quickly as possible with the least effort and smallest number of titles involved. Just-in-time document access is an alternative to acquisitions which should be as carefully planned as other library acquisitions, or collections and service will suffer.

Before a title is cancelled or rejected for acquisition, the affects of alternative access should be considered. The following questions must be answered:

- From whom is the title available—other libraries or commercial services?
- How current are "current only" holdings?
- Are multiple volumes available for delivery; or, are there limits on the number of articles or volumes which can be supplied at one time?

- In what format will the material be supplied?
- How will the material be shipped?
- How long can it be kept?
- What are the delivery charges of potential suppliers?
- What is the average turnaround from potential suppliers?
- What are the prospects for continued access?

ASSESSING CURRENT DOCUMENT DELIVERY SERVICES

The ARL/RLG ILL Cost Study showed that the average cost to "borrow" a document from another ARL library is $18.62 and the average cost to loan an item or send a photocopy between ARL libraries is $10.93—just under $30 for the complete transaction between all involved institutions. (Roche 1993, iv) With this data in hand, there is a tendency to jump head-long into commercial access with a logic that says if an article is acquired commercially for about $19, this is as cost-effective as acquiring it for "free" from a reciprocal library. The $19 covers the indirect costs of the requesting library to process an ILL transaction. Unless internal proce-dures are simplified when ordering from a commercial supplier, the cost to process the request may be about the same, with the added cost of the document itself. If internal processing costs are not reduced, then com-mercial suppliers need to provide value-added services to woo libraries from their long-time interdependence.

Libraries that do not match the profile of an ARL library may do their own cost study, using Roche's methodology or that of other cost studies. Libraries primarily using BLDSC as a supplier may prefer the model described in *Modelling the Economics of Interlibrary Lending* (British Library Board 1990). Several statewide library networks use the model described by Dickson and Boucher (1989). Donna Runner Ruda (1990) supplied a model used successfully by Curtin University in Western Australia. An older model used in a previous ARL study may be still appro-priate for ILL/DDS units that are not automated (Palmour, et al. 1972).

A thorough assessment of in-house document delivery services and costs must consider products and systems with document delivery poten-tial that are in use throughout the library system (not just in ILL/DDS). Fully assessing what a library already has usually leads to some surpris-es. Often CD-ROM or online databases were acquired piecemeal, or var-ious departments have separate accounts to access the same online data-bases, many of which provide document delivery services. Commercial document suppliers may be available via "gateway" services provided by vendors with whom the library may already have an account. A docu-ment delivery option may have been developed by a database producer after one of their products was originally acquired.

A database producer may have developed a document delivery option after a product was acquired. Staff members may simply need training is the use of new features in an upgraded product. Such was the case with libraries using OCLC or RLG's cataloging and ILL subsystems when document delivery options were added to FirstSearch and Eureka databases. Libraries in the CARL Corporation Network may be using the UnCover database for document delivery, but may not be using other databases on the network that provide document delivery or full-text retrieval, such as the databases from BLDSC; *Journal Graphics Online, Expanded Academic Index, Magazine Index, Business Index, and ASAP,* and the full-text journal *Online Libraries.*

Databases acquired for use by one department may have broader applications, but other units may not realize their availability. Acquisitions may use a subscription agent's database from EBSCO, Faxon, or Swets. The same services provide options for ordering single articles and single issues as well as annual subscriptions. Acquisitions may sometimes order a replacement issue or volume, but ILL/DDS might make fuller use of such a service. Reference may acquire full-text subject databases, but if the titles available are not listed in the library's catalog, requests for articles in those journals are probably being turned into ILL to acquire from other libraries. Envision $30 lost by two or more libraries—yours and potential lenders—when this happens. Multiply this by all databases in the library with document delivery potential, especially databases with annual licensing for unlimited use, to compound the costs of under-utilized in-house document delivery resources. There may be costs associated with expanding existing services to other departments, but once one knows what is available, a cost assessment can be made.

Determining Document Delivery Service Needs and Funding Resources

After assessing what is already in-house and current costs, consider future needs and service objectives. Users' needs may soon surpass system capabilities if only the bare minimum is acquired. It may cost more to upgrade later than to include extra features at start up. Is the objective primarily to provide access to cancelled journals? If core journals, rather than duplicates and marginal titles, are cancelled, rapid access may be vital. Is direct patron access a key component?

Does the library wish to be an active participant in resource sharing networks or less involved as a net lender? Resource sharing agreements with net borrowers are costly. Net lenders can stretch operating budgets by using commercial document suppliers for their own needs and get out of unbalanced reciprocal agreements. Revenues generated through lending fees to nonreciprocal libraries may cover the cost from commercial suppliers.

The value of resource sharing agreements, however, may be based on more than a balanced borrowing:lending ratio. Consider other consequences before cancelling such agreements if reciprocal ILL is just one component. Budget restrictions are forcing resource sharing agreements into the limelight. The focus is on cooperative acquisitions and/or serial cancellations. In years past, this type of agreement got more lip service than actual practice, but technological improvements make sharing collections feasible.

While copyright legislation favors publishers, libraries can legally "share resources" as long as royalties are paid when "fair use" guidelines are exceeded. When consortia are formed exclusively for resource sharing purposes, guidelines on what constitutes "fair use" are much tighter than when reciprocal ILL is just one of many benefits provided by a local network. The "rule of five" applies in the latter case, while royalties may well be owed from "copy one" in the first instance, if the same titles are regularly accessed rather than acquired.

An analysis of library serial cancellations undertaken at five midwestern libraries covering fiscal years 1988/89 and 1989/90 confirmed that journals most likely to be cancelled are English language science titles costing over $200 that fall outside the scope of high use, core serial titles (Chrzastowski and Schmidt 1992, 2). A study by Ann O'Neill at the University of North Carolina at Chapel Hill indicated that use, rather than cost, is more likely to be the deciding factor in deciding whether to cancel a subscription (Alexander 1993, 90—summarizing a presentation by O'Neill).

While publishers may lambaste resource sharing agreements as leading to cancellations of serials and limiting book purchases, O'Neill's study indicates that such agreements may actually protect titles that otherwise might be cancelled if renewals were based strictly on use by one specific library rather than group needs. The publishing trade created the necessity for resource sharing with serial pricing strategies over the past thirty years that milked the library market as far as possible; but publishers are not going to sit idly by and lose a good market. Commercial document delivery provides the marketing strategy most likely to curb excessive resource sharing when libraries cannot afford to buy books or maintain subscriptions.

As buying power shrinks, more library administrators acknowledge the need to divert part of the materials budget into alternative access. From the bureaucratic point of view, managers find it easier to divert funds into commercial document supply than into ILL. Funding for ILL is normally from the operations budget; the materials budget applies only if items acquired are added to the collection. Funding *only* commercial access could be shortsighted, however, because many libraries are well equipped and staffed to provide services comparable with many commercial document suppliers.

Auditors, nervous about the idea of paying for temporary access as opposed to permanent acquisitions, seem less likely to question payments to commercial suppliers than for access through ILL, possibly because many commercial document vendors are familiar subscription agents and publishers. Tim Collins (n.d., 17) of EBSCO describes the status of a subscription agency in this new market: "The future role of the subscription service agent will be to provide a mechanism for libraries to search the information available via a current awareness service and select articles from multiple publishers for purchase. This mechanism must provide easy accounting and produce reports and services related to the management of a library's periodical information access budget. The future role the subscription service agent will also be to provide publishers with a way to provide access to, and sell, their information by the article to librarians and to collect revenue from multiple libraries on one check."

Review of Document Delivery Products and Services Available

For ILL/DDS staff members, the phrase "commercial document supplier" usually brings to mind trade names or acronyms (where every supplier is more likely known by a symbol) of well-established document providers, such as UMI or IOD. Before thinking about whom to use, it's best to determine what to use. Most of the published literature describing the various types of commercial document delivery services and products, treat each type as a separate entity with a specific purpose. So much overlap exists among all of the options that the best course is to apply the same criteria (as much as possible) across the whole array to determine which product(s) or service(s) best meet the needs of a particular library. The main purpose of such a comparison is to determine the most cost-effective method of document supply to meet one's specific requirements. It does away with the illusion of library vs. commercial supplier, and simply considers what is available from all suppliers where a cost for delivery is involved.

In this chapter, therefore, "commercial" carries a broad interpretation that includes *every* form of document procurement other than that from reciprocal libraries. This holistic approach to document delivery will undoubtedly be criticized by purists who believe that everything is either black or white. The variety of delivery options and the overlap among fee-based document suppliers is more like a kaleidoscope of color, constantly evolving and changing!

Available products/suppliers consist of:

- Online citation services with full-text document ordering facilities
- Database- or product-specific document suppliers

- Information brokers or full-service information providers (IPs), for whom document delivery is only one aspect of their business
- Clearinghouses and consortia for document supply (The latter is an admittedly grey area. Many consortia are "not-for-profit libraries for libraries," but because their use requires a large budget commitment, their document delivery services are best compared with "commercial" suppliers.)
- Full-text databases—CD-ROM or online (another format with a mix of "free and fee")
- Fee-based information services within libraries (another grey area; these often operate on a cost-recovery basis as opposed to "for-profit" commercial IPs) to serve a library's nonprimary clientele, but often providing rush service to other libraries and information brokers as well
- Electronic journals and electronic libraries (some of which are only available on subscription, while others are free through research networks)
- Nonreciprocal libraries with lending/copying fees (the greyest area of all!) [This particular category is not discussed in depth, but included here as a reminder that all suppliers should be evaluated on speed, cost, dependability, and quality of service. The same criteria apply to library suppliers, whether they charge or not: the key issue is the best overall service to the patron, regardless of supplier.]

Many vendors offer a variety of products, or the same product may be available from different vendors or in various formats. Coverage options and costs also vary widely. Ann Okerson describes this multiplicity, as it relates to electronic and print journals, in her introduction to the *Directory of Electronic Journals, Newsletters and Academic Discussion Lists*: "Some electronic serials are electronic only, but various of them either index or review paper publications, and others move between electronic and more traditional formats. Some electronic journals produce paper or microform spinoffs and some paper journals appear selectively in electronic form. Various paper publishers are beginning to produce tables of contents or abstracts in advance electronically. The electronic preprints phenomenon is of increasing interest, particularly in certain scientific disciplines." (Strangelove and Kovacs, 1993, i-ii)

Comparative shopping and consumer awareness are keys to discovering the most effective document delivery services. While formats, methods of access, ordering, and shipment may vary, in each instance the requester will only acquire and pay for the portion of a document which is actually desired—not a whole issue or full subscription, much of which might go unread.

The following sources provide a starting point for determining what is available and from whom:

1. Association of Independent Information Professionals' *Membership Directory* (1994). A list of information providers or information brokers who support the AIIP's Code of Ethical Business Practice, which calls for members to accept only "legal" projects. The directory is available from the association's current officers, whose address is listed in the annual *Encyclopedia of Associations*. It is also available on Compuserve as an electronic bulletin board. AIIP membership ranges from corporate giants in the industry (Dynamic Information, Information on Demand, the Information Store, for example) to small home-office operations. Too often, anyone with a computer, modem, an account with an online bibliographic database vendor, and access to a major library can decide to open shop as an information broker—particularly "specializing" in document delivery from local libraries. A firm's membership in AIIP offers some assurance that it is legitimate (Quint 1992, 85-86). An ethical information broker will be registered with CCC or other RROs to pay copyright royalties, be registered as a legitimate business organization in the local community; and be registered as a bona fide patron of all local libraries used to fulfill clients' requests (and, one hopes, a supporting member of "Friends of the Library" groups for those libraries that provide the broker's "stock").

2. *BiblioData Fulltext Sources Online: For Periodicals, Newspapers, Newsletters, Newswires and TV/Radio Transcripts* (Orenstein 1994). An excellent resource to discover whether an article is available in full-text online. This alphabetical directory lists almost 4000 titles in all subjects available from 18 major vendors. Entries include all databases, dates of coverage, file names or numbers, and whether the coverage is "cover-to-cover" or "selective." Cross-references to title variations are also included.

3. The *Burwell Directory of Information Brokers* (Burwell and Hill 1994). An international directory of information providers, many of whom offer document delivery services. Entries are alphabetical under state or country. It includes a subject index and a service index with a key to document delivery providers. [Description based on the 1984 edition with the former title *Directory of Fee-based Information Services*, compiled by Helen P. Burwell and Carolyn N. Hill. An ad for the 1994 edition indicates it will provide more information for international document delivery—language, country, and foreign database experience of the brokers listed.]

4. DIALOG Information Services' "Yellowsheets" provide a list of seventy-five (as of June, 1994) document suppliers for various Dialog databases.

Requests could be sent to suppliers by some means other that "DIALORDER, but DIALOG provides a training database, ORDERTRAIN, to assist those new to online ordering. Many suppliers are not database-specific, but it can be cost-effective and fast to order documents directly from a database producer because the documents on the database are usually in stock, especially if stored electronically. DIALOG also provides a free guide, *DIALOG Full-text Sources Alpha List*, an alphabetical list of full-text journals in DIALOG databases; a subject list is also available.

5. *Directory of Electronic Journals, Newsletters and Academic Discussion Lists* (Strangelove and Kovacs 1993). The title describes the scope—each of the three sections contains an alphabetic listing with information regarding date of first issue, price, how to subscribe, how to access back issues, and who to contact for more information. There is even a style manual (contributed by the National Library of Medicine) on how to cite electronic resources. This directory is also available electronically (ARL.CNI.Org).

6. *Directory of Fulltext Online Resources* (Kessler 1992a). As a "directory" this source is less complete than other references listed, but it is outstanding as a good "how-to" guide for the novice accessing full-text online resources. It also serves as a directory of directories (many online) to a variety of full-text services.

7. *Directory of Online Databases and CD-ROM Resources for High Schools* (Parisi and Jones 1988). This directory is not specifically devoted to full-text or online document retrieval products, but serves as an excellent (though dated) guide to "user friendly" databases within the price range of most libraries.

8. *Document Retrieval: Sources and Services* (Erwin 1987). An international directory of commercial document suppliers, including information brokers, fee-based DDS in libraries, and information centers of nonprofit organizations such as trade associations or research institutes which offer DDS for nonmembers, it provides four indexes—the standard subject and geographic indexes, plus a copyright compliance index and an index by online ordering access capabilities. Rates and average turn-around are also given. [Description is based on an earlier edition—Champany and Hotz 1982, iii-v].

9. Electronic discussion groups, such as FISC-L Fee-Based Information Service Centers in Academic Libraries (contact: <FISC-L@NDSUVM.BIT-NET>) and ILL-L Interlibrary Loan Discussion Group (contact: <ILL-L@UVMUM.BITNET>). Listservs are excellent for referrals from librarians using commercial suppliers. The listservs are particularly helpful when trying to locate foreign suppliers or suppliers for an esoteric subject. Other electronic lists and bulletin boards in specific subject areas may also be helpful.

10. Exhibits at professional conferences—particularly good for demonstrations of vendor-specific products .
11. The FISCAL *Directory of Fee-Based Research and Document Supply Services* (Coffman and Wiedensohler 1993). The Fourth Edition (formerly, the *Directory of Fee-Based Information Services in Libraries*) is expanded in scope to include international coverage and commercial document delivery services *not* in libraries. Each service profile includes basic directory information, OCLC codes, e-mail and Internet addresses, the time zone for the service plus the number of hours different from GMT, services offered and price ranges, ordering and delivery options, research specialties, online services, and billing terms. Besides a broad subject index, it includes an "Areas of Special Expertise" index which could lead one to possibly the only company providing document delivery in certain fields. The "Services Offered Index" includes organizations with document delivery services—a category further subdivided into organizations which only use their own collections and those that also use outside sources. Other indexes include geographic locations, online services, bibliographic holdings codes (mainly OCLC codes), and service names. Service profiles are alphabetical under the name of the parent organization.
12. *The Information Broker's Handbook* (Rugge and Glossbrenner 1992). This source is not recommended solely as a "directory," but some chapters include lists of information brokers, providers, or suppliers in specific areas, some of which are not easy to locate elsewhere.
13. The "Yellow Pages" of telephone directories, particularly for large cities where major libraries are located. "Library Research Service" is one heading under which such services are located, but headings may vary.

Library-specific Criteria for Selecting Commercial Suppliers

After assessing current library services and future document delivery needs, and after reviewing the variety of products and services available, you can design your library-specific criteria for a coordinated, funded document delivery program. Such criteria include:

- maximum cost limits
- minimum turnaround time required
- general versus subject-specific suppliers
- single comprehensive supplier or multiple suppliers and/or formats
- patron-directed or library-mediated ordering and receipt.

Library-specific criteria can then be matched with possible commercial document suppliers or systems, eliminating those that fail to meet your needs. Eliminating a supplier does not always equate with eliminating a document delivery product or system, because many document suppliers lease services from others. That is why some of the same databases may be available from several database vendors such as BRS and Dialog. Options exist even for seemingly producer-specific systems, such as ADONIS (a CD-ROM biomedical database). That product can be leased directly from its producer, which might be the preferred route for services that require large volumes of biomedical articles almost instantaneously, but articles in the ADONIS database are also distributed by other document suppliers, such as EBSCO, UMI, and Faxon, on a single article basis (Ashton 1994—telephone conversation).

Critique the remaining contenders against general factors:

- scope of inventory and/or subject expertise
- turnaround time
- shipping methods, including document format(s) available
- delivery costs
- copyright compliance
- ease of accessing citations and ordering documents
- staff requirements or user-friendliness
- equipment needs, including supplies and maintenance
- accounting and billing options
- value-added services
- references from other library clients. [Check on quality of documents supplied; percentage of unfilled requests or requests not filled within advertised time frame; customer service; accuracy of supplier's catalog; and percentage of requests requiring follow-up.]

Scope of Inventory

The main factor in selecting a document delivery supplier is the scope of the inventory. Scope includes subject areas, type of materials included, years of coverage, languages included, and limits on place of publication. Special libraries may find one supply source to handle all of their commercial document delivery requests (Davis et al., 1992, 186), but less specialized ILL/DDS units will likely need more sources. Diversifying among suppliers is particularly wise while evaluating performance. Consider the following issues to determine whether the scope of a supplier's inventory meets an organization's needs.

What titles are available from the supplier? Most commercial suppliers can provide a list of titles available. How many titles are currently owned by the library and not likely to be used except for replacement issues or rush requests? The size of the inventory (i.e., number of titles available or records in a database) is less important than the relevance of what is included. Do the titles match the target audience and collection development guidelines? How does the list compare with titles frequently borrowed through ILL? How many titles duplicate those easily available via memberships in consortia, networks, or the Center for Research Libraries? How many titles being considered for cancellation, previously cancelled, or never acquired due to budget constrictions are on the list? How does the supplier determine titles to drop or add in the future? Are clients notified when titles are to be dropped, and are there options for retrospective purchases?

Does the stock correspond to titles covered in indexes, abstracts, and bibliographies available within the library? Find out whether the inventory is limited to titles indexed in a specific database or to materials actually owned by the supplier. Many commercial suppliers own their own stock of current materials, but rely on third-party suppliers for retrospective stock of titles listed in their catalog, which may slow turnaround and raise costs. If the supplier does not have a private inventory, which libraries and/or online sources does it use? Are the vendor's resources equally accessible through ILL with comparable turnaround times and/or lending fees? Does an information broker have access to databases to which your library does not subscribe or to libraries which normally do not provide ILL service or charge high lending/copying fees?

What time span is available? One to five years is common among commercial suppliers. This may not be a factor if one's objective is to provide temporary access to articles in journals recently cancelled or temporarily unavailable in one's own collection, or if the majority of requests is for documents published within the last five years. It is a major consideration if retrospective materials are regularly required, or if access rather than acquisitions is to be long-term. If retrospective stock is available, are the conditions for delivery the same as for current journal articles? If the same database is available in various formats or from various vendors, which provides the best coverage and which is most frequently updated?

Is the stock limited to periodical articles only? Will users need serials, proceedings, patents, standards and specifications, books, government publications, scores, dissertations, or AV materials?

Are foreign publications available? Are documents in foreign languages included? If so, are titles, abstracts, or translations available in English? Is the language of the document clearly indicated?

Case Study 1. Arizona State University (ASU) compared the holdings of three online document delivery databases: Faxon XPress; Inside Information (BLDSC's document delivery service through RLG); and UnCover. After comparing a sample section (the "A's") of the vendors' title lists, the ASU study *projected* that 18,586 titles (56.7 percent) were unique to one database or another. Faxon held 52.6 percent of the total list and

14.4 percent of the unique titles; Inside Information held 56.3 percent of the total list with 22.7 percent of the unique titles. UnCover held 51.6 percent of the combined list with 19.6 percent of the unique portion. Together, Inside Information and UnCover held 85.6 percent of the combined list; Faxon and Inside Information held 80.4 percent; while Faxon and UnCover held 77.3 percent (Brownson 1993, n.p.).

Case Study 2. Woodward Library of Austin Peay State University made a similar comparison of CD-ROM databases—InfoTrac, *Wilsondisc*, and IAC's *Academic Index*. This study was more concerned with bibliographic access to titles owned by the library than with document delivery, but both InfoTrac and Academic Index have document delivery possibilities. InfoTrac covered the largest percentage of library-owned journals, 338 of 1,186 titles, but the hit rate of users finding citations for library-owned journals was more than twice as high for *Academic Index*. *Wilsondisc* covered the lowest number of library-owned journals and also had the lowest use rating in this particular study (Buchanan, Berwind, and Carlin 1989, 10-14).

Libraries should closely compare coverage options among database producers. Many large CD-ROM databases exceed the capacity of a single disc. When this happens some vendors sell the whole database as a packaged set, while others may offer the same database in separate segments. Arrangement may vary even among segments—some divided by subject; others, chronologically. A subscription may include the current year and a specified backfile on the same disc, or backfiles may be available separately. When such subscriptions involve "roll offs," the oldest year is lost with each revision. Some vendors also require the return of old discs as new ones are received, or the library must pay extra to retain coverage. (Davis 1993, 69) "Rolls offs" are particularly critical if the library cancels any of the publications included or withdraws backfiles, because they may have a gap in holdings as years go by if the electronic database only covers current years. Libraries considering a document delivery system based on CD-ROM products may wish to consult Peter Jacso's *CD-ROM Software, Dataware, and Hardware: Evaluation, Selection, and Installation* (Jacso 1992) for detailed instructions on comparing CD-ROM databases.

Full-text electronic retrieval is available on CD-ROM and online databases, and more recently from electronic journals. Most of the commercial online full-text databases are derived from bibliographic databases available through services such as Dialog, BRS, Nexus, etc. Document delivery can be in two forms—actual retrieval of full-text articles online, or retrieving the bibliographic citation from a database and then placing an order through one of the document suppliers for that database. The document can be sent online for full-text retrieval at a later time or shipped by mail, fax, or courier.

Databases that offer full-text document retrieval require close comparison. The major concern is whether the full-text edition is complete. Ruth Orenstein (1994, iii-v) explains that "the term `fulltext' . . . implies that an article is provided in its `complete form'—not as an abstract or summary." Complete articles are found online, but coverage is not "cover-to-cover." The entire *text* of an article is usually available, but other pertinent data may be excluded. Coverage policies differ widely from one periodical to another and from one database vendor to another. Illustrations, photographs, tables, charts, and graphs are frequent exclusions, as well as letters to the editor, editorials, book reviews, obituaries, stock quotes, legal notices, wire service stories, syndicated columns, fillers, announcements, meeting notices, calendars of events, classified sections, advertisements and indexes to advertisers, corrections, or information for authors. "Selected full-text coverage" may have other exclusions including publisher-imposed limits, such as the length of an article, the importance of the article (in the judgment of the database producer), or lack of permission from the author.

While several sources in professional literature laud online full-text retrieval as a means of document delivery (Bjorner 1990, 109; Gillikin 1990, 28; and Hawley 1992, 45-48), others disparage the reliability of this medium for regular document retrieval. Everett (1993a, 23-24) found a low number of hits (only 3.3 percent over a year-long study at Stetson University) due to limited subject coverage and limited backfile coverage, making full-text retrieval a poor choice for document delivery in a general academic library. Findings at Arizona State University validate Everett's assessment (Walters 1993a, 7). Even with the growing number of full-text databases available, it is doubtful that many libraries could rely solely on full-text retrieval as the primary means of document delivery.

Online retrieval may be the only means of acquiring some documents published electronically on research networks—particularly those published by noncommercial entities. Commercially produced electronic publications, such as e-journals, are more likely to be available in more than one format. Some are also available from commercial document suppliers, or as offline prints delivered by mail or fax from the publisher. At this time, more noncommercial e-journals are more plentiful on research networks than on commercial publications. As subscription and copyright issues are resolved, the number of commercial e-journals can be expected to rise, since the format provides almost instantaneous delivery.

D. Scott Brandt (1992, 17-18) defines an e-journal:

> In its broadest definition, an e-journal is some grouping of information which is sent out in electronic form with some periodicity. Thus an e-journal could be a listserv discussion sent by e-mail; the electronic "edition" of a jour-

nal which is also commercially printed, sent out on disk or through the internet; or a unique entity created and distributed solely electronically.

This last definition is the one usually referred to in the academic world. And it is usually implicit that the focus is on scholarly e-journals—those which require some review of the contents of each issue and are of research value. Thus a scholarly e-journal is different from a listserv discussion because of its content, and usually different from commercial editions in the way it is distributed.

The biggest problem related to the scope of electronic documents is that the content of the document can change and earlier "editions" may or may not be archived. This is no different from any other publication that goes out of print, but the chances of locating an "original" edition for a requester are probably nil.

A major publisher of full-text electronic documents is the U.S. government. Since the Office of Electronic Dissemination Services was formed to assist agencies in publishing electronically, major portions of documents collections have been converted to various machine-readable formats. The diversity includes census data; the Government Printing Office's *Monthly Catalog*; the *Congressional Record*; the Department of Commerce's *National Trade Data Bank* (NTDB); the U.S. Navy's Paperless Ship Project; the Marine Corps University's collection of war-fighting materials stored on optical discs; and a prototype multimedia collection developed for the public, the Library of Congress' American Memory Project. Many of the publications from government agencies have no print counterpart.

Unfortunately for library staff and for end-users, little standardization exists for government-produced documents in electronic formats. Many libraries may not have the budget to acquire the hardware required to access some of the files. It is also difficult for staff to master the search protocols for the full array of available software. Many libraries may opt to use a commercial supplier that specializes in U.S. government documents to provide rapid document delivery, rather than trying to develop a full-text electronic government depository collection.

Turnaround Time

Once you have narrowed the choice of suppliers down to those with the coverage desired, speed of delivery becomes the next major consideration in selecting commercial suppliers. Many ILL departments are willing to pay a commercial supplier or lease a CD-ROM product or network licensed document delivery system to provide timely alternative access to materials not owned. Those departments experience considerable discontent, however, if commercial suppliers or products fail to provide documents listed as available in a timely fashion when a library could be acquiring the same materials from reciprocal libraries. The following are considerations which relate to turnaround.

What is the supplier's standard method of shipment? If it employs the postal service, a commercial service will not likely be any faster than ILL. If fax, electronic transmission, or full-text electronic retrieval is the standard method of document supply, then fast turnaround should be achievable. In-house systems provide the fastest delivery of all. Systems such as UMI's ProQuest or ADONIS systems are as close to instant delivery as one can find.

What rush services are available? What subject expertise does the information broker's staff possess? Is the provider staffed so that it can provide high-volume rush service on a regular basis? Does the supplier's definition of rush and high volume match your expectations? (If a contractual arrangement or large deposit account is a requirement for service, ask for references from clients that regularly require high volumes of rush requests.)

Does the supplier respect "need before" dates and automatically cancel requests if unable to fill within the time frame? The "need before" date should allow for the supplier's advertised delivery time frame and the "maxcost" given should equal the supplier's advertised rates for delivery within the requested time frame. If material is available after the stated time frame, does the supplier check first before shipping?

How are requests sent over national ILL networks handled? Does the commercial supplier honor system protocols? The supplier must either fill the order within the designated time frame or consider the order cancelled if the system automatically forwards a request after the designated time frame expires. The supplier must also update records accurately and in a timely fashion. Can other lenders be listed in the lending string, or does the commercial supplier insist on exclusive or multiple listings (a sign that requests are unlikely to be filled regularly within the system's designated time frame)? If commercial suppliers do not honor the protocols of an ILL system, then using the ILL subsystem for commercial requests is likely to delay rather than speed the order process.

How are requests for retrospective materials filled? Many commercial suppliers use remote storage or off-site libraries for retrospective stock and may require several extra days to deliver articles from such stock. Are titles requiring additional delivery time identified in the supplier's catalog?

Shipping Methods

Most commercial suppliers still use first class mail as their standard shipping method, relegating telefacsimile, couriers, and overnight express to higher priced options. Newly established firms are more likely to use fax as a standard shipping process. A few suppliers employ the electronic research networks to transmit documents, either as digitized or bit-mapped images, using software such as RLG's Ariel or FTP processes to send files. The format currently used by most full-text online databases is ASCII, which reproduces only standard text. This limitation is unacceptable to patrons who need illustrations, scientific symbols and formulas, or articles in non-romanized languages.

The fact that commercial document suppliers can employ a large number of shipping methods puts them ahead of those ILL departments that primarily use first class mail for photocopies and "Library Rate" for shipping book loans. Such libraries use fax primarily for rush requests, but that is changing daily. Telefacsimile machines have not only improved considerably in the last few years, they have also become much more affordable. The capability to use Ariel or otherwise transmit documents via the Internet is also becoming more common in libraries.

In addition to the importance of the method of shipment, it is also important to know how the supplier will identify materials shipped, so that the receiving library can quickly match the material with the patron's request. Material shipped without a copy of the request or with no mention of the patron's name or a request number, slows its processing upon receipt.

Delivery Costs

Cost is the reason libraries most often give for not using a commercial supplier. Obviously, if one's criteria states a maximum cost of $15 and a vendor's minimum cost is $25, that supplier will not be used. Internal processing costs should be examined along with document costs in comparing commercial and library suppliers.

Determining cost in advance is difficult with some commercial suppliers. Most have a basic delivery fee, but what is "basic" varies widely. "Basic" usually includes shipment by first class mail within an advertised time frame. Fax or electronic transmission is "standard" for a few suppliers, but most of them reserve those methods as rush shipping options, applying a surcharge based on the delivery time requested—usually ranging from shipment within 30 minutes to 48 hours after receipt of request. The rush fee may include shipment by courier, fax, or other electronic means, but many suppliers impose an additional surcharge based on mode of shipment, especially if optional shipping modes are available for "nonrush" requests. Some services add a "processing" or "handling" charge to their "basic" delivery fee for each request—a practice which can only be considered a ploy to make their "basic" fee appear to be lower than actual!

Information brokers usually charge extra for verifying incomplete or inaccurate citations; they also usually assess a service charge for processing requests that they cannot fill. Charges may be added when a request is filled "off-site"—rather than from a supplier's own stock. The off-site location is usually a large library where the supplier sends a "runner" or has staff located. In some cases, the commercial service sends the equivalent of an ILL request to another fee-based document delivery service within a library. Thus, a library could be paying a middleman to acquire a document from a library to which an ILL request could have been sent directly! (This isn't necessarily bad, but it should be done knowingly, based upon the staffing and resources available to one's own ILL/DDS unit.)

In comparing suppliers, find out if prices are reduced when they do not make delivery within the requested time frame. For example, consider a request that specifies 48-hour, rush delivery by fax, authorizing a "maxcost" equal to the advertised rate for those options. The supplier misses the 48-hour deadline, but faxes the document anyway—possibly after the library has already acquired the document from another supplier. If the material is accepted, does the supplier eliminate or reduce the rush and/or fax fees because the material was shipped late? Will the supplier drop all charges if the library no longer requires the item after the stated "need before" date?

Single article document delivery is so new that pricing has not stabilized. Suppliers are still trying to determine marketing strategies. Original prices are often based on intuition. Articles expected to be used frequently may be provided as inexpensively as possible, while articles with low use expected may be priced higher to recover initial costs. Some publishers price articles by well-known authors higher than those by lesser-known writers. Authors have little say in the pricing of their writings, an argument which many academic institutions have used to encourage faculty not to publish in journals with inflated institutional subscription rates.

Costs are particularly hard to predict when retrieving articles from full-text online databases, unless the searcher is familiar with a particular file and knows the length of the article. Downloading the file and printing it after logging off lowers costs, but it is easy to go over one's budget in expensive databases. More database producers are lowering connect time charges and raising display and print charges to capitalize on changing searching patterns with microcomputers. Most full-text document retrieval databases have built-in per item costs covering per page print charges and copyright royalties. Volume discounts may apply, especially if several departments in the library are accessing the databases under one account.

Group discounts through consortia or professional memberships are also available, and educational discounts to academic institutions are commonplace. Discounts are not always automatic, so be sure to ask the service representative about every possibility for which a library may qualify. Discounts are also available for searching at nonprime times. Some European databases give North American searchers a discount when

European users are likely to be offline, and some American vendors do the same for foreign users. Make use of any free time offers from database vendors to become familiar with the contents and search protocols on seldom-used databases (Basch 1992, 16-18).

Some advocates of full-text databases cite low average costs for full-text "delivery" (Bjorner 1990, 110; Gillikin, 1990, 28; Hawley 1992, 48), but the rates cited often exclude the cost of searching and identifying the citation prior to the actual retrieval or delivery process (particularly personnel costs if librarians are involved in retrieval). The delivery costs may indeed be relatively low, but the total cost of acquiring the document online as compared to acquiring it by other means is the true test of whether full-text retrieval is cost effective.

Curry and Olsen (1985, 6-7) compared search times and costs of verifying and ordering documents over Dialog versus using the RLIN/OCLC ILL subsystems. They found that the fill rate and turnaround times from four Dialog suppliers did not warrant the extra expense on "easy" requests, which they defined as "those which have an easily identified periodical title with a verifiable citation." Their results might have been considerably different had they compared online verification and ordering of "difficult requests," which they defined as "publications with unclear titles or issuing bodies and citations which are difficult to verify" such as conference proceedings, publications of corporate institutions and foreign publications."

There are usually several other delivery options for easily verified documents, many more cost-effective than online retrieval or document ordering. Fewer options exist for "difficult" requests, however, so online verification or document ordering from database producers and full-text retrieval can be a cost-effective delivery mechanism.

An institution must decide when it is more cost-effective to acquire an on-site document delivery product or system (a full-text CD-ROM database, for example) rather than pay per document charges for delivery. The initial costs to acquire CD-ROM or network-licensed databases are high—hardware with massive storage capacity, laser printers, as well as the databases, must be acquired. The volume of use must be significant before a library reaches a break-even point compared with paying on a per document basis for online full-text retrieval or document ordering (Leach and Tribble 1993, 363). Although the per document costs may be higher with a "just-in-time" approach, overall costs may be significantly lower and a library will have more flexibility in using a greater number of suppliers. Cost is not the only factor in such a decision, of course; patron satisfaction, speed of delivery, and ease of use all count.

When a library wishes to offer document retrieval initiated by its patrons, it must decide between acquiring the CD-ROM version of a database or the magnetic tape version to load onto the local OPAC. That decision is usually based on the number of simultaneous users expected, whether remote access is required, the number of sites to be served, and the number and sizes of the databases to be acquired. Either choice represents large financial and service commitments. On CD-ROM workstations, probably no more than ten users can simultaneously access the same disc without excessively slowing response time. There are also limits to how many workstations can be linked through a LAN and to the distance a LAN can cover (Lippert 1992, 132).

A prime consideration is determining whether to purchase or lease a product—if a choice is even offered. It's difficult to understand why so many database producers lease rather than sell products. Subscribe to the print counterpart of a database and the library owns the product; subscribe to the electronic format, and the library may or may not own it. A cancelled print subscription does not require that you return back issues (with a few rare exceptions), whereas a database lease not renewed creates all sorts of consequences—all negative for the library.

Retrospective coverage is of particular concern with leased subscriptions. Database prices may include pre-established backfile coverage or only the most current years. Subscribers may be required to return older files when new files arrive. A library may have to pay extra each year to retain backfiles—basically, paying twice, once to lease access to the database and then again to keep any portion not included in the next year's lease for "current files." Subscribers may be able to acquire backfiles separately through a one-time purchase of an archival set, but publishers may charge to continue the software license and customer support if the current subscription is cancelled. The cost of updates could be included in the subscription rate or disguised as a shipping charge. Some databases are divided into separately priced segments, which can represent a cost savings for libraries that do not need a complete set. Advertised prices do not always reflect added charges or possible discounts. Discounts are often available for multiple copies shipped to the same address; for subscribers to associated journals, related databases, or equivalent print products; or for libraries with low budgets or small collections.

Maintaining a subscription to the print equivalent of a database or to journals cited is required by some database-specific document suppliers. In a recent study by Cahners, 57 percent of 608 libraries surveyed carried the print version of databases to which they also subscribe on CD-ROM. (Berry 1992, 46) The same percentages probably apply to libraries that have licensed magnetic tape versions of databases. Maintaining print subscriptions may be acceptable when a full-text CD-ROM or network-licensed database primarily supplements traditional access, but this would not be a viable option for any library that uses online retrieval as an alternative to subscribing to expensive, infrequently used journals. The one advantage to the confusing array of pricing options on licensed databases is that almost everything is negotiable; unfortunately, it must all be renegotiated every time the lease expires.

Copyright Compliance

Publishers have relied for years on subscription agents to sell journals, and one would expect that publishers would want as many commercial suppliers as possible to be selling single articles. Unfortunately, publishers are much more territorial when it comes to granting distribution rights to single articles within their journals, a fact beyond the understanding of most librarians and library users. One-stop shopping is difficult when titles bounce in and out of document delivery catalogs and databases as fast as tennis balls at Wimbledon! The many shifts in distribution rights are undoubtedly linked to the collection and payment of copyright royalties.

Copyright royalties represent a cost factor that is difficult to determine in advance. The fees are not always listed in supplier's catalogs and royalty costs vary remarkably. Fluctuation in royalty fees plays havoc with document delivery budgets. Try to select vendors who list the total cost up front.

Commercial suppliers' policies regarding royalty payments vary, and a financial advantage may accrue in selecting a vendor based on copyright fee policies, assuming other charges are equal. Some charge a flat royalty fee based on average rates ($2 to $10 per copy); others include royalties in their base rate. Most commercial suppliers allow publishers to set their own royalty rates, which currently ranges between $7 and $10 on services such as UnCover (Long 1994—letter).

Publishers set the royalty rate independently for articles in each journal. It's not unusual for articles within an issue to have different fees. Variations between articles are most often based on the number of pages, but some publishers try to estimate demand based on the author's reputation or the subject. Such factors do not cause variations in per issue rates or subscription rates, so it is annoying that publishers can't settle on a standard royalty fee regardless of whether the articles they publish are best sellers or best forgotten!

Maurice Long (ibid.) estimated that with an average subscription rate of $225 for the BMJ Publishing Group's journals and with royalty fees for the three most recent volumes at $7.50 per article, BMJ would need to receive at least thirty royalty fees to recoup the cost of one cancelled subscription. Since BMJ (*British Medical Journal*) reduces the royalty fee to $2.50 for articles published more than three years hence, it would take even more single article sales to recoup costs if this did not occur in the first three years. Long goes on to state that if the fee were to slip to a nominal one dollar, it would take 225 royalty fees. He estimated that the average journal issue has 10 articles, so it would take approximately 22.5 copies of each article to recoup costs through royalties alone. That level of activity may be expecting a lot, since a typical article has no more than twelve readers (Dyer, 1992, 33-35). Even if all of the ten articles per journal used were from copies rather than originals, chances are that most of the copies fell within current fair use guidelines, so few royalties would have been paid. These statistics, coupled with current U.S. CONTU guidelines, explain why a few publishers try to recoup the cost of an entire subscription with every article sold, and why many publishers are wary of depending on royalties from single articles ranter than annual subscriptions.

On the other hand, single article sales plus revenue from royalties for copies acquired, are likely to increase when delivery charges *and* royalties are within the budget of the average reader. It seems unlikely that royalties will ever "slip to a nominal one dollar," but if they did let's surmise that it occurred at the same time that "fair use" required payment of the nominal fee on every copy (no more "rule of five")! Using Dyer's typical twelve readers per article and Long's average of ten articles per issue, a monthly journal would average 1,440 articles read per year. Even considering that some of the articles would simply be read, not copied, the publisher would recoup the cost of a $225 subscription five to six times each year from royalties alone. If most of the copies were acquired from commercial document suppliers with distribution rights granted by the publisher, the publisher would clear even more based on annual licensing agreements with each document supplier.

The average ILL/DDS unit might cringe at the thought of budgeting one dollar per copy based on annual statistics. But libraries that invest in just-in-time access to seldom requested journals would be more fiscally sound than those libraries that adhere to a just-in-case acquisitions policy. It is a scenario that comes much closer to satisfying the needs of publishers, libraries and readers than current practice produces.

A major difference between commercial suppliers and library suppliers involves just when copyright royalties are due—up front with commercial suppliers, or after "fair use" guidelines are exceeded with library suppliers. When materials are acquired from libraries, the receiving library is responsible for maintaining records and paying copyright royalties when appropriate. Conversely, commercial suppliers must pay royalties on every copyrighted document provided to clients; publishers do not give commercial suppliers the right to distribute their stock unless royalty fees are included. They are especially leery of the possibilities of copyright abuse where electronic document delivery and remote access are available.

Understaffed libraries, those not inclined towards record-keeping, or those that get hassled by the accounting department over small invoices, are best advised to use commercial suppliers. Paying at delivery is also more cost-effective if royalty payments are likely for the majority of documents acquired—often the case for corporate libraries. Libraries that charge back fees to requesters may also prefer to have the royalty included in the delivery fee. On the other hand, libraries that rarely exceed the "rule of five" may find that annual delivery costs are raised unnecessarily by acquiring documents from commercial suppliers with royalties automatically assessed. It is more economical (if all other costs are equal) for such libraries to use suppliers that allow copyright payments to be made by the library. Other than fee-based services in libraries, however, few commercial suppliers provide this option, which is why so many libraries still prefer ILL.

Ease of Accessing Citations and Ordering Documents

The methods by which a commercial supplier's inventory can be accessed and the methods by which document requests are initiated are vitally important. CD-ROM databases often have different search interfaces, which can frustrate users and staff. Centrally mounted databases, on the other hand, are more likely to use the same interface as the OPAC.

The user should be able to readily identify the online database in use. Sometimes access is so "seamless" that patrons think they are in one database when they are not—particularly at public access terminals where users fail to return the terminal to the OPAC's "home screen." Users should know whether a database is primarily used for alternative access, rather than as an index to materials likely to be in the library. Ideally, patron-directed document delivery services should be integrated with a library's own holdings, offering options to order from the commercial suppliers offered or directing the request to the library's ILL/DDS service to acquire elsewhere or deliver from the library's own collections. This should be done without the user having to switch between the database and the library's OPAC; the end-user should be able to capture citations and automatically produce either an ILL request or a commercial document delivery order. If database contents are not full-text, enough information should be given about the document to allow the end-user to assess relevancy, either through the provision of abstracts or subject headings.

Online systems should be easy to use, requiring little staff intervention or prior instruction, although help should be readily available. Search commands should be appropriate for the users' level of expertise. Instructions for searching and ordering online should be clear—either menu-driven or with help screens. Instantly accessible help screens at the point where help is needed are preferable to those in separate tutorial files. Help screens should provide more than definitions of terms; they should also offer examples and solutions to problems.

Other concerns afflict searching of online systems. Will free-text searching provide relevant hits? Is the text of documents searchable or only certain "fields?" Author, title and subject (word) searches are standard, but other search fields vary considerably. Databases that use ASCII full-text may have searchable text, but no software yet developed can search text from bit-mapped page images (Leach and Tribble, 1993, 363). Is a thesaurus available online if a "command language" is required? Is proximity or Boolean searching or truncation possible? How can searches be narrowed or limited? Sort and retrieval speeds should be fast. Can searches be saved or downloaded? Can search results be manipulated by other software programs?

What is the system's track record for "downtime" due to updating or malfunctions? Will the database be accessible at all hours that the library is open or only when library staff is available? Is remote access available? Is a password required?

Examples of some systems that provide remote access to document delivery databases include the CARL Corporation Network, Faxon Xpress (Faxon Research Services), OCLC's ArticleFirst and ILL Link, RLG's CitaDel, and IAC's Expanded Academic Index. Examples of on-site access systems include UMI's ProQuest MultiAccess Systems and ADONIS (Leach and Tribble 1993, 359-362).

Other considerations come into play with staff-mediated document ordering. ILL/DDS order options should fit into the unit's daily workflow. If a library regularly uses OCLC, RLIN, or UnCover for search and verification, then document suppliers on those systems may be preferred, especially if other department processes, such as statistical reports, are automatically generated. Conversely, libraries seeking to reduce overall OCLC or RLIN costs might prefer commercial services with order options requiring less search and verification. Even libraries that do not subscribe to the services of the major bibliographic utilities can still use such services for document delivery purposes, since most can be reached by TELNET via the Internet. Access and delivery via the Internet is certainly a consideration when selecting a commercial supplier.

If document suppliers accept requests by telephone or fax, is there a toll-free number available for ordering 24 hours per day? Are there surcharges connected with phone or fax ordering? Does the supplier prioritize requests based upon the method by which the request is sent? Is there a required format for sending requests? If so, does it vary from the format used to send other ILL/DDS requests?

Staff Requirements

A service that provides for patron-directed ordering should be considerably user friendly. What type of instructional materials, orientation, or assistance will users need? What type of staffing is required to assist patrons either to use the system or maintain equipment (adding paper or toner, changing printer ribbons or discs, rebooting or troubleshooting when the system goes down). Installing CD-ROM or OPAC databases usually frees reference staff from mediated computer searching while expanding the number of users who have access to databases. In turn, increased access adds to public service duties related to assisting end-users and equipment maintenance. While some such duties require professional staff, others are more appropriate for technicians or office assistants. If the requests volume is high, selecting a document delivery system that requires mediated document delivery means extra staffing. If new equipment is required to process requests, how much time will it take to train staff? Using a service with familiar search and order processes is preferable to one requiring extensive staff training and "refresher courses" to maintain proficiency. Look for vendors that offer product training and support services.

Equipment Needs

Preferably, the document suppliers selected will be those from whom accessing, ordering, and receiving requires equipment already owned— but that may not be the case. Access to a supplier's inventory may entail checking a printed catalog, a national online bibliographic database, a database on a CD-ROM workstation, or even using TELNET to access a catalog on the Internet. Many commercial suppliers, particularly those that use CD-ROM or online databases or networks, have vendor- or software-specific requirements. A variety of options may exist, including site licensing of vendor specific software, access via the Internet, or gateways through other vendors. Some CD-ROM document delivery systems require a LAN "jukebox" system. Each system providing patron-directed document delivery will have specific requirements for equipment in public areas based on the number of simultaneous users. If all document delivery requests are to be funneled through ILL/DDS, then less equipment is needed. If access to files on the Internet is needed then a microcomputer and an Ethernet cable or other type of connection to the institution's avenue to the Internet will be required.

System-specific software may also be needed to receive documents. For example, documents ordered over RLG's CitaDel system cost less if the supplier transmits the document over the Internet, but the receiving library must have RLG's Ariel software. A library does not have to be on the RLIN system to use Ariel, however, and the same software can be used to receive and send documents to any other Ariel library. To do so requires the latest version of Ariel, which operates in a environment requiring at least a 486/25MHz PC with 4 expansion slots, 8 megabytes of RAM, an 80-MB hard disk or better, an Ethernet connection, and a compatible scanner and laser printer. Send-only and receive-only versions of the software are available for libraries with multiple workstations, requiring only a scanner or a printer respectively (Coleman 1994—letter).

These equipment requirements are typical for receiving documents electronically regardless of using Ariel or another vendor-specific software package, especially with multi-tasking workstations. Nearly all electronic retrieval systems originated with lower requirements for hard disk and RAM capacity have proven too slow, deficient in storage capacity, or disruptive to other programs being run of multi-tasking workstations. Be wary of a system using anything less than the requirements cited above without a guaranteed, money-back trial period.

Libraries that wish to acquire materials via fax need the most affordable fax machine designed for high volume use (although price is not synonymous with quality). A plain paper machine with a laser printer and large paper bins is a minimum requirement. Is an easy-to-use computer interface option available to expand the memory capacity? A computer interface may also be necessary if one wishes to receive documents in the library and forward them to requesters. Is the equipment designed for library processes (not office procedures)? Before purchasing an expensive product-specific computer interface, be sure that the vendor includes staff training, customer support, and a trial period.

Customer support from the supplier is crucial regardless of which automated document delivery products are acquired. A total customer support package would include help in installing equipment, staff training, well-written user manuals, tutorials, site visits from customer reps, plus a toll-free help line. Maintenance contracts should be available with repairs guaranteed within a designated time. Equipment loans should be available if the system is likely to be down long enough to have a detrimental affect on service or workflow. References should come from other libraries, copy services, or document delivery services, not from the general business community.

Accounting and Billing Options

Almost as important as the cost of individual documents is the method of payment. The irony behind low-cost document delivery is that the payment transaction may cost the institution more than the actual document. As far as the authors are aware, no ILL cost study has taken into consideration the internal costs for paying by check (other than staff time), but various accounting departments have suggested that the average cost to "cut a check" ranges from $30 to $50, which includes personnel, and bookkeeping and bank charges. That is why libraries, when dealing with other libraries, tend to favor reciprocal agreements or coupons rather than an exchange of funds. Colbert (1979, 78) stated that "the most highly favored of any payment option is one that allows a user to order material and be billed later. The obvious advantage is that there is no delay in processing the order."

The most common billing and payment options of commercial document suppliers consist of:

- payment upon receipt—invoice accompanies document (preferred) or follows shortly thereafter
- invoicing (monthly or quarterly)
- deposit accounts
- prepaid coupons/blocks/units/vouchers
- payment in advance
- licensing fees
- credit cards
- annual membership fees (which may or may not include all document delivery costs for the year).

Payment upon receipt is most common when dealing with library suppliers. Libraries prefer to know the total cost at the time they order a document, or at least when they receive the item. Knowing the costs is especially important when the patron is expected to pay. Even when "invoice follows" is indicated on the paper work accompanying the document, a supplier that indicates the item cost is preferred over those that do not. This type of billing allows a library to try a supplier without making a large financial investment. It also provides flexibility in tracking down elusive material from specialized suppliers.

Monthly/quarterly invoicing is another method of payment upon receipt, allowing small payments for single items to be merged into one payment—a definite convenience when suppliers are used regularly. Invoices should itemize each transaction, providing at least an order number and cost for each one. Suppliers that give discounts based on volume of requests often do so based on a month's transactions. Some document suppliers are better known for their turnaround time than for their accounting, so invoices do need to be checked thoroughly. Common errors include billing twice when a poor quality copy was replaced; billing for items requested, but never shipped; and billing for rush shipping when the item was not received within the time requested.

Some vendors require deposit accounts—in some cases, minimum amounts apply. Per-item delivery fees are often less with deposit accounts, but it is probably safest to pay the higher item-by-item fees for a trial period, until a library is ready to commit to regular use of a supplier. Frequently, deposit account clients receive extra services. (The Genuine Article, for example, provides an assigned customer service representative and waives additional copyright royalties for documents over 10 pages long). Even with a deposit account, a supplier should provide a record of the per-item cost at the time of shipment (unless a standard fee is charged) and an itemized monthly statement. Online document suppliers, such as

UnCover, Dialog and BRS, often provide online account reports which the library can pull up, as needed.

Prepaid coupons, order forms, or "blocks" of searches and/or documents comprise another type of creative accounting. While managing the coupons (filing, remembering when/how to use them, distributing to end-users) can be a nuisance, this type of payment offers advantages, especially when end-users are to be allocated portions of the document delivery budget. RLG's Citadel databases and OCLC's FirstSearch databases offer "block" purchases as a method of payment—the more blocks purchased at one time, the lower the per item cost. BLDSC utilizes coupons on mail requests.

Coupons are often used as a method of payment for international transactions to avoid problems with currency exchange rates. Coupons can protect against inflation, although if rates change dramatically, as when national currencies are devalued, some suppliers may simply require two coupons instead of one—inflation at its worst! International reply coupons (acquired at main post offices at very reasonable rates) are often accepted as payment from international suppliers (particularly libraries). The International Federation of Library Associations (IFLA) recently initiated a voucher scheme of payment, similar to that long used in Australia. Reusable vouchers can be purchase from IFLA. Net borrowers buy as needed, while net lenders may cash in surplus vouchers, thus eliminating most of the accounting and currency exchange problems related to international lending and document delivery (Gould 1994, n.p.).

Payment by coupons is problematic in some libraries. Purchasing agents and accounting clerks tend to question their use. The coupons are easily lost, forgotten, or even thrown away by new staff who think they are cleaning up "clutter," because most coupons do not resemble script. Staff opening up the mail must be trained to recognize coupons as "currency," not junk mail.

Payment in advance is the least favored method, because it adds days and internal costs to the delivery process. Except in small libraries where bills may actually be paid within ILL/DDS, pre-payment often requires hand carrying an invoice through the entire accounting process to get a check to enclose with the order request. Prepayments must often be justified to accounting departments (who have their own priorities and are unlikely to be concerned with one patron's request). Only a few library suppliers (e.g., the Library of Congress Photoduplication Service, New York Public Library Photoduplication Service, and some national libraries that are international document supply centers) require prepayment. Advance payments are almost always required whenever an ILL/DDS department deals directly with a publisher, and frequently when using commercial suppliers on an irregular basis.

Licensing fees come into play when a CD-ROM product or online data-base is acquired. Usually, all searching is free, and even some full-text retrieval may be included in the licensing fee. More likely, additional per page print charges and copyright royalties will apply. An annual licensing fee may, therefore, not cover the entire document delivery fees for the year.

Other suppliers, particularly those designed for end-users, may require payment via a credit-card (usually Visa, MasterCard or American Express). Departmental credit cards, which could eliminate numerous small checks and international bank notes, are often anathema to comptrollers who fear misuse. For patrons printing or downloading documents on library equip-ment, coin boxes may need to be installed. Maintaining coin boxes and security add to overall delivery costs.

Some suppliers sell "debit cards." (A few accept "universal" multifunc-tion debit cards, others issue the cards only for document delivery.) Debit cards have been known to cause problems when the item requested costs more than the card's available balance. The order for the document will be taken, and all available funds are deducted. The patron is then given an item number to use to re-order the document after adding additional funds to his or her debit account (Stromquist 1993, n.p.). It would be sim-pler to just stop the transaction due to insufficient funds rather than start-ing the process with only partial payment available.

Only a few suppliers, such as WISE for Medicine ($100) and Universal Serials and Book Exchange ($150), charge annual membership fees. Both of these services also charge a quite reasonable per-item delivery fee. The Center for Research Libraries, a resource sharing consortia, has an annual membership fee that varies depending on the type of membership and a formula based on the number of volumes owned and a library's materials budget. CRL fees range from a mean of $896 for "User Members" to a mean of $217,693 for "Voting Members." (CRL 1992) The fee provides, among other things, "free, unlimited" document delivery from a stock of hard-to-find titles, seldom owned or loaned by other libraries. Nonmembers may acquire up to ten documents per year from CRL by pre-paying a flat per-item delivery fee.

Value-added Services

Various commercial suppliers have "value-added services" designed to put them ahead of their competition. They may offer discounts for high volume usage, a set copyright fee, dedicated customer service represen-tatives for libraries with deposit accounts, copyright clearance services, customized billing, or drop shipping (shipped directly to the library patron, billed to the library). The possibilities are too numerous to list. If the service is indeed beneficial, then value-added services may be the deciding factor in your choice of suppliers.

References from Other Library Clients

Check with other libraries before opening large deposit accounts, pay-ing licensing fees, or acquiring expensive document delivery equipment. When checking references be sure that you are not comparing apples with oranges. Performance seems to vary considerably for different libraries using the same suppliers, depending on volume of requests, turnaround requirements, whether a deposit account was established, distance between the library and the supplier, methods of sending requests and receiving materials, and how long ago the service was tried. Many libraries use the ILL-L Listserv as an excellent resource for current comparative information. If the conditions in one's library are similar, published cur-rent comparisons may also be helpful.

Preparations

Once a library determines which commercial suppliers best meet its needs, it can begin the actual in-house preparations to use the selected services. How involved this preparation is depends entirely on the formats in which requests will be sent to and materials received from commercial document suppliers.

Establish Relations with the Commercial Document Supplier. Find out the name of your service representative and get a direct tele-phone number (preferably toll-free and by-passing automated call routing systems). Explain your service objectives in using the company or its doc-ument delivery products. Determine if requests must be submitted in a vendor-specific format. Openly discuss any negative references to cir-cumvent similar problems. Ask for a free or discounted trial period, espe-cially if staff must be trained on vendor-specific equipment or techniques.

Open an account with the supplier, setting up any required deposits. The amount of the deposit may affect delivery fees, but unless the firm has a well-established track record, a minimum deposit is advised.

Acquire, install, and test equipment required for the service(s) selected. Sometimes standard office equipment and supplies are all that is needed, but you may need to add or upgrade equipment. Delays are not unusual if installation must be subcontracted. For example, so many offices need Internet connections that telecom crews in some areas are up to six months behind in "hard wiring" buildings. Delays may cause war-ranties on new equipment or "introductory discounts" to expire before the total installation is ready to test.

Train staff on new equipment. Revise work procedures to incorpo-rate new technologies or processes. The order process with commercial suppliers may be different from an ILL order. Learning what works and what doesn't will impact service.

SUMMARY

Selecting appropriate commercial document delivery alternatives to complement a library's interlibrary loan program is complex. Choices consist of article clearinghouses; information brokers, which may include large private sector companies of full-service information providers with employees strategically located around the world; fee-based information retrieval and document delivery services within libraries; and small home-offices delivering resources from a few local libraries. Online document delivery services are offered in various formats by bibliographic database producers or vendors that can be accessed by users at single terminals or simultaneously over electronic networks. Service may be bought directly, leased, or used via gateways and dial-up access, or acquired on a per-item basis from secondary suppliers that have access to systems or products beyond those in one's library. Some products are directed at end-users, others best-suited to library mediated document retrieval, and others for either environment. There are suppliers for specific subject areas and generalists with huge collections. Virtual collections of full-text electronic documents available from publishers and libraries are increasing. Each library must decide from the diverse assortment which commercial services meet its needs for alternative access.

Chapter 5

Evaluating Document Delivery Service

"Information industry, meet your new part-
ners—or your successors." (Quint 1993, 74,
quoting an introduction of a panel of librarians
at the Information Industry Association's annu-
al meeting, December 1992)

Finding a quote to begin this chapter was not easy. It was tempting to use the *Field of Dreams* aphorism, "If you build it, they will come." Document delivery is now such a driving force of the information industry that almost everyone is either developing a product or service or looking for one to acquire; but, once it is built and they come, will they stay?

The veterans of the document supply industry no longer enjoy a monopoly, and no one in the industry can afford to sit back and hope their service or product is good enough to attract and hold customers. Libraries form a huge potential document supply market, vital to the many companies competing for the business. Those companies that treat libraries as partners and are eager to develop products that are user-oriented and fairly priced will prosper, while those that are less service-oriented may soon lose out.

EVALUATIVE CRITERIA

What are the options if the chosen product or document supplier turns out to be a nightmare instead of a dream? The previous chapter dealt with evaluating a service or system prior to acquiring or using it. What does one do to monitor, evaluate, or tweak a system or supplier after it is in use? All

117

document delivery services can be evaluated, regardless of whether it is an online or CD-ROM document supply product acquired for patron-directed document delivery, a commercial information broker, a document supply clearinghouse, or another library (reciprocal or charging). To some extent, the same criteria used to select a document supplier can be used to determine whether a vendor is meeting expectations. Most performance evaluations revolve around the following issues:

- turnaround time
- cost of documents—usually features the average cost/item delivered, but may include overall costs to use service or make it available to end-users
- fill rate
- quality
- reliability
- user satisfaction
- value-added services.

The same criteria can be adapted to evaluate internal document delivery services such as a fee-based document delivery service within a library, or a document delivery system or network between branch libraries or member libraries in a resource sharing consortium. Cost evaluations, in such cases, are based on an internal cost analysis, whereas the actual document cost is more significant when comparing external suppliers.

Turnaround Time

Turnaround time measures the elapsed time between the initiation of a request and its fulfillment. When to start and stop the count, however, depends upon who is counting and the purpose of the data. A requester measures turnaround in the broadest sense—the total time from when a request is submitted until the material is in hand (weekends and holidays included). Total turnaround is what most libraries strive to improve, but in the eagerness to show improvement, there are myriad ways to control measurements.

For starters, an ILL/DDS unit will likely define turnaround as the time from which the request was logged in until the item is received, possibly including patron notification, but probably excluding the time until pickup. Units that deliver to end-users rather than relying on client pick-up are more likely to include the delivery phase in an appraisal. The receiver will evaluate a specific supplier's turnaround by measuring from the time the request was sent until the material arrives, while the supplier will count from the time the request was received until the material is shipped (or ready to ship). Nonbusiness days may or may not be counted, depending upon the tracking system employed. Add the quirks of automated tracking systems and differences in time zones, and it's fairly easy to add or subtract days in an amazing number of ways—all being totally accurate variants of turnaround!

Automated ILL tracking systems are particularly adept at providing deceptive turnaround data. Libraries that collect requests all day for batch processing may begin the batch log-in process at the start of the next business day. To the patron, it is Day 2, but to an automated system, it is Day 1. On the day that a supplier updates a record to "shipped," the sender considers the request filled even if it is Friday afternoon and the item doesn't leave the premises until Monday! A document faxed on Friday after 2:00 p.m. from a West Coast supplier probably won't be counted as received by an East Coast library until Monday morning; on the other hand, documents transmitted electronically from Australia to the United States are often received the day before they were shipped, thanks to time zone variations!

Some ILL record-keeping software packages do not account for multiple lending strings or resending unfilled requests to additional suppliers. For example, a library on the OCLC ILL subsystem may list five potential suppliers in a lending string. A request filled by the fifth lender in the string might be "automatically" assigned a turnaround that included the days in which the request was floating between the four previous non-suppliers—up to twenty days.

A request that went unfilled by all five suppliers could then be sent to five new potential lenders. If the first supplier in the new set filled the request, turnaround might be "automatically" recorded as low as one day. Whereas twenty or more days would reflect true turnaround for the request, it would not be an accurate assessment of the supplier's performance—one to four days might be correct for the document supplier, but would not reflect total turnaround for the request!

In the same example, if the software treated the "re-send" as a new request, then it would record that the request was filled by the first supplier, rather than the sixth. (An added controversy among libraries regards how to count requests automatically forwarded to multiple libraries. If mailed to each of the six libraries, six sends would have been tallied, but with automatic forwarding, the request was only handled twice, so an automated system will likely count it as two sends.) Software which downloads from other systems, such as OCLC or RLIN, may not include data from requests filled in-house or requests filled by suppliers outside that ILL system, such as mailed, faxed or e-mailed requests, unless such data is input separately.

Some libraries need detailed statistics, others do not. No department should do more record keeping than is necessary for efficient management. Automated record keeping can free staff for other tasks, but be aware of the quirks when interpreting automated turnaround data! Judge suppliers only on the aspects of the process which they control, although it may be difficult to totally eliminate all other factors, except under controlled experimental conditions.

If the tracking system employed doesn't recognize the vendor's place in the lending string, some other method will have to be devised to accurately reflect turnaround. If a copy of the request is sent with the material, the actual dates when the supplier received the request and shipped the material, and the date of receipt, can be accurately derived. An alternative is to randomly assign all vendors to various places in the vendor string to balance out over time—some online ILL systems such as DOCLINE automatically do such "load leveling."

When in-house processing time is included in comparative results, either the actual time spent processing each request has to be carefully controlled, or the study must cover a long time period and high volume to account for differences between easy and difficult requests. Comparisons should also indicate if weekends and holidays were counted, or only work days.

Some libraries use commercial suppliers, particularly full-service information brokers, only for difficult, unverified requests. Suppliers that accept unverified requests are not as likely to have consistently fast turnaround as those that only process requests for materials verified as available in their own catalog.

The method of transmitting requests directly affects on turnaround, as does the method of shipping the material.

The following methods are those most commonly used:

- Electronic transmission and/or delivery, including online transmission, telex, and e-mail
- Mail service
- Telephone
- Telefacsimile
- Special delivery services—commercial delivery services with regular and overnight services (e.g., UPS, Purolator Courier, Emery, Federal Express), and numerous local delivery and courier services, plus same-day postal services.

Online ordering over OCLC, RLIN, WLN, ARTEL, DIALOG, BRS, etc. provides almost instantaneous transmission, although receipt depends on how often the supplier downloads requests and updates records. "A major advantage of using electronic mail systems linked to online database producers is the elimination or reduction of errors which naturally occur when a cited reference is copied or recopied." (Allen and Alexander 1986, 39)

Electronic transmitting speeds requests, but then timely processing is up to each supplier. Some ask to be listed more than once in the lending string—a clue that service may not be consistently fast. If understaffing is the reason behind multiple listings, that supplier's turnaround evaluations will almost always be slow. In many cases, multiple listings are requested because much of a supplier's inventory is off-site as, for example, libraries in a multibranch or multicampus system. Turnaround is usually within the network-prescribed time for materials available from the main library or main campus, but more time may be needed to get documents from other libraries in the system. (Turnaround inconsistencies sometimes relate to the subject being acquired.) Less validity exists for multiple listing among commercial suppliers who advertise average turnaround of one to four days.

When a commercial supplier requires multiple listing, it indicates understaffing or that it depends on off-site locations (usually large research libraries) to acquire materials. In some cases, current stock may be available within one to four days, but retrospective material is held off-site and requires more time. Valid reasons for requesting multiple listings are when a supplier serves as a referral center for other libraries in a network or when it accepts unverified citations with no known potential supplier ("blind" requests). If possible, blind requests should not be included in turnaround evaluations as they skew results; if included, the evaluation should indicate the inclusion of difficult, problem requests.

Some studies indicate that commercial suppliers do not respect the protocols of national online ILL systems, such as OCLC. Records may be updated to "will supply" almost as soon as they are downloaded, thus preventing automatic forwarding to the next supplier if not filled in the allocated time; but then the supplier may fail to ship the document or provide a timely status report. Other commercial suppliers tend to ship materials after requests have been forwarded to other suppliers in the lending string, sometimes after the item has been received elsewhere (Leath 1992, 4; Walters 1993a, 6). If staff members can send requests to commercial suppliers as well as libraries on the same system, sending requests is simplified, but those vendors selected that do not follow the same protocols as library suppliers are likely to have poor performance evaluations.

The slowest method of transmitting requests and delivering materials—postal service—remains the most heavily used by libraries and commercial document suppliers around the world. No alternative exists for library suppliers in some areas, but commercial document providers should offer a wider range of options for receiving requests and sending materials.

"Rush" requests can be transmitted quickly by telephone (assuming few busy signals or unanswered calls), but that method is expensive if long distance business rates are involved. Suppliers that offer a toll free line will rate higher in comparisons. Telephone requests are subject to spelling and enunciation errors. Many libraries do not have the staff available to accept document delivery requests over the telephone; this gives commercial suppliers an edge, but some suppliers charge extra for telephone service.

Faxing a request is usually faster and more accurate than a telephone request. Modern fax machines can store requests throughout the day and transmit them in a batch during nonbusiness hours when telephone rates are lower. A supplier with 24-hour telephone/fax service will rate higher than hose inaccessible due to varying business hours in different time zones. Telefacsimile is an efficient way to transmit requests to any supplier not linked to a major online bibliographic utility, so the library should evaluate the availability and effectiveness of this service option.

Suppliers that ship material electronically, either by fax, via the Internet (using Ariel, for example), or over other online networks, usually have fast turnaround. If not, the cause may as likely be on the receiver's end and, if so, should be taken into account in turnaround evaluations.

Books, microform and media materials can only be shipped by mail or courier. Most studies show that courier services, such as UPS, Purolator, and Federal Express, transport materials faster than the U.S Postal Service, but delays may occur at both the supplier's and receiver's end if such delivery modes are not part of the daily routine. More paperwork must be done at the supplier's end and the receiver must have staff available to receive the courier delivery. For example, if overnight delivery were requested and an item is shipped on Friday, the library may not be staffed to receive materials after standard business hours. The item might not be re-delivered for several days, or pick-up may be required. Besides adding needlessly to shipping costs, it inconveniences the supplier when a library requests rush handling but is not prepared to receive the material expeditiously. If it foresees any problems, the library should include instructions for the supplier advising when the receiver can accept rush deliveries.

The distance between the requesting library and the supplier affects speed, unless electronic transmission of requests and materials is the standard method of operation. Libraries in fairly close proximity often form a network using local couriers and/or electronic transmission to speed turnaround. Many networks give priority processing to member libraries.

Another factor affecting deliveries relates to the supplier's inventory source. When documents have been electronically scanned and stored and the receiving library is set up to receive scanned images electronically, delivery time can be very fast. If the supplier must pull and photocopy an item prior to sending, turnaround is slower. If the supplier depends upon materials stored in an off-site location, particularly one that is not in a controlled "closed stacks" environment, turnaround will be even longer. If the borrowing library supplies the lending library's call number (including location symbols), time may be saved on the lending end. Paying attention to location symbols in the call number of a potential lending library will alert the borrowing library to materials unlikely to be supplied, such as books in reference or special collections. Select a different supplier, thus averting an unfilled status and the necessity to resend to another supplier.

The actual speed of delivery may be less critical than consistent and predictable delivery time. Just knowing that a patron can expect to have a document arrive within a promised time may be sufficient in many libraries.

Cost of Documents

Suppliers' fees should be competitively priced for the value given. The service chosen may not necessarily be the least expensive, particularly if it offers added value in terms of convenience, rapid delivery, and customer satisfaction. Cost factors include the following: order costs, supplier's handling charges, transmission or delivery costs, copyright royalties, internal processing costs, and start-up or annual fees for some document delivery services, memberships or licensing agreements.

If a certain supplier or mode of service requires dedicated equipment for its use, then those costs should be factored into overall comparisons, but pro-rated based on the equipment's expected service life. Personnel costs should be considered only if the time involved in processing requests varies greatly between document suppliers, or if different staffing levels are required to process requests. Personnel costs are a factor if trained professional staff are required to process document delivery requests from some suppliers (such as online full-text retrieval from a database vendor), while student assistants or classified staff are able to process requests sent to other suppliers.

In the ARL/RLG ILL Cost Study (Roche 1993, 4), the average cost for two research libraries to process an interlibrary loan request is almost $30 ($18.62 for the borrower and $10.93 for the lender). If a commercial document supplier can provide an item faster and/or at lower costs, then it benefits both the borrowing library and potential lending libraries to use a commercial service instead of ILL. By doing so, the potential lending libraries are saved all indirect costs. The borrowing library may not save indirect costs unless the method of accessing holdings and ordering from a commercial supplier requires little internal processing. Thatis what should be compared.

When to use a commercial supplier rather than a library is not an easy question to answer. Delivery fees are most often based upon the following costs: pulling and photocopying (or scanning) articles, making copies (staff and paper costs), preparing items for shipment, and shipping charges. Indirect costs—online fees, costs, equipment, reshelving, wear and tear on the collection, etc.—are normally absorbed in the interest of "resource sharing" among libraries, but commercial suppliers include all indirect costs in their pricing structures.

Copyright royalties on items sent between libraries are paid only when the amount of material exceeds "fair use" guidelines. The number of articles acquired in excess of fair use is relatively low. Some libraries do not even process requests that exceed copyright guidelines, and most journal issues are rarely requested more than once or twice each year, as shown many times in copyright records kept by U.S. libraries, by commercial suppliers, and in international networks.

Even the subject may affect the cost of items acquired from both commercial and library suppliers. Some subjects tend to have higher internal processing costs because they are more difficult to verify and locate among lending libraries. Commercial suppliers who can readily supply "problem requests" at a reasonable cost are worth their weight in gold. Some subjects almost always incur a lending fee from nonreciprocal libraries—another case where it may be beneficial to use either an information broker specializing in subjects for which one's library is regularly assessed lending fees.

Another option is to acquire access to a full-text CD-ROM or online document delivery system that covers the journals most frequently requested. For example, a library affiliated with an institution that does not have a medical library, may still receive a large number of health-related requests. The library is unlikely to have many reciprocal agreements with medical libraries, so ILL lending fees would be incurred regularly, and the chances would be higher of exceeding fair use and owing additional royalty fees. Many medical libraries give priority to ILL requests from other medical institutions which slows turnaround on requests from a nonmedical library. Faster and less expensive service might be available commercially—either by ordering single articles from a document supplier such as Wise for Medicine, or by investing in a CD-ROM product such as Adonis.

The number of pages in an article may affect the cost of copy requests. The number of pages covered in a supplier's "basic" fee varies widely. Most commercial suppliers charge extra for anything above ten pages, while reciprocal libraries may provide as many as fifty pages at no additional charge. Suppliers using fax or other types of electronic transmission may lower the number of pages included in basic fees.

Many suppliers lower turnaround times and costs by using local couriers to deliver materials between nearby libraries. While national carriers are usually more expensive than library rate postage, fees vary considerably. Local couriers with routes restricted to a relatively small delivery area can bid quite competitively for regular deliveries between libraries in a local network or system. It pays to check around. Even large national and international carriers may be competitive when bidding on a daily delivery contract.

Another cost factor in using couriers is insurance. Most couriers have limited automatic coverage that is unlikely to cover the loss of an entire daily shipment. Additional coverage costs more and requires added paperwork (raising internal processing costs as well as shipping costs), but a library's regular insurance may already cover loss or damage in shipping. In ILL transactions the requesting library is always at risk for material en route, coming and going. Commercial suppliers are more likely to bear this risk in a transaction. An evaluative comparison of various providers would note how often materials were lost or damaged en route.

Other factors abound that may either raise or lower costs. Volume discounts, membership privileges, deposit accounts, educational rates, current subscriptions, and licensing agreements may reduce document delivery costs. On the other hand, most libraries overlook follow-up costs in their cost analyses. If follow-up requires a toll call, correspondence, or excessive staff time, those costs should be considered in evaluating supplier performance.

Fill Rate

Fill rate is defined by Waldhart (1985, 314) as "the relationship between the total number of interlibrary loan requests and the number of requests successfully completed." Fill rates can be analyzed in the following ways:

1. To measure the borrowing unit's efficiency in verifying requests, selecting appropriate suppliers, and utilizing appropriate modes for transmitting requests.
2. To measure, to a certain extent, the efficiency of the lending unit in filling requests for materials in your library. The lending unit, of course, has no control over "blind" requests received for materials not owned; nor is performance at fault when items are already in use or noncirculating. The reason behind the "unfilleds" is more important than a lending unit's fill rate!
3. To evaluate suppliers' performance, particularly when a title was listed as available in a catalog or union list. Suppliers with a consistently low fill rate should only be used as a last resort. Library suppliers can be expected to have a higher rate of unfilleds, due to items being noncirculating or in use, or due to restrictions on copying (copyright or preservation limitations), but commercial suppliers are expected to produce what they advertise as available. If a commercial supplier's stock is limited to only certain years of a journal run, the supplier's catalog should so indicate. If distribution rights to a title are withdrawn, the supplier's catalog should be updated as quickly as possible to reflect the changes. When libraries withdraw materials or cancel serial subscriptions, the library catalog and all union lists in which the library participates should be updated as quickly as possible.

Quality

Product quality pertains to the readability and clarity of the delivered document. Missing pages, blurred margins, edges cut off, and light copies are all factors to evaluate. Text, illustrations, graphics, nonroman languages, and scientific symbols should all be evaluated. In the case of a faxed or electronically transmitted copy, the equipment on the receiving end affects quality: do not downgrade a supplier on quality if the problem is at the receiving end.

System quality pertains to factors such as the ease of using the supplier, both to access citations and to send orders. How easy is it to process materials once they are received? Are costs known in advance or at the time of receipt? Do the invoices provide required documentation for your accounting system? Are they itemized and accurate (reflecting what was delivered, method of delivery, and the time span)? How often is follow-up necessary and by what means?

Reliability

Reliability is based primarily upon availability of documents purportedly in the supplier's inventory, delivery within the supplier's advertised time frame, and no mistakes related to either the document supplied or billing. Reliability also involves the responsiveness of the supplier. If unable to supply, are prompt reports issued? By what means? Are reasons given for unfilled requests? Is service consistent and dependable?

User Satisfaction

User satisfaction pertains to turnaround time for delivery, the quality of the material received, and the cost of the service. When evaluating a patron-directed service, ease of locating the citation, ease of ordering, and ease of paying are additional considerations, as well as any problems encountered. The same factors apply with mediated document supply, but the staff involved would evaluate aspects related to ordering, paying, etc. When document delivery is in a mediated environment, user satisfaction might include the ease of determining whether the library owned the articles found and, when necessary, the ease of submitting a document delivery request to ILL/DDS. Measuring user satisfaction in a comparison of various document delivery suppliers may not always be possible, but the chances are that if users are *not* satisfied, someone will hear about it. Disgruntlement with traditional ILL bears witness!

While patrons are often quite vocal about dissatisfaction with ILL turnaround, every ILL staffer will confirm that most patrons are not willing to pay for improved service. Most library users perceive document delivery as a "right." While access to online databases has increased demand and expectations of users, patron-directed document delivery service has made them more aware of the cost of information. Many library users show more appreciation when ILL/DDS staff manage to get documents "free" (to the patron) and to library administrations that subsidize access. Faculty and students on campuses and taxpayers in publicly supported institutions are becoming more appreciative and supportive of information suppliers.

Value-added Services

Value-added services are intended to make one supplier stand out above another when product lines are similar. For example, online vendors of full-text databases might have considerable overlap in titles available and basic fees might also be very close. If only one supplier or product is to be used, however, value-added services may be the deciding factor. Such services should only be compared if the library actually intends to use them. Otherwise, it is like paying extra for a washing machine with a delicate cycle while always washing delicates by hand! Indeed, some "value-added" services should really count against a supplier if they make more work or duplicate services already incorporated into regular routines.

Examples of extra services that may distinguish one supplier from another include a copy of the request or adequate paperwork sent with documents for easy matching with requests, cost information in advance or with material, customized invoices with enough information to match with requests, a customer rep who stays abreast of clients' needs and standards of service, and search and verification services. A value-added service of mixed value is the practice of following up a rush fax copy with a mailed paper copy. This procedure might be a plus for a library with a poor quality fax machine—the supplier would be providing speed with the fax and quality with the paper copy. A library with a high-quality fax machine and prompt delivery to the end-user, however, might have no use for the second copy and might waste time trying to identify "orphan documents" with no request "in process." Prepaid copyright royalties may be a plus for a corporate library likely to owe royalties on most articles, but it adds to costs for libraries that rarely exceed fair use. A management report may be a plus if only one or two major suppliers are used, but libraries using multiple suppliers may prefer to produce a standardized comprehensive report.

Although usually thought of in terms of commercial suppliers, many libraries also offer value-added services that set them apart from other library suppliers. These include being part of the same online ILL subsystem for transmitting requests and checking holdings and union lists, the ability to transmit documents electronically over the same networks, or willingness to give priority service to reciprocal partners.

COMPARATIVE CASE STUDIES, 1985-1993

The following chronological summary compares various document suppliers. The selected studies do not all report data of the same type, thus limiting across-the-board comparisons, but they do chronicle the slowly turning wheels of change prior to the 1990s. Much of the lack of progress may be due to the landmark survey *Document Delivery in the United States*, by consultants Richard W. Boss and Judy McQueen (Information Systems Consultants 1983). Their work gave the impression that although much was wrong with document delivery, since no one seemed to care at the time, it wasn't worth trying to improve! The survey quelled document delivery progress for a number of years, but Boss and McQueen were right. At that time too many other things required attention first. The development of better online catalogs, national bibliographic and full-text databases, and electronic research networks, combined with improved equipment for transmitting scanned images, helped to lay the foundation from which rapid document delivery systems evolved.

In 1985 Gary Dorman Wiggins ([1985] 1990) wrote an influential dissertation that revived interest in document delivery. Although change was still slow, the works listed below answered Wiggins' plea for additional study and reopened the dialog on document delivery. By the '90s the surveys chronicle the struggles of meeting increased demand and higher expectations from patrons and library administrators. The methodologies employed in the various studies may serve as a "how to" or "what not to do" for those who wish to compare their own suppliers.

Authors: Jean Currie and Jan Kennedy Olsen (1985).

Compared: The time required to locate holdings and send requests to commercial and library suppliers, as well as turnaround and costs.

Type of library: Special, academic—The Albert R. Mann Library, Cornell University's agricultural library.

Suppliers: Chemical Abstracts Document Delivery Services (CAS), the Commonwealth Agricultural Bureaux (CAB) (in the U.K.), Information on Demand (IOD), OATS [hereafter referred to as The Genuine Article (TGA) for consistency with later comparisons] from the Institute for Scientific Information (ISI), and various libraries.

Time frame: September 1984-March 1985.

Sample: 124 "straightforward" requests for periodical articles published 1975-1985.

Methodology: Each request was sent to 3 sources: 1) an ILL library supplier over RLIN or mailed ALA forms; 2) a publication specific source—CAB over DIALORDER on DIALOG, CAS, TGA; and,

3) a commercial information broker, IOD using ORDERITEM on DIALOG.

Method of shipment: Not consistent among suppliers—CAS, IOD,and TGA used first-class mail; CAB, air mail; libraries, UPS.

Average Turnaround: Varied from 6.04 days to 20.53 days, with CAS being the fastest, followed by TGA, the libraries, CAB, and IOD.

Average Costs: Search and verification costs were decidedly lower over RLIN as compared to the DIALOG databases. [The figures are severely outdated now by changes in the price structures and technology for all of the suppliers.] Document costs were also much lower from libraries (most were supplied by reciprocal libraries). Commercial suppliers average costs ranged from $8.72 to $13.25, with TGA being the lowest, followed by CAB, CAS, and IOD.

Fill Rate: 121 citations were verified on RLIN or OCLC, 119 of which were supplied by library sources; 82 citations were verified on DIALOG databases with 68 actually supplied by the online vendors or IOD.

Authors: David B. Allen and Johanna Alexander (1986).

Compared: Potential turnaround and costs of commercial suppliers and conventional ILL suppliers.

Type of library: Academic—California State College, Bakersfield(CSB).

Potential Suppliers Meeting CSB's Criteria: British Library Document Supply Center (BLDSC); Data/Courier Inc.; Information on Demand (IOD); The Information Store; Institute for Scientific Information (ISI); Los Angeles Public Library; Predicasts Article Delivery Service; University Microfilms International Article Clearinghouse (UMI). Included were two fee-based services in libraries, two information brokers; three online database producers; and one clearinghouse. Seven of the services were located in the USA, one in the UK. Two suppliers (Data/Courier and Predicasts) offered subject specific coverage, the others offered coverage in all subject areas.

Time frame: 1 year, 1985.

Sample: 1,053 periodical ILL requests—64 percent published within the prior 1 to 5 years; 365 published within the prior 6 to 10 years.

Methodology: This study focused on selecting potential suppliers based on the above sample of requests and did not actually involve using the suppliers. Steps taken to determine potential suppliers included review of CSB's ILL service; analysis of patrons' external resource demands; literature review; performance evaluation of the document delivery aspect of CSB's ILL service; survey of document delivery suppliers listed in *Document Retrieval: Sources and Services*, (see Erwin 1987) which identified sixty-nine suppliers potentially capable of meeting CSB's document delivery needs. A questionnaire (included in Appendix B of Allen and Alexander's work) was sent to the sixty-nine suppliers. By comparing the answers on the questionnaire with the sample of requests, the authors determined that the eight suppliers could meet CSB's pre-established criteria (Allen and Alexander 1986, 59-60).

Turnaround: Forty-six percent of the requests were filled in five to fifteen working days from within the California State University system. Average turnaround from libraries was compared with potential time frames provided by commercial suppliers' responding to the questionnaire.

Cost Comparisons: Actual supplier costs; how much patrons had indicated they were willing to pay, plus the amount the library was willing to subsidize; and the number of pages included in each article were all compared against the information provided by the vendors. Eighty-two percent of the requests were 10 pages or less; 96 percent were 20 pages or less. Seventy-seven percent of the ILL users had indicated a willingness to pay (although the study does not indicate how much they were willing to pay). An average ILL cost range of $6 to $16 was compared with the vendor's prices quoted in the questionnaire, based upon the same delivery method actually used to fill the request.

Author: John Budd (1986, 75-80).

Compared: Turnaround time for phases of an ILL transaction on the OCLC subsystem. Budd also summarizes similar studies.

Type of library: Academic—Southeastern Louisiana University.

Suppliers: Libraries using the OCLC ILL subsystem.

Time frame: One year (1985).

Sample: 493 transactions—280 photocopies (56.8 percent) and 213 loan requests (43.2 percent).

Methodology: Tracked the time required to complete each phase of an ILL transaction from receiving the request through fulfillment or cancellation.

Average Turnaround: From receipt of request to sending request—6.46 days; from sending of request to shipment by supplier—4.58 days; from shipment to receipt by library—8.21 days (mean figures cited). Budd further breaks down the data to reflect turnaround for both loan and copy requests.

Authors: Douglas P. Hurd and Robert E. Molyneux (1986, 182-185).

Compared: Supplier turnaround (from the time the request was sent until the material was received) and costs.

Type of Library: Academic, special—University of Virginia Science and Engineering Library.

Suppliers: Center for Research Libraries (CRL) via OCLC. (CRL was backed up by BLDSC's Journal Article Service (a document delivery service provided at that time through a joint effort of CRL and BLDSC); University Microfilms International (UMI) via OCLC; and various libraries with requests sent over OCLC, mailed ALA forms and Telex.

Time frame: April, 1985.

Sample: A total of 212 requests were ordered from various suppliers; 41 (19 percent) were available from UMI, including pre- and post-1977 articles.

Methodology: The final comparisons were based on twenty-six identical documents received from both ILL suppliers and UMI.

Turnaround/Costs: UMI delivered higher quality copies faster (averaging 10 days) than conventional ILL sources, although the costs were higher and service was inconsistent. [The study does not indicate how the twenty-six items received from conventional library resources were originally requested—OCLC, mail, or Telex—which would have affected total delivery time; average turnaround/costs is not given for the library suppliers.]

Fill Rate: Of the orders sent to UMI, 13.9 percent went unfilled even though listed as available in the UMI catalog. [Fill rate was not provided for the library suppliers.]

Authors: Susan B. Ardis and Karen S. Croneis (1987, 624-627).

Purpose: Determine whether commercial suppliers could provide rapid document delivery as an alternative to acquisition.

Type of Library: Academic, special—University of Texas's five science libraries (Chemistry, Engineering, Geology, Life Sciences, and Physics-Math).

Suppliers: Selected for strong holdings in the sciences—Chemical Abstracts, (69 requests); CTIC (68 requests); NTIS (66 requests); and Global (only 4 requests).

Time frame: 3 months.

Sample: (See "Suppliers.")

Methodology: Only scientific and technical titles were included in the study. Requests for materials not owned were accepted from all UT students, staff and faculty.

Average Turnaround/Costs: CAS—7.3 days, $9.62; CTIC—8.4 days,

$9.04; Global—10.2 days, $32.50; NTIS—14.4 days, $8.45.

Fill Rate: Received 196 of 212 requests; 188 were received by the first supplier selected, eight by the second supplier.

Results: Alternative access was deemed highly successful, with plans to expand the number of commercial suppliers.

Authors: Anthony K. Kent, Karen Merry, and David Russon (1987).

Compared: Scope of inventory and overlap of holdings.

Type of libraries: International document supply centers.

Suppliers: British Library Document Supply Centre (BLDSC)—U.K.; Chemical Abstracts Service (CAS)—U.S.; Centre de Documentation Scientifique et Technique (CDST)—France.; National Library of Medicine (NLM)—U.S.; and Online Computer Library Center (OCLC) member libraries—U.S. (although members are worldwide).

Time frame: Varied slightly depending upon the fiscal year for each institution, but basically 1983/84-1985. BLDSC and OCLC utilized a sample from their huge annual volume rather than analyzing requests for a full year.

Sample: BLDSC—61,946 (a sample); CAS—51,650; CDST—250,588; NLM—99,162; and OCLC—54,870 (a sample), representing requests from the following number of unique journal titles: BLDSC—18,465; CAS—7,734; CDST—10,324; NLM—11,853; and OCLC—20,031.

Results: A core list of 500 most frequently requested titles was devised, as well as a "minimum core list" of the forty titles that were common to the top 1,500 titles requested from all of the organizations

---

Author: Mary Williamson (1987).

Compared: Turnaround, costs, fill rates, and coverage.

Type of Library: Clearinghouse for statewide ILL network—Wisconsin Interlibrary Services (WILS)

Suppliers: Chemical Abstracts Service (CAS) and ISI's Original Tear Sheet Service, [now called "The Genuine Article" (TGA)] and various libraries.

Time frame: five consecutive work days.

Sample/Methodology: Twenty requests were sent to libraries and ten each to the two commercial suppliers. Although only twenty requests were actually sent to the suppliers, the survey checked on potential availability of 153 scientific/technical article requests.

Fill Rate: Seventy-seven of 153 articles were available from the commercial services; 73 of the 77 were also available from libraries. Of the 20 requests sent, the commercial services filled 15, the libraries filled 14.

Author: Kathleen F. Halsey (1988).

Compared: Coverage, fill rate, turnaround time (from the time the order was sent until receipt of material), document quality, and direct cost per article were evaluated for a statewide ILL clearinghouse and a commercial supplier.

Type of Library: Academic—University of Wisconsin-Stevens Point.

Suppliers: UMI's Article Clearinghouse and Wisconsin Interlibrary Services (WILS).

Time frame: October 21 through December 11, 1987 (thirty-seven consecutive workdays), covering both heavy and light usage periods.

Sample: Five hundred and two ILL requests were checked against holdings; 124 requests available from both WILS and UMI were used. Coverage included a broad range of subjects, but excluded "requests for photocopies of transactions, proceedings, tables of contents, sections of books, Educational Resources Information Center (ERIC) materials, technical reports, and government documents." (ibid., 6)

Methodology: Replicated, in part and on a larger scale, Williamson's 1987 study. The study includes criteria for selecting a commercial supplier. One hundred and twenty-four identical article requests were ordered almost simultaneously over OCLC from both suppliers.

Fill Rate: UMI filled 60 percent; WILS filled 100 percent.

Turnaround/Method of Delivery: Fax delivery was requested on twelve articles. UMI advertised [and still does] same day processing and transmission on orders received before 1:30 p.m. (EST). Seven orders were placed before the deadline, but only two arrived the same day, although UMI did fax all twelve as requested; mean turnaround for the fax requests was two business days. WILS faxed only four of the twelve requests, each arriving on the second day. Turnaround was almost identical with both mail and fax requests.

Costs: Article costs were higher from UMI and there were several billing problems; billing from WILS was reliable.

Other Results: Both services had a high percentage of quality control problems with the faxed documents, but UMI edged WILS on copy quality overall. WILS was more likely to copy two pages together, whereas UMI copied each page on a separate exposure. [Separate exposures can be considered a "value added" service, as many suppliers will try to cut processing time and paper/toner costs by copying two pages or making back-to-back copies whenever possible.]

Comments: Halsey (ibid., 33) states: "The Article Clearinghouse guarantees 48-hour in-house turnaround time for all available articles published within the last 5 years. If the deadline is missed, the company fills the order at no charge." That guarantee may be one reason why UMI figured in so many comparative studies; current documentation from UMI still indicates that 48-hour processing is "standard service," but makes no reference to free delivery if it fails to meet the "standard." WILS also claimed a 48-hour, in-house turnaround time.

A separate study of the 378 "excluded" requests would have made an interesting comparative study of more difficult requests. Halsey (ibid., 6) states that the exclusions were "unlikely to be available from a commercial document delivery service." This may be true since UMI was the only such supplier tried, but many commercial providers specialize in the very types of materials excluded. Halsey's thesis summarizes other comparative surveys prior to 1988.

Authors: Connie Miller and Patricia Tegler (1988, 352-366).

Compared: Commercial document suppliers with library suppliers for turnaround, cost, and fill rate.

Type of Library: Academic—University of Illinois at Chicago.

Suppliers: Libraries in the statewide ILL library network, ILLINET; various OCLC libraries; commercial suppliers available through DIALOG's electronic ordering system. Five database-specific suppliers were selected according to their advertised specialties in the subjects required, and six general suppliers were chosen—two of which were "local" suppliers in Chicago, the remaining four in different geographical regions of the United States.

Time frame: One quarter.

Sample: A total of 186 photocopy requests, consisting of 165 requests filled the previous quarter by OCLC libraries or the ILLINET library network and twenty-one previously unfilled copy requests that were sent to the commercial suppliers. All of the requests were for journal articles, proceedings, papers, and technical reports in the subject areas of science, technology, or mathematics, requested by graduate students or faculty.

Methodology: Each request was sent to at least two commercial suppliers—preferably one subject specific and one general. If no subject-specific supplier was available, requests were sent to one local and one nonlocal general supplier. General suppliers were assigned to documents randomly.

Method of transmittal: All of the requests sent to commercial suppliers were ordered electronically over DIALOG's Dialorder when able to verify online, or ORDERITEM when unable to verify in DIALOG. A librarian with many years of online experience handled all verification and ordering on DIALOG, but this proved complex, especially determining an appropriate database, even with the subjects restricted.

Method of shipment: OCLC libraries shipped by mail; ILLINET libraries used couriers; commercial suppliers used "standard" shipping methods. Rush methods were not used.

Average turnaround: ILLINET suppliers were fastest, averaging 9.8 days (median time 12.5 days) with seventy percent arriving within eight days. Database-specific suppliers were next, averaging 12.8 days (median time, 14.5 days) with fifty-two percent arriving within eight days. OCLC libraries averaged 13.2 days (median time, 14.5 days), with thirty-six percent arriving within eight days. General commercial suppliers were slowest. Nonlocal suppliers averaged 17.8 days (median time, 25.0 days) with thirty-one percent arriving in eight days. Local suppliers averaged 24.3 days (median time 28.0 days), with only fifteen percent arriving in eight days.

Average costs: Online order costs were less using ORDERITEM, but suppliers often added verification fees, making document costs higher for unverified documents. Document costs for library suppliers were negligible, coming mostly from reciprocal libraries. Of the commercial suppliers, database-specific suppliers cost the least, averaging $12.11, with forty-four percent of the requests under $11, while providing the fastest commercial service. General suppliers averaged $12.64 with thirty-eight percent under $11. Local suppliers averaged $14.85 with only seven percent under $11, thus costing the most while being the slowest.

Fill Rate: Although Miller and Tegler did not calculate the fill rate for the library suppliers, the combined fill rate for both network and OCLC libraries would have equaled 88.7 percent. The general nonlocal commercial suppliers had a fill rate of 93.3 percent; followed by the database specific suppliers with a fill rate of 89.9 percent; and the general local suppliers once more came in last at 83.6 percent. Engineering requests were the most difficult to fill for every supplier.

Other Results: Date and language had little effect on delivery time or delivery cost. Technical reports and the unfilled library requests took a little longer and cost slightly more, but the samples of those two groups were too small to be definitive. Agricultural requests sometimes took longer to fill because they often had to come from overseas.

Authors: Janice S. Boyer and John Reidelbach (1990, 7).
Compared: Commercial document suppliers with ILL, comparing cost, turnaround, fill rate, and user data.
Type of Library: Academic—University Library, University of Nebraska at Omaha.
Suppliers: Information on Demand (IOD), Institute for Scientific Information (ISI), and University Microfilms International (UMI).
Time frame: Four months, 15 February-15 June 1989.
Sample: Complete information not provided, but fifteen requests went to ISI; ninety-four requests to UMI; and IOD was "the most heavily used vendor." (ibid., 7)
Methodology: Not provided.
Average Turnaround: IOD had both the fastest (same day via fax) and slowest time (forty-five days by mail). ISI's fastest time was same day (fax); its slowest, eleven days. UMI's fastest time was one day (fax); it's slowest, forty-two days.
Average Costs: ISI—overall, $16.26; by first-class mail, $13.15; by fax, $20.92. UMI—overall, $16.67; by first-class mail, $16.38; by fax, $17.75. IOD—overall, $32.60; by first-class mail, $27.24; by Federal Express, $38.44; by fax, $49.03.
Results: Determined that holdings of UMI and IOD were too limited and service was slower than expected.

Author: David P. Gillikin (1990, 27-32).
Compared: Full-text article retrieval using online bibliographic databases as an alternative to ILL. Searching and retrieval methods; coverage and costs were compared. Delivery within twenty-four hours of the request having been submitted was the goal.
Type of Library: Academic—University of Tennessee, Knoxville.
Suppliers: Bibliographic Retrieval Services (BRS) and DIALOG.
Time frame: 9 January-17 March, 1989.
Sample: Thirty requests, pre-screened as available in a full-text database, ordered a total of ninety-one times distributed over the following databases: CCML (5) and TSAP (4) files in the BRS database, and from DIALOG files 625 (41); 635 (2); 647 (2); 648 (31); and 675 (6).
Methodology: If possible, the same article was retrieved from both data-

base hosts and/or printed twice on DIALOG (immediately while online or ordered for next day online delivery via DIALMAIL and downloaded and printed the next day) to compare costs.

Average Costs: Gillikin cited an overall average cost of $3.03 per article—including the cost of searching for eight unfilled requests. [Gillikin's average costs may have been skewed since forty-two of the eighty-three articles retrieved were from *American Banker* (average cost, $1.54 per article) and forty-two were through DIALOG File 625, the least expensive file used, particularly since the sample of databases and volume of requests were small. Using Gillikin's data, the average cost per article in each database was calculated: CCML—$3.42 (including one unfilled); TSAP—$4.63 (including one unfilled); 625—$1.43; 635—$9.76 (including one unfilled); 647—$4.38 (including one unfilled); 648—$4.02 (including three unfilleds); 675—$4.71 (including one unfilled), all of which are reasonable per article costs for document delivery, although they exclude internal processing costs.]

Fill Rate: Ninety-one articles were ordered; eighty-three were retrieved. Unfilleds were primarily due to patron error (bad citations) which added to online searching costs (as in file 635) and journals with limited full-text availability (most prevalent in BRS).

Other Results: Using *American Banker* (annual subscription $525) as an example, Gillikin estimated that the subscription cost alone would have paid for approximately 350 articles. [That estimate excludes internal processing costs, but such costs are probably no more than the library's internal costs for checking in, binding, storing, reshelving, and maintaining the printed edition of the title.]

Author: Richard E. Raske (1991, 61-69).

Compared: Commercial article delivery suppliers with library suppliers, comparing costs vs. turnaround.

Type of Library: Corporate library.

Suppliers: Not named, but codes are given for four commercial suppliers.

Time frame: One year (1989).

Sample: Not provided.

Methodology: Not applicable; this is more of a "how to" guide for selecting and evaluating suppliers, rather than a definitive research analysis.

Average Turnaround: Varies between seven and sixteen days for commercial suppliers with the overall average being ten days; library suppliers averaged seventeen days.

Average costs: Commercial suppliers averaged between $6 to $24; library charges averaged $4.

Author: Frank L. Davis, et al. (1992, 185-187).
Purpose: To determine how hospital libraries use independent information brokers.
Type of Library: Special—hospital libraries in four southeastern and two midwestern states.
Suppliers: Not named specifically.
Time frame: Autumn, 1989.
Sample: Surveyed 223 hospital libraries.
Methodology: Questionnaire.
Results: Of the total libraries, 13.9 percent used independent information brokers for document delivery, filling an overall average of 10 percent of the total document delivery requests (the range was from one to forty-five percent among the libraries). 70.9 percent confined their business to one independent information broker; seven libraries used two, and two used three or more. Most routine requests were sent to libraries. In some cases, information brokers were used mainly to verify and fill incomplete citations or to obtain rare materials—the implication being that librarians were willing to pay for the specialized service. For hospital libraries, rapid turnaround was most important in selecting a supplier. Cost was the most important economic influence, followed by inclusion of copyright royalties, low referral fees, and volume discounts.

Author: Janis Leath (1992).
Compared: Commercial article delivery services with ILL.
Type of Library: Academic—University of Wyoming Libraries.
Suppliers: Fourteen commercial suppliers, including some vendors with broad subject coverage and some specialized suppliers. Some suppliers had in-house collections; others had access to major research library holdings. Chemical Abstracts Service—20; DataSearch—9; Dynamic Information Corporation—1; Engineering Societies Library—4; GeoRef—6; I-Med Information Services—12; Information on Demand—10; The Information Store—19; ISI's The Genuine Article—126; Medical Data Exchange—9; Michigan Information Transfer Source—10; Sport Information Resource Centre—2; UnCover—105; University Microfilms Article Clearinghouse—163; various libraries—140.
Time frame: Phase 1, 19 August-4 October 1991; phase 2, 21 October-22 November 1992.
Sample: 636 requests, disbursed as indicated under suppliers. Method of

transmitting: Various means—OCLC, fax, telephone, and Federal Express.

Method of Shipment: Mail, fax, and Federal Express. The most frequent problem was suppliers failing to fax upon request.

Average Turnaround: Libraries and vendors averaged about the same delivery speed; overall average was 9.7 days, 12.4 days by mail and 2.8 days by fax. The fastest, UnCover, averaged 1.8 days; the slowest, Data-Search, averaged 36.0 days. Mailed documents averaged five days en route.

Average Costs: Excluding reciprocal requests, the average cost per request from commercial suppliers was $9.87 overall; by mail, $9.14; by fax, $11.52. Libraries averaged $6.07 overall; $5.81 by mail and $9.50 by fax. Excluding ESL which failed to bill, the lowest costs were from I-Med—$8.59 overall, $8.50 by mail, $9.50 by fax; the highest costs were from IOD—$27.96 overall, $25.19 by mail and $50.15 by fax.

Fill Rate: Commercial suppliers averaged 84.1 percent; libraries, 89.3 percent. Only UnCover stood out in terms of fulfillment success and delivery time. Three vendors—Dynamic, GeoRef, and SIRC—filled 100 percent (all with very small samples); UnCover was next with 95.2 percent; ESL had the lowest fill rate with 25 percent (also a small sample).

Other Results: Overall, nineteen percent of the requests from commercial suppliers had problems—most often long delivery times or not shipped by fax as requested. Several vendors supplied the article twice or did not update OCLC. Twenty-two percent of the libraries had problems.

Comments: The author concluded that commercial suppliers did not provide a clear alternative for filling article requests, but made librarians more aware of available options and the need to balance time and expense in document delivery.

Authors: Arizona State University (ASU) Libraries Journal Document Delivery Task Force (1992a; 1992b); Sheila Walters (1993a; 1993b).

Compared: Commercial suppliers with a select group of reciprocal libraries known to use fax, ARIEL, or couriers, and a third group of "traditional" library suppliers. Primary goal was to improve turnaround.

Type of Library: Academic—Arizona State University Libraries.

Suppliers: DIALOG full-text databases, The Genuine Article (TGA), UnCover, and University Microfilms International (UMI) were the commercial suppliers in Phase 1. WISE for Medicine (WISE) was added in Phase 2. ASU was striving for a maximum turnaround of four days from the day a request was logged in until the requested item was received, at a maximum cost of $25 (reduced to $15 in Phase 2). The criteria for selecting commercial suppliers centered on speed, cost, method of transmitting requests, and method of delivery. Commercial suppliers were expected to ship articles within 48 hours by fax and/or ARIEL. Suppliers available on systems used regularly by ILL/DDS (UnCover and OCLC, primarily) were preferred. Only two of eleven suppliers listed on OCLC's Name/Address Directory met the price limit for the turnaround required, UMI and TGA. TGA was borderline because copyright royalties were not known in advance and could push fees over the limit. Fee-based document delivery services in academic libraries were considered but rejected because most did not meet the cost limit for 48-hour turnaround; most charged for unfilled requests and extra verification if needed; and ILL/DDS had access to collections in the same libraries, many of which were reciprocal partners. DIALOG was selected for online full-text retrieval as it had the widest selection of full-text journals and newspapers available.

Library suppliers included a select group of eight libraries known to provide rapid delivery by courier, fax, or ARIEL. No formal network or contract for priority service existed between these libraries, but seven were reciprocal academic libraries within the Southwest or Rocky Mountain region. The Center for Research Libraries was included in this group, since it also provided "free" delivery, but we considered it to be on the cusp between a commercial supplier and a library supplier due to our annual membership of $35,000. Other libraries (both reciprocal and nonreciprocal) and other commercial suppliers were used when none of the suppliers in the first two groups could fill a request, but any analysis of individual suppliers in this group was based on very small samples.

Time frame: Phase 1, January-June, 1992; Phase 2, January-June,1993.

Sample: Phase 1: 1591 filled photocopy requests from faculty and graduate students. After the success of Phase 1, the scope was expanded in Phase 2 to include undergraduate requests and book loans. [Except where noted, the data below primarily covers Phase 1.]

Methodology: After being logged in, the daily batch of photocopy requests was immediately checked against the commercial suppliers' online or print catalogs. When found, the articles were ordered with no further search and verification. Requests unavailable commercially were searched on OCLC and ordered immediately when available from among the select group of libraries by listing one to five of these libraries in the lending string. Requests not available from any of these suppliers went through traditional ILL processes to verify citations and locate potential suppliers, preferably other reciprocal libraries.

Method of Transmittal: UnCover requests were sent direct over UnCover on the CARL Corporation Network; requests to UMI, TGA, and the select group of libraries, via OCLC; WISE for Medicine, via fax; some requests were sent by mail, fax, or RLIN to other suppliers, but that was a small sample.

Method of Shipment: Photocopies from commercial suppliers, via fax; from out-of-state libraries in select group, by fax or ARIEL; select in-state libraries, by daily courier or fax (rush only during the projects, but a later project involved fax and/or ARIEL between the state university libraries). Book loans were not available from commercial suppliers used. They were shipped by daily courier from the two other state university libraries; library rate mail from others.

Average Turnaround: UnCover in Phase 1 and 2 and WISE for Medicine in Phase 2 were the only commercial suppliers to consistently meet the goal of a four-day maximum turnaround (from request being logged in until receipt of materials). UMI averaged 7.4 days; frequently meeting the four-day goal, but inconsistent service caused the higher average. [By working with the customer service rep, UMI's average improved and UMI now ships by ARIEL.] The select group of libraries averaged eight to eighteen days (depending on where the supplier was in the OCLC lending string). TGA averaged fourteen days. Overall, average turnaround was 16.7 days, but 37 percent of the copies were received within one week; 66 percent had arrived within two weeks; and 79 percent of the filled requests were in house by the end of the third week. Twenty-one percent took more than three weeks. Prior to the project only 21 percent of the filled copy requests were received within one week, and only 60 percent had arrived within two weeks. There was no change in the percentages received within three or more weeks, so using commercial suppliers and a selected group of first-rate library suppliers made a dramatic improvement. (The overall average turnaround remained high due to a few problem requests which skewed overall averages).

Average Costs: WISE for Medicine, under $9.00/article; followed by UnCover, $9.60/article; TGA $18.45/article; and UMI

$19.65/article. The average cost/article from all suppliers was $4.40 (compared to an average cost of $0.92/article in the previous six months when few commercial suppliers were used). Copyright royalties of $1413.50 were paid on 514 articles acquired commercially, although CONTU guidelines were exceeded on only twelve journals (fifty articles); royalties for only the fifty articles would have been about $150. Subscriptions to the twelve journals exceeding copyright would have cost $1,890, while document delivery costs, including royalties, for the fifty articles was $503.

Fill Rate: Overall costs were low because only 32 percent of the 1,591 articles were filled by commercial suppliers (approx. $7,500 expended). Five percent were filled by nonreciprocal libraries—a category in which ASU included national libraries that serve as international document supply centers, such as BLDSC. A few commercial suppliers used only occasionally were also in this group, as were used primarily as a last resort rather than for rapid document delivery. ($3,000 expended to nonreciprocal suppliers). The remaining 63 percent were supplied by reciprocal libraries. (An additional $9,000 was expended for book loans and photocopy requests filled outside the scope of the project in Phase 1.)

Other Results: Follow-up was required on 18.5 percent of the requests. Error rates regarding the quality of copies received was about the same for all suppliers—commercial and library. Commercial suppliers had more errors due to failure to update records on OCLC, shipping after the request had been forwarded to another OCLC library, shipping twice, not providing a copy of the request or price information with the request, and billing errors. Full-text retrieval proved the most difficult process to work into daily routines, primarily due to lack of staff trained to search DIALOG or Internet resources prior to the projects. Only sixteen full-text articles were retrieved during the span of both project phases, but few conclusions could be drawn as the sixteen articles were acquired from almost as many different databases!

Author: Robert T. McFarland (n.d. [1992], [115-eoa]).

Compared: Commercial document suppliers as an alternative to acquisition; suppliers were evaluated on cost, turnaround, cost efficiency, fill rate, reliability, and vendor responsiveness.

Type of Library: Special, academic—the Chemistry Library, Olin Library System, Washington University, St. Louis, MO.

Suppliers: Nine commercial suppliers, including fee-based services in libraries. Pilot project began with only Chemical Abstracts Services, but after six months, expanded to include John Crerar Library Photoduplication Service at the University of Chicago (Crerar); Linda Hall Library (LHL); ISI's The Genuine Article Service (TGA); UMI's Article Clearing House (UMI); Engineering Information (Ei); Engineering Societies Library (ESL); American Mathematical Society MathDoc Service (MATHDOC); and American Geological Institute's GeoRef Service (GEOREF).

Time frame: March-October, 1991.

Sample: One hundred thirty different titles predetermined to be available from CAS were requested in the original pilot project.

Methodology: Over eighteen months, the Chemistry Library moved away from ownership of materials to access, primarily by online and fax. Over 90 percent of the requested documents were expected to be available from CAS. After a six-month trial period, eight additional document delivery vendors were selected for a comparative study, using requests from five science disciplines: biology, chemistry, earth sciences, mathematics, and engineering. Four documents each were ordered from each of five vendors within a given subject area, except for mathematics in which five documents were requested from three different vendors. Two documents each were requested to be delivered by regular or first-class mail and two to be delivered "rush" by express mail or fax.

Turnaround: Overall average turnaround was seven days by mail and two days (mean delivery time) for rush. MathDoc and UMI were fastest for regular mail; LHL and Ei were significantly slower; Ei was also the slowest for rush. LHL had the largest turnaround differential between regular and rush; UMI the least.

Reliability: Determined by the consistency of documents being supplied within advertised time limits or requested time limits. UMI was the least reliable for regular requests, meeting advertised or stated time limits only 50 percent of the time; the other suppliers met stated times 80 percent or greater. For rush requests; UMI scored lowest meeting advertised times limits only 25 percent of the time; Ei and ESL met advertised time limits 50 percent; the others all met advertised limits 80 percent of the time or better. Crerar and CAS were "exceptional" in providing timely notification of problems. CAS stores documents older than ten years in remote storage, but always notified the library that the request might require extra time. Other suppliers tended to send notification of problems by the same mode in which a document was requested; thus, mailed requests required several days before the library found out there would be a delay.

Average Cost: ESL was the most expensive with average costs for regular requests at $23 and rush averaging $40. Crerar and LHL had the lowest cost differential between regular and rush—about $5 more for rush requests; Crerar had the lowest rush costs ($15) and the fewest "add-on" charges. The total spent during the pilot project was $2,789.50. Overall average turnaround was seven days by mail and two days (mean delivery time) for rush. MathDoc and UMI were fastest for regular mail; LHL and Ei were significantly slower; Ei was also the slowest for rush. LHL had the largest turnaround differential between regular and rush; UMI the least.

Cost Efficiency: Cost efficiency measured which vendors could supply documents in the least amount of time at the lowest cost. Crerar, LHL, and GeoRef scored lowest; UMI highest for regular mailed documents. Crerar, LHL, and MathDoc scored highest for rush service; Ei and ESL scored lowest for rush. Overall, rush was more cost effective than regular mail.

Fill Rate: A total of 192 requests were received. Fill rate was 90 percent or above for all vendors. LHL had the broadest scope, supplying documents in all five subject areas; CAS, Crerar, and TGA were appropriate for four of the five subject areas. Not surprisingly, GeoRef, MathDoc, ESL, and Ei proved to be subject specific.

Other Results: John Crerar became the vendor of choice for biology, chemistry, and earth sciences, and second choice for mathematics (MathDoc was first); second choices were TGA for biology, CAS for chemistry, and GeoRef for earth sciences. For engineering Ei was the first choice and LHL second. TGA was the second choice for biology based on the scope of its inventory, but had limited availability of retrospective literature, numerous add-on charges that complicated ordering, above average prices, and inconsistent reliability. CAS scored high in every category except "ease of ordering." The best rates were obtained when documents were ordered online over STN, after 5:00 p.m. local time, with the CA accession number included; CAS had a number of different combinations by which documents could be ordered, each one with a different rate. (It took staff six months to figure out the best ways to order with the best results.) GeoRef had excellent reliability and customer service, and documents could be ordered easily, but its charges were extraordinarily high, particularly for rush requests; Crerar matched GeoRef's service with much lower rates. UMI was not recommended as either a primary or secondary source for science libraries as the scope of its inventory did not match the requests as well as the other vendors. Crerar was the only supplier that did not provide a copyright cleared service, leaving it up to the requesting library to pay copyright royalties when it exceeded fair use.

Comments: The report demonstrates both the frustrations that librarians frequently encounter as they begin using commercial services as well as the importance of establishing good vendor relationships and learning what works best with a particular supplier. This was an excellent study. Even though the results are based on a small sample from each vendor, it demonstrates the many ways that vendors can be compared.

Author: David Everett (1993a, 22-25; 1993b, 17-25).

Compared: The viability of full-text online as an alternative to ILL for document delivery.

Type of Library: Academic—Stetson University.

Suppliers: Various online full-text database vendors. The requests were not actually ordered, only compared for potential fulfillment.

Time frame: One year.

Sample: A total of 1,896 photocopy requests, plus a list of periodicals indexed in five of the Wilson periodical indexes.

Methodology: Compared photocopy requests received with journals available full-text online as listed in *BiblioData Fulltext Sources Online* (see Orenstein 1993) and *DIALOG Full-text Sources* (see DIALOG Information Sources 1992). Periodicals indexed in *Readers' Guide, Business Periodicals Index, General Science Index, Social Sciences Index*, and *Humanities Index* were also compared against the list of full-text databases.

Average Costs: Since the articles were not actually retrieved online, no comparison of average costs was made. However, Everett noted sixty-three articles would have been retrievable online, and costs would have included copyright royalties for all. A survey of the 1991 ILL borrowing records indicated that of the 562 articles acquired, 96 percent of the journals were requested four or less times; twenty-two titles reached the CONTU limit of five requests, and only nine titles exceeded CONTU guidelines (Everett 1993b, 20). Regular online full-text retrieval would thus add excess royalties to document delivery costs.

Fill Rate: Only sixty-three titles (3.3 percent) were available in full text from the fifteen databases surveyed. Had the databases carried additional back files, another eighty-four titles could have been acquired. Three databases—Westlaw, DIALOG, and BRS—contained the majority of the sixty-three titles, but some unique titles were in databases that had few hits overall—LEXIS, Nikkei Telecom, and STN. Articles in business databases were more widely available in full text that other subjects in the Stetson study. Sixty-two percent of the periodicals indexed in *Business Periodicals Index* were available in full text, compared with 33 percent from *Readers' Guide*, 20 percent from *General Science Index*, 7.6 percent from *Social Sciences Index*, and 0.3 percent from *Humanities Index*. (Everett 1993a, 23)

Author: Peggy Richwine (1993).

Compared: Commercial document suppliers used to "augment" ILL in acquiring obscure materials, such as for issue supplements or conference proceedings, for "rush" requests, and for materials unavailable through OCLC libraries—in other words, the "tough stuff."

Type of Library: Special, academic—Indiana University School of Medicine Ruth Lilly Medical Library.

Suppliers: The Genuine Article (TGA), The Information Store (TIS), UnCover, University Microfilms International Article Clearinghouse (UMI).

Time frame: Three months, January-March 1993; data for UnCover covered six months—October 1992-March 1993.

Sample: A total of 48 requests—UMI (34); TGA (6); TIS (3); UnCover (5).

Method of Transmittal: OCLC for all but UnCover, for which fax or electronic transmission over the Internet was used.

Average Turnaround: UnCover, one day; TGA, seven days; UMI, eleven days; TIS, fifteen days.

Average Costs: UMI, $7.75 (group discounted rate; standard rate would have been $12.50); UnCover, $10.50; TIS, $12.50; TGA, $15.25.

Fill Rate: TGA and TIS, 100 percent; UMI, 88 percent; UnCover, 80 percent.

Although these comparative studies may not be a true test of how a particular supplier will work for your library, they do provide a basis for identifying potential problems that can be discussed ahead of time with suppliers' customer service representatives. Many of the older, established document suppliers have only recently felt the need to improve services in response to increased competition.

Many libraries on the OCLC and RLIN ILL subsystems use the monthly statistics from those networks to compare the performance of library suppliers (as well as commercial suppliers on those systems). Other national and international ILL networks provide similar data. From such data, one can easily determine which libraries are used (or abused) either as a lender or a borrower. Some of the reports provide data only on filled requests, but even limited data makes it easy to determine which libraries might become appropriate reciprocal partners and, conversely, which reciprocal relations need adjusting or canceling.

Libraries that are net borrowers should monitor their borrowing patterns, spreading out requests among reciprocal partners. Net lenders should try to use net borrowers as much as possible to balance reciprocal relations. If the net borrowers do not provide good service, the best move may be to get out of the reciprocal relationship. This decision may generate some ill will, especially if commitments are based on regional resource sharing agreements. Every library, however, has an obligation to its own patrons and to the ILL/DDS staff to concentrate activities with suppliers who share a mutual commitment to excellent service.

All comparative studies more than ten years old indicate that commercial suppliers had few requests for "rush" shipments and that they had few complaints with an average turnaround of ten to fifteen days. In those times, however, libraries primarily relied on commercial suppliers as a last resort, not as primary suppliers. Check to see what technological or staffing changes suppliers—both commercial and library—have made to improve turnaround or quality, or what new online products have been developed which may appropriately serve end-users. If a supplier meets your library's established criteria, then try it to determine whether it is worthy of continued support.

Chapter 6

Document Delivery and the Librarian/Publisher Relationship: Changing Roles, Evolving Products

The long history of good relationships between librarians and vendors goes back to the days when book salesmen brought news and samples as they made their rounds among libraries. Books were the principal items for sale, a publisher's reputation was the principal sales tool, and most of the salesmen were not librarians (Katz 1987, 174).

The relationship between librarians and publishers is changing, largely in response to technological possibilities that "facilitate the dispersion of traditional roles and blur the boundaries between the formerly discrete functions of the generation, publication, dissemination, and archival storage of knowledge." (Battin 1988, 1) Broad, theoretical questions about scholarly communication and the roles that various parties play in that process are being raised in the professional literature of both groups.

More immediately, the circumstances surrounding the publication of scholarly journals, particularly to the academic librarian, make the issue of serials pricing one of increasing moment. The problem affects not only "acquisitions," but also the growth and development of the flip side of that process, "access," of which document delivery is one manifestation. As library budgets undergo contortions to maintain essential subscriptions, alternatives to subscription are increasingly becoming part of the practice of libraries' information delivery services. Technological advances support the development of these alternatives, driving a reexamination of the traditional roles of publishers and librarians.

In the 1960s, when prices of scholarly, particularly scientific, publications began to be less affordable to individual purchasers, libraries remained a reliable market. "Libraries were at the center of publishers' sales universe. . . . Virtually every reputable publication . . . found an archival home." (Hunter 1984, 75) Libraries still might aim for some definition of completeness in subject or type of holdings. As early as the 1970s, however, a perceptible shift began in library acquisitions, with monograph purchases subjected to greater scrutiny than the more "passive" journal subscription continuations (Carrigan 1992, 139). More recently, even the previously untouched serials collections have suffered dramatic changes.

JOURNAL PUBLISHING TODAY

The prices of journals—notably in the areas of scientific and technical publishing—increased faster than inflation during the 1970s and 1980s. Journal prices experienced a 13.5 percent annual rise during those two decades (Cummings, et al. 1992). Scholarly publishing now, rather than an activity of academic or society presses, is largely in the hands of increasingly monolithic commercial concerns that operate with expectations of continuing profits. Gherman and Metz (1991, 317) have pointed out that scholarly publishing represents what amounts to a monopoly: "It is clear that many scientific and scholarly journals are exempt or nearly exempt from the pressures a perfect market would exert to limit their prices. Each journal, and certainly each prestigious journal, is a monopoly by nature. . . . The monopoly enjoyed by each journal exists not only for the journal as a whole but for each article." Ironically, while much of the research reported in scholarly journals is publicly funded, the research institutions themselves are increasingly unable to afford to purchase it in its published form.

"In the U.S. . . . research has doubled in the last 15 years, but investment in libraries has increased by only 50%. . . . Of all the scientists who have ever lived, half are alive today." (Cameron 1993, 23) Compounding the problem of higher per-title costs for serials is the explosion in the number of titles. Particularly in technical areas, topics once subsumed within larger areas of inquiry (and published in the journals within those disciplines) are being splintered off, with new publications of more limited readership and thus higher subscription prices devoted to them. Among faculty, the length of one's publication list continues to be currency for tenure and promotion, encouraging research to be reported in fragments, in multiple sources.

Librarians Respond to Serials Prices

Librarians have responded to serials price increases in a number of ways, but only lately with critical subscription cuts. Reallocation of acquisitions money in an effort to maintain serials titles led to disproportionate spending on serials, and resulting in what Eldred Smith (1991, 232) identified as distortion of long established acquisitions and collection development programs, and imbalance among disciplines. A Mellon Foundation document (Cummings et al. 1992) reported that the institutional data in one study suggested that science journals accounted for 29 percent of serials titles, and 65 percent of the serials budget.

Initially, cancellations of titles were proposed only as a last resort. These cancellations did not cut to the core, but focused primarily on duplicate subscriptions and then on easily obtainable or arcane, low-demand titles. Today, drastic cuts have become so commonplace that it has been suggested that libraries rethink the archival aspect of their mission, and purchase journals for current rather than future use. In the *ARL Annual Statistics 1990-1991*, sixty percent of surveyed libraries planned cuts averaging $150,000 in subscription dollars; eighteen libraries planned to drop more than $250,000 worth of journal subscriptions. Ann Okerson and Kendon Stubbs (1992, 22) reported that "last year, U.C. Berkeley's $400,900 cancellation project was a high water mark; in 1992 at least six members plan to exceed it." The library at the University of California at Davis canceled 2,762 subscriptions—$650,000 worth (Nicklin 1991, A29), and the picture is not improving. In his introduction to the *ARL Statistics 1991-1992*, Kendon Stubbs (1993, 9) wrote that "ARL members paid $15.7 million more for serials in 1992 than in 1991, but had an average of about 600 fewer subscriptions per library, or around 60,000 fewer serials among all ARL academic libraries."

Some institutions cancel journal subscriptions to make statements underscoring growing dissatisfaction with serials prices. Princeton University Librarian Donald W. Koepp notified Pergamon Press of that university's cancellation decision this way: "It was agreed to cancel sufficient Pergamon journals so that the total cost of them in 1992 will be no more than the cost was in 1991 plus the percentage of inflationary increase in the allocations for each of these departments for this fiscal year. Consequently, you will find that almost one fourth of the journals in the September 1 invoice have been canceled." (Okerson 1992, 1) Other libraries have made similar subscription cuts to maintain budgetary stability.

In some institutions, faculty are urged to recognize the cost implications of the shift of scholarly publishing to commercial control, and to alter their publishing, reviewing, editorial board, copyright and other relationships with the most offending publishers. Academic librarians have had mixed results in broadening the discussion of the serials crisis to include scholarly communication in a larger forum. Faculty continue to demonstrate "no elasticity of demand for access to the publications that are the source of libraries' problems, giving librarians little latitude for developing leverage with publishers by threatening cancellation of subscriptions as prices rise." (Dow, Hunter, and Lozier 1991, 522) The mutual benefits derived from the publication of scholarly articles—cost-free text for the publisher, professional "credit" for the scholar—are central to the apparent invulnerability of this system.

Intensified resource sharing arrangements with other libraries and, more significantly, the use of electronic products and commercial vendors for single article delivery, are providing patrons with access in the face of subscription cuts. Some libraries have made a direct link between subscription cancellations and document delivery services, a link that supports both the fact and the perception of unimpeded access. At Virginia Polytechnic Institute and State University, some of the savings generated through major cancellation projects were earmarked for article delivery services. University Librarian Paul M. Gherman (1991, A36) wrote, "I expect this fund to continue to grow until by [the year] 2000 a majority of our periodical information will be purchased this way . . . although we will probably continue to subscribe to some heavily used journals in paper. . . . By using the on-demand article suppliers, we can offer wider access to information without the costs of cataloging, binding, and shelving." And, at Arizona State University's Hayden Library, the 1992 serials cancellation project was also accompanied by the trial of a rapid document delivery service.

How Document Delivery Affects Serials Subscriptions

What relationship exists between journal publishing, serials prices, and document delivery services? Kingma and Eppard propose that the availability of good quality, low cost photocopying services in libraries discourages individual subscriptions to journals in favor of inexpensive and convenient (to the user) article copies. By raising photocopy prices, libraries would, in effect, support an increase in individual subscriptions which "might result in publishers' decreasing library subscription prices." (Kingma and Eppard 1992, 533) This suggests that ready availability of single articles in libraries contributes to the high cost of subscriptions.

The provision of single article copies to individual users does impact the reprint sales for articles, and leads to the perception on the part of publishers that subscription cancellations are a result of article availability. Royalties collected for photocopying are not filling the revenue breach left by canceled subscriptions. Journal publishers are finding "a mismatch between the fall-off in journal subscription revenue on the one hand, and the revenue we are getting from photocopying and what we are likely to get from document delivery royalties on the other." (Cameron 1993, 24)

The proposition that the availability of single article delivery encourages libraries to drop subscriptions is debated by library professionals. A continuing discussion on the ILL-Listserv in 1992 yielded the statement that some libraries find it "more cost effective to borrow articles than to maintain an expensive subscription." Other librarians responded to the contrary: "Our records over the years seem convincing that there is a genuine market for single issue document delivery services and that ILL is definitely not a substitute for a subscription since the bulk of the requests are for single issues for a single patron." Another respondent's comments further that argument: "I really haven't seen any research that concludes that it is more cost effective to ILL than to buy (unless it is to compare commercial document delivery prices this way, which conveniently leaves out the overhead of staff to place these requests, contact patrons, re-request if the first supplier can't fill, and etc.). I think that collection managers are still canceling subscriptions based on low use, and based on which titles they think their users will miss the least."

Irene B. Hoadley (1993, 194) cites a 1968 Center for Research Libraries study that determined that "unless a title is used more than about six times per year, it is less expensive for the library to acquire a photocopy of articles from it when needed than to maintain its own subscription and file." She suggests a real need to determine whether that result is still valid. The ARL/RLG Interlibrary Loan Cost Study found that a research library spends an average of $18.62 to borrow a document or article or to purchase a photocopy, and $10.93 to lend a document. Weighing these costs against the cost of subscription, staff time in processing the subscription and in handling the issues, and space, would provide a truer cost comparison of access and ownership.

Serials cancellations are not the only factor affecting the demand for single article delivery services. Broader bibliographic access through online databases also intensifies the demand. While access to their journals is enhanced by inclusion in online databases, publishers do not necessarily see any profit in this, certainly not enough to make up for diminished subscription income. Both the online searching costs as well as the article delivery, either through interlibrary loan or other means, may call upon funds otherwise dedicated to journal and monograph purchases. Library budgets are now divided to support nonprint and one-time-only expenditures, which may likely come out of an "access" (formerly "acquisitions") budget.

DOCUMENT DELIVERY:
CONSIDERATIONS FOR PUBLISHERS

Publishers responded in different ways to this climate in libraries. Journal prices continue to increase. Bert Boyce (1993, 272) points out that "publishers who have already decided what they expect the bottom line from a particular journal must be, simply increase their prices if volume falls." Based upon copyright considerations, publishers have tried to control library photocopying and the development of electronic document delivery systems that bypass their interests. Publishers have also sought to develop new products and to market new services in order to more aggressively retake some of the perceived lost profits. Some are expanding into the document supply business with the idea that, if they do not, someone else—librarians, vendors, brokers—will fill that need.

The publisher must first decide whether to market articles through vendors or other document suppliers, or to provide them directly. A study of the document delivery needs of the United Kingdom library market found that from the librarian's perspective "the two most significant requirements of a publisher's document delivery service . . . are the ability to check availability and the price of documents, and the provision of a comprehensive range of documents from a single source." (Brown 1982, 77) Such features often are more readily offered by brokers' or vendors' document delivery services, representing a broader list of titles and with good mechanisms in place for communicating prices and holdings. It is no small task for the librarian to maintain current information on which publishers offer direct supply of articles, which supports a tendency to look to the comprehensive vendors' services first. Nonetheless, one publisher pointed out that "the publishers who stand the greatest chance of profiting the most are those who have the power to provide delivery services themselves. . . . It will be one of the publisher's great assets to control the promotion of and purchase price for the documents it creates." (Abbott 1992, 204)

A host of related decisions accompanies the choice to offer article delivery directly, such as in what format to provide access and what mechanisms to employ for ordering and for delivery. Both the amount of fees and the method of collection, including alternatives such as deposit accounts, per item charges, and licensing, must be established. Some document delivery services require an annual subscription fee as well as a per article charge. Another option is two-tiered pricing, one for journals held by the library and another for titles not held. Robert Campbell (1992, 217) suggests the possibility of a range of charges based upon "the age, length, and nature of the article." Images of a commodities market come to mind as, according to Clifford Lynch (1993, 11), "it is easy to imagine publishers applying information technology to vary prices of articles over days or weeks, based on usage levels, topic interest, citation analysis, or media coverage." The publisher must also consider rights: What kinds of reproduction will be permitted, for what purposes, and by whom, and, in an increasingly networked library world, what kinds of shared access will be tolerated, supported, or encouraged?

When publishers address not only the library but the end-user market, they should provide greater flexibility in access, ordering, delivery, and payment. For example, the acceptance of credit card orders for single articles might appeal to noninstitutional users. (CARL Uncover statistics show that more than half of their article orders are from individuals, purchased with credit cards.) Unlike journal publishing, the cash flow of document delivery occurs after the fact. Payments are made upon receipt rather than in advance of delivery (or even production) in the case of journal subscriptions. Karen Hunter's excellent article, "Document Delivery: Issues for Publishers," in *Scholarly Publishing Today* 1, no. 2 (March-April 1992) provides a veritable checklist of such considerations from the publisher's perspective.

According to statistics on royalty payment through agencies such as the Copyright Clearance Center, publishers are not profiting from the millions of photocopies shared among libraries annually in this country. It is estimated that between 2.7 and 9.4 million photocopied documents are delivered annually without payment, at a cost of $9.5 to $37 million in income unpaid to publishers (Feldman 1992, 33). Whether this copying is within fair use limits is unclear. While fledgling document delivery services are not recapturing dollars lost to serials cancellations, they do provide a way for publishers to gain some income and to retain some amount of control over their publications.

Articles on Demand: Publishing Alternative

Some publishers and vendors are turning to facsimile technology to provide articles on demand. Fax machines, and the fax boards and modems that enable them to work with personal computers, have become standard equipment, particularly in academic and corporate settings. While some document delivery suppliers have long offered fax delivery as an option, a number of new services and products have emerged offering fax delivery as their core method. Some suppliers integrate scanning, electronic storage, and other process or delivery enhancements as well.

CARL's Uncover is one of many emerging citation and article delivery services. The service is based upon faxed delivery of articles from the Uncover table of contents index. The user can search the database by keyword or scan an individual title's tables of contents, check availability and fill out a request for delivery. The request is routed to CARL staff members in libraries, who retrieve the articles, scan them, and transmit them to an optical storage device on the CARL mainframe from which they are faxed to the requester, usually within twenty-four hours. The turnaround time is faster than most interlibrary loan operations, and the average cost is less than that of an interlibrary loan request. Copyright royalties are included in the price. Deposit accounts or credit cards may be used for payment. Registering a profile enables the user to batch requests made at one time to expedite the patron information forms.

The UnCover Company has launched another service, UnCover Reveal, that provides current table of contents delivery for the journals included in UnCover directly to users by e-mail. Users select and register the titles of interest to them; the tables of contents are mailed to their electronic addresses at about the same time that the printed journal issues appear in their libraries. Articles of interest could then be retrieved either from their own library or through UnCover article delivery service. For further information, contact Martha Whittaker (mwhittaker@carl.org) or the UnCover Company (303/758-3030). The UnCover services provide a good example of the trend toward integrated document delivery services, combining a current awareness product with article delivery.

A fax delivery service is also offered by Meckler Corporation, one of a range of products and services in their "complete electronic storefront." A table of contents database including both periodicals and monographs published by Meckler is available on the Internet, and may be browsed free of charge. Not only articles but book chapters may be ordered electronically and shipped via fax or mail within four hours of order receipt, at a fixed uniform price.

Several news organizations are now providing fax document delivery services in different formats. In 1989, the *Hartford Courant* began offering subscribers "Fax Paper," a single-page summary of news and business information available a day before the regular publication; subscribers away from their designated fax could even have their copies faxed to any temporary location. Nonsubscribers could also use this "Fax-It-Back" service to receive single copies of the publication when a topic of particular interest was in the news. The *Globe and Mail* of Toronto is also offering a faxed edition, transmitted within five minutes, to buyers anywhere in the world (Rosenberg 1990, 29).

Specialized, industry-oriented, fax-based document products are also emerging. For example, Demand Publishing, Inc. offers the on-demand publishing service for legal opinions, which provides access to the full texts of opinions summarized in their client publishers' legal publications. The documents are scanned and stored, and are faxed upon request via touchtone telephone. Paul Jones attributes the success of this service to its link to the current awareness publications, and to its immediacy of delivery (Jones 1991, 15).

Suppliers face problems in relying upon fax for delivery of articles. Interruptions to transmission occur often. The paper capacity of the receiving fax machine is a concern, and, of course, if fax machines are turned off for the night the transmission will not succeed. Susan M. Stearns of Faxon Research Services has also pointed out that the quality of the faxed product is ultimately out of the hands of the supplier, but instead depends upon the transmission medium and the quality of the end-user's fax machine. Stearns (1991, 21) cautions that "several issues including the technology of facsimile, the speed and reliability of telecommunications (including the Internet and NREN), and document storage requirements must be understood and addressed by each community—vendor, publisher, librarian, and individual user—to insure widespread marketability of fax-on-demand for document delivery."

Electronic Document Delivery: Another Alternative

Electronic document delivery encompasses the use of computer technology for the identification, location, request, processing, delivery, and/or record keeping/accounting aspects of document provision. While electronic communications technology is increasingly used for request and verification, perhaps the most eagerly anticipated refinement of the document delivery process, electronic delivery of documents among libraries, is only beginning. This aspect, the breaking of the traditional interlibrary loan time barrier, will have the most revolutionary effect on the definition of access.

Publishers are incorporating some form of electronic document delivery into many of their services. Electrocopying, the scanning of documents for storage, access, and transmission, may be used in conjunction with a variety of delivery systems, such as CD-ROM storage, network distribution, and faxing. This process provides images which are accurate representations of a page including graphs, charts, and drawings, although not searchable without complicated conversion; indexes are often provided for machine-searchable access. In other formats, the text is machine-searchable, but formulas and other figures are not.

ADONIS, a full-text, CD-ROM storage and retrieval system, is perhaps the most widely known example of an electronic document delivery product initiated and developed by a consortium of publishers. Its roots extend to 1982, when six international scientific publishers, concerned about the widespread photocopying of their materials, developed an optical disk system to create journal archives for sale of single articles on demand. By joining together, the publishers were able to provide a comprehensive service for frequently requested titles in the field of biomedicine and the life sciences, and do so in a manner that was acceptable to their library market. ADONIS was launched in 1991, after ten years in development, funded by participating publishers who pay for inclusion, consisting of Blackwell

Scientific Publications, Elsevier, Pergamon, Springer Verlag, and others, with support from the European Economic Community and The British Library. By 1993, over 550 journals from more than forty-two publishers were available.

In the ADONIS process, journals are scanned as soon as they are published. The machine-readable images are then stored on CD-ROMs, which are shipped to customers soon after the print publication date. These disks are mounted in the library on jukeboxes, each holding 100 disks, which can then be linked to increase capacity and accessibility. The annual subscription fee for the service is about $16,000 U.S. dollars currently, and a per-article royalty on printouts set by the individual publishers averages $5 to $6 (U.S. equivalent). Printing is automatically monitored on site in libraries, and a quarterly report generated and sent to ADONIS on floppy disk to be decoded for per-article billing. ADONIS is very popular in Mexico and Latin America, where document delivery is difficult: for $16,000 a library can have instant access to titles that would cost over $250,000 for individual subscriptions. UNESCO supports ten installations around the world, primarily for document delivery in developing nations. Most subscribers in the U.S. are medical school libraries, commercial document suppliers, and pharmaceutical companies, which, although the fewest in number, have the highest volume of usage.

TULIP (The University Licensing Project) is a cooperative venture involving several major institutions, among them Carnegie Mellon University, Cornell University, Georgia Institute of Technology, Massachusetts Institute of Technology, all campuses of the University of California, and Elsevier Science Publishers. Begun in 1991, the project's intent was to examine networked delivery and use of journals produced only in print format. The service provides electronic files for forty-two material sciences and engineering journals from Elsevier and Pergamon to the universities, which can make them accessible in a range of local prototype or operational networked environments, according to their own needs and resources.

The TULIP files, updated biweekly, are maintained in an archive by Engineering Information, which provides Internet access to the participating universities. The universities will receive the electronic full-text for those journals to which they hold print subscriptions, at no additional charge, and will have bibliographic access to those to which they do not subscribe; an article-on-demand service will provide copies of those as requested.

At Carnegie Mellon, the full text of the journals will be accessible through its Project Mercury electronic online campus library. Keyword searching of the journal database is not available, but programmers have included table of contents pages that permit browsing and instant retrieval of desired articles from those pages (Krumenaker 1993, 11). Other TULIP libraries will rely upon on-demand delivery.

A range of configurations and implementations is evident among the various institutions, from single-sites to multiple campuses sharing one server, to regional network distribution. "Access tools and distribution systems on campus will also include a wide range of alternatives, from high resolution images sent directly to desktop workstations to print-on-demand of individual articles and of locally sold subscriptions." (Zijlstra 1993) It is hoped that alternative pricing and cost structures will emerge from the project, providing points of comparison with existing document delivery models; information on user behavior will be collected and analyzed. For further information contact Jaco Zijlstra, Project Manager, Elsevier Science Publishers Group, 655 Sixth Avenue, New York, NY 10010. (J.ZIJL-STRA@ELSEVIER.NL). Call 212/633-3757.

Red Sage, a project initially of the University of California at San Francisco Medical Libraries, AT&T, and Springer Verlag, is intended to examine issues associated with the electronic delivery of primary journals directly to scientists. Articles from twenty-four journals in the areas of radiology and molecular biology are scanned and provided to the scientists according to individual user profiles. Their precise use, including information on what they view, what they print, what they store, even whether they read summaries first, will be captured by Right Pages, a software package that manages, measures, and analyzes network activity. The libraries involved have agreed not to cancel their subscriptions to the journals and not to transmit the electronic journals over the Internet to other libraries.

The CORE (Chemistry Online Retrieval Experiment) project of Cornell University, said to be the largest body of electronic text of its kind, contains the full text and graphics of twenty American Chemical Society journals, some from 1980 onwards, with associated abstracting and indexing information (Barden 1992, 435). CORE represents another collaborative effort, though this time between a publisher (The American Chemical Society), Cornell University, OCLC, and BellCore, the research organization of Bell Telephone. The publisher provides the journals in machine-readable format and Bell Core translates the text into Standard General Markup Language. OCLC verifies the tagging and loads the database, which is sent over the Internet to Cornell. Two storage modes are employed—one of the tagged ASCII text files which is searchable, and the other of the full-page bitmapped graphics—in order to accommodate Cornell's multiple user interfaces. When all of the journals are on the system, ACS may offer it commercially (Krumenaker 1993, 10-11).

The Vendor and Document Delivery

Document delivery is increasingly offered by the serials vendors that have established working relationships with the library market and that seek to capitalize upon their existing access to journals. Linked with table of contents indexing and current contents products, article supply enables the vendors to recoup some of their losses due to serials cancellations. The vendors serve as intermediaries, working with the publishers to interpret varying pricing, copyright, licensing, and other issues, and maintain records and generate management information reports for their customers. These services are piggybacked upon established accounts and in-place communication channels. The service relieves librarians from some of the burden of locating sources for articles, and of negotiating individually with publishers about permission and royalty. With more comprehensive document vendors, we may in the future find something close to "one-stop shopping" for articles from many publishers. Vendors may offer flexible delivery options, including, as do UnCover and Faxon Research Service, delivery to the end user, if the library so desires.

For journal publishers, the inclusion of their titles in the indexing and current awareness tools of these vendors makes market sense. The vendors may continue to distribute the publishers' wares, though "unbundled" into smaller units, and they enforce copyright, in effect, by incorporating royalties into their fees. Vendors can also provide valuable market information to the publishers, reporting on what articles are being purchased and by whom. The cost to vendors of providing document delivery services can be high, particularly when the stock of journals is not already held as part of other operations (as with UMI and ISI), and each request must be located, reproduced, delivered, perhaps stored, and recorded.

One vendor, Faxon Company, has responded to issues of access and document delivery by combining table of contents indexing (Faxon Finder) and article delivery (Faxon Xpress) under the umbrella of the Faxon Research Service. The service is not intended to undercut the company's primary product line. "We see FRS as a service designed to facilitate access to the content of material itself and to do that quickly and inexpensively in a way that is complimentary to, rather than as a substitution for, the current library subscription holdings." (Hawks and Alexander 1992, 94)

Faxon Finder contains the table of contents information from over 10,000 journals, including all citable references, editorials, book reviews, and conference proceedings, as well as research articles. A search profile capability is planned to permit the saving of a search strategy for future automatic current awareness searches. The system alerts the user to new search results at log-on time, or sends the results to an electronic address. There is a single, fixed, annual licensing fee. Faxon Xpress is the delivery component of the service. Faxon CEO Rowe believes that this service is only an intermediary process, to be replaced when publishers, working in electronic format, begin "actually releasing the articles at the time that the editor says the article is ready to go." (ibid. 1992, 95)

UMI, another vendor, offers a variety of document delivery-related components in a line of products called ProQuest MultiAccess Systems, which enables the library to customize a system or to purchase a turnkey package, the ProQuest Article Delivery Network. *ABI/INFORM & Business Periodicals Ondisc* is the business database that includes citations from 900 publications and full text images for about half that number of titles. *Periodical Abstracts & General Periodicals Ondisc* is the general periodicals citation and article image database, available in three editions (depending on the number of titles included), that provides full-text articles from over 300 indexed titles. Article delivery system components include a CD-ROM jukebox with a 240-disc capacity to store the full-text articles; the ProQuest Network IMAGEserver, which receives article requests, locates the documents in the CD-ROM jukebox, and routes them; the Remote Image PRINTstation, which prints the requested documents; and the FAXserver, which automatically sends articles to fax machines or PCs with fax cards. An electronic link to the UMI Article Clearinghouse (document delivery service) is planned to provide supplemental access to the full text of articles from 12,000 journals.

Electronic Publishing

Electronic document delivery may be "only an interim stage before full electronic publishing takes off, and could become technologically redundant in the longer term." (Brown 1982, 87) Full-text documents in electronic format are already increasing in number and will continue to do so. The trend in both online and CD-ROM databases is shifting from the predominantly bibliographic to periodical or other full-text content (Tenopir 1993, 54). While industry experts do not foresee the replacement of print with electronic media, they can envision electronic communication as a parallel channel, perhaps seated in the noncommercial or scholarly environment.

Scenarios for electronic publishing include a range of applications of technology: the use of computers in the writing and editorial processes of traditional paper documents; the electronic delivery of printed materials; article-on-demand publishing from a centralized electronic database; and the completely electronic journal, with no print counterpart. Farther along the continuum are those published products that are either made possible or greatly enhanced because of the technologies that permit collaborative endeavors, that encourage interaction, and that incorporate hypertext and hypermedia in their presentations. Electronic publishing offers not only a variety of product alternatives to the publisher, such as articles on demand, current awareness services, and customized publishing, but also a number of pricing alternatives, such as licensing for network access, subscriptions for a set number of articles, pay-per-use (for the copying and retaining of a document), even pay-per-view (for the read-only access to a document).

Librarians are already used to electronically published and accessible bibliographic products such as indexes and abstracts, which textually mirror print counterparts but with the enhancement of computerized access and retrieval. Many librarians access the full text of some of the over 3000 journal, newsletter, and newspaper titles that are now available through online hosts such as DIALOG and BRS. BiblioData offers a twice-yearly directory of these publications, *Fulltext Sources Online*. Journals published solely or primarily in electronic formats, "e-journals," are less familiar, distributed commonly over the national and international networks rather than through the online vendors. *The Directory of Electronic Journals, Newsletters, and Academic Discussion Lists*, by Michael Strangelove and Diane Kovacs, published by the Association of Research Libraries, catalogs existing electronic journals and newsletters available over academic and commercial networks, and is itself electronically accessible. For further information contact Strangelove by e-mail (441495@ACADVM1.UOTTAWA.CA). We expect to see growing acceptance of these publications as their inclusion in bibliographies and indexes brings them to the attention of library patrons who, in turn, bring requests for articles to the library staff.

Electronic Journals on the Networks

Journals in electronic format have the advantage for the publisher of incorporating production and distribution activities, and of offering a variety of marketable format options, from paper journal to unbundled single article delivery to custom packages compiled to a user's profile to networked delivery. They also provide publishers with a means to recapture the individual subscription market, and to take advantage, through the networks, of two-way communication with their readers. On the other hand, access is not yet universal; a publication solely in electronic format has a more limited market. Parallel publication in both formats is not economical, with print costs per unit rising when the volume drops. Advertising as a source of revenue will have to be reworked by publishers for the electronic environment.

Some electronic journals have print counterparts and maintain some of the structural elements that define a periodical. *The Chronicle of Higher Education* has been broken into separate sections in its electronic version, *Academe This Week*. Although out of context, these separate sections are still issued according to a publishing schedule: advertisements for academic employment, for example, are available weekly on the Internet, as well as mounted locally on institutional networks. Another e-journal, QUANTA, "has even gone so far to make itself look like a paper journal as to provide dotted lines marked `cut here' at the start and finish of each release, assuming that readers or libraries will print the journal, cut away

its 'electronic' header information, and bind it like any other periodical."
(Stoller 1992, 655) Other publications take full advantage of the flexibility
of the medium, with single article "issues" published on no particular
schedule; electronic mail may serve as the means to announce the publication of another issue, with users retrieving the files when available
(Bailey 1992, 31). Alternate delivery formats, such as print or CD-ROM, are
sometimes offered in tandem with an electronic journal subscription.

For the user, a primary advantage of electronic journals is speed. They
are often published article by article, breaking free of the issue structure
and delivering information with an immediacy that is impossible with a
quarterly or monthly publication schedule. Electronic journals also offer
greater searching capabilities. With electronic publications, the user can
search large quantities of information, and can compile and assemble
documents into personalized packages. A range of personalized features,
such as current awareness and SDI, also enhances the electronic format.
Access through resource sharing is another advantage. Even texts held
remotely can be read locally, without the delay of traditional interlibrary
loan arrangements and the uncertainty about the potential relevance or
usefulness of a requested, long-awaited document. Sound, images, and
even movement are being incorporated into electronic "publications,"
although the slickness, color, and graphics package of the print medium
is not yet paralleled in electronic journals. Access to electronic journals,
however, is not as familiar, nor archiving as reliable, as in
the print collections of libraries.

OCLC has been at the forefront of some of the collaborative electronic
publication efforts. It introduced *The Online Journal of Current Clinical
Trials* on July 1, 1992. A joint venture of OCLC and the American
Association for the Advancement of Science, the work is the first peer-
reviewed, medical research journal produced solely in e-journal form. The
speed and flexibility of electronic publishing are particularly relevant to
this discipline; the latest research findings can be made available to physi-
cians almost immediately upon peer review. For ordering information, call
AAAS Subscriptions Department at 202/326-6446.

OCLC, under its OCLC Electronic Journals Online service, offers several
other online journals that combine full text and graphical images. *The
Online Journal of Knowledge Synthesis for Nursing*, a peer-reviewed e-journal of critical reviews of nursing research literature, provides current
research information to the medical profession. This is also a joint venture,
involving OCLC as the publisher and distributor, and Sigma Theta Tau
International Honor Society of Nursing for the editorial content. (Sigma
Theta Tau, 317/634-8171.) *Electronics Letters Online*, the electronic version
of *Electronics Letters*, is a biweekly journal of research in the field published by the IEE. The journal provides, via links to the INSPEC database,

access to the abstracts of other articles cited in the journal. For more information, call Michele Day at IEE, 908/562-5556.

Current Cites, a current awareness citation journal in the field of library and information technology, illustrates the migration of a print publication to electronic format. Begun in 1990, *Current Cites* was a printed list of citations and abstracts for the internal use of the staff at the Library of the University of California at Berkeley. At first a mailing list provided distribution to other libraries; in 1991, an ASCII text version of each issue was distributed through MELVYL, the University of California online catalog system. Later, it was posted as part of the PACS-L electronic discussion group and became primarily an electronic journal, with a paper version almost as a byproduct. Most recently, *Current Cites* was incorporated into *Computers in Libraries*, a printed monthly, as well as continuing to be available and archived electronically through PACS-L, and it is delivered to direct subscribers in printed form as well. Through its various formats, *Current Cites* counts over 11,000 subscribers and readers (Robison 1993, 26).

Electronic publishing will present challenges to library practice as we now know it. How do we know that an electronic publication exists, and how shall we evaluate its usefulness? Electronic products need to be announced by their publishers and reviewed by the profession, along with their print counterparts, in order that their "acquisition" be based not on the cachet of format alone. They also need to be included in traditional indexing and abstracting sources, and on bibliographies.

How can these publications be provided to users? Does the library download, print, mount locally, or offer network access? How are they to be cataloged? Some libraries are adding electronic journals to their serials lists and catalogs, but this involves cataloging something that is not "in hand" that also may not have any of the recognizable elements by which printed documents are described. Not only must the cataloging indicate that this is an electronic publication, it must also help the users access that publication (Thorburn 1992, 15). How are titles "counted" when they are unowned but accessible? How are they purchased, and for how much? How and for how long should libraries keep and provide access to back issues of electronic publications?

How is an electronic article cited, and is the citation reliable? For example, Sandford Thatcher writes of *Surfaces*, a comparative literature journal from the University of Montreal that "intends not only to publish online originally but to allow authors to go back in and to revise articles over time as they get feedback." (MacEwan and Chamberlain 1993, 221) Problems of bibliographic verification and documentation and delivery will abound when the documents themselves are in a state of flux. How are they shared? When something is available only electronically, will nonsubscribers be able to borrow articles through traditional channels from other libraries, or will licensing arrangements preclude this? A raft of copyright considerations will attend the development and dissemination of electronic publications.

CONCLUSION

Some grim visions pervade the relationship between publishers and libraries in this electronic future. Some persons fear that the on-demand availability of single articles will result in market-driven scholarship, with publishers fashioning their products to the interests of the moment; indeed, one vendor foresees a "keyword-of-the-month" club, where the hottest keywords will be planted in every article to ensure its retrieval and purchase (Hawks and Alexander 1992, 95). And, in an article-on-demand system, will an article's value be measured not by its use and citation by other authors, but by its mere purchase, even unread, which can easily be monitored? If libraries forswear their archival roles in favor of access, will the library's role become "an electronic switching centre for requested items"? (Barden 1993) As electronic documents, hot off the wire, come directly to end-users, will the librarian's vital role in the scholarly process be subverted to that of keeper of dusty print collections?

Librarians must continue to choose among available formats and methods of delivery. The evolving information repackaging and "de-packaging" brings the ability to select and use the precise information that is wanted, just when it is needed, weighing the advantages and disadvantages. A subscription is an investment with potential for future use; while a single article, delivered only once, is a disposable product that does not even remain within the library. Yet the costs involved in the maintenance, handling, control, and storage of that subscription may not justify its purchase over the article ordered "just in time." These considerations certainly challenge both library "ecology" and economy.

To users, the immediacy of a paper journal in-house may not be matched by the delivery of a single article within twenty-four hours; but the downloading of electronic text on the spot may well satisfy. Whether the article ordered remotely will fit the bill cannot now be foretold until the charge is already on the tab. Both librarians and patrons will increasingly depend upon the development of access tools that serve as selection aids, and rely upon their descriptions for purchasing decisions. Librarians are challenged to integrate the variety of possibilities into a cohesive and efficient system, and to select the most appropriate tool for each document and patron, considering price, format, technology, time frame, and patron preference.

Chapter 7

Document Delivery: Copyright Issues

As a lawyer specializing in copyright matters, I am continually struck by how few precise answers exist to so many issues that librarians and other information users confront daily in relation to their use of copyrighted materials (Rich 1992, 105).

Copyright law, its interpretation and functionality, has emerged as the focal point in the increasingly complicated relationship between librarians and publishers. For libraries seeking balance at the point where increased bibliographic access collides with diminishing financial resources, a variety of emerging systems offer information delivery solutions as alternatives to purchase. Innovative resource sharing arrangements and networks, commercial document delivery services, facsimile transmission, and electronic full-text databases are increasingly seen as a means of providing what was formerly, in some cases, on our own library shelves. These systems may also be, intentionally or not, circumventing the established publishing/marketing structure.

Not surprisingly, libraries, for long a primary target market for published information and partners to the publishing community in marketing information to readers, find now that their attempts at resource sharing encounter publishers' more vigorous efforts to control the means by which products are disseminated. In the midst of general confusion over copyright law interpretations and difficulty with enforcement, together with technological changes that are redefining the possible, publishers are striving through legal and economic avenues to arrive at clear and mutual acceptance of more restrictive definitions and practices. And while technology seems to be on the side of users, easing both access to and reproduction of information, court rulings appear to support copyright holders in their attempts to control access and replication (Rothman 1992, 103).

The primary documents that define copyright, and a plethora of explication, commentary, and interpretation in the legal and library literature, have not yet really helped the library community establish clear policy and practice. As John Garrett (1991, 24) wrote: "Despite elaborate rules, real behavior in the real world will fall frequently into the gray areas between what is definitely permitted and what is unquestionably forbidden." Libraries appear willing to comply when mechanisms are available, convenient, and comprehensible. Noncompliance, however, in the view of the copyright holders, is not excused by confusion about the law and the inefficiency of royalty collection systems.

COPYRIGHT IN THE UNITED STATES: A BRIEF HISTORY

The first federal copyright law in the United States dates to 1790 and closely resembles the Statute of Anne, in English law. That law, in 1710, extended copyright to authors themselves; also, by setting a time limit on the term of a copyright, with renewal available only to the author, it supported the concept of a public domain. The U.S. law was also based upon Article 1, Section 8 of the U.S. Constitution, which gives Congress the power to "promote the progress of science and the useful arts, by securing for limited times to authors and inventors the exclusive rights to their respective writings and discoveries." Thus authors were encouraged and rewarded and the public benefits resulting from their efforts were recognized and assured.

The Copyright Act of 1909 extended to copyright holders the right to make copies of their works. That law, enacted when the technology for photoreproducing was just being developed and refined, "did not create an exception for copying by or within libraries and did not manifest any anticipation of new technologies of document reproduction." (Tepper 1992, 343) Later, libraries incorporated use of this new technology into their procedures, under the vaguely defined doctrine of "fair use."

In the years following 1909, photocopying increased under the nonstatutory provision of fair use. Some voluntary and cooperative efforts among libraries and publishers addressed growing concerns about the effects of these practices upon copyrights. The parties reached agreement about which of these practices and to what extent they were, indeed, to be considered fair use. The Interlibrary Loan Code of 1935—adopted with some revision by the American Library Association in 1940 as the ALA

Reproduction of Materials Code—was the "gentlemen's agreement" that established guidelines permitting libraries to photocopy parts of books and journals for researchers rather than loan the original materials (ibid., 346). This early definition of document delivery was conceived jointly by librarians and representatives of publishers' trade associations.

A later effort to respond voluntarily to the issue of reproduction and copyright was a survey conducted in the 1950s by the Joint Libraries Committee on Fair Use in Photocopying. The conclusion drawn was that "photocopying did not cause measurable damage to publishers or other copyright owners." (ibid., 348) Documentation of economic damage resulting from document delivery arrangements continues to be a focal point in copyright discussion today.

Fair Use

Fair use, that is, use of a copyrighted work without infringement, was codified in U.S. statutory law for the first time in the Copyright Act of 1976. During more than a century before, even "exclusive rights" as defined by statute in the U.S. were "subjected to a judge-made equitable rule of reason ultimately known as fair use." (Latman 1986, 239) Most early questions of fair use involved quotation or excerpt, rather than duplication.

A test of the doctrine of fair use as it applied to library photocopying practice had reached the court in 1973. Williams and Wilkins, publishers of medical journals, charged that the article photocopying practices of the National Library of Medicine and the National Institutes of Health constituted copyright infringement. Although the Supreme Court ultimately affirmed a close Court of Claims (4-3) ruling that no infringement had taken place, the split decision failed to provide conclusive direction for the drafting of 1976's Copyright Act. Some opinions hold that the 1976 enactment statutorily reversed the judicial conclusion of the Williams and Wilkins case. They express the belief that the photocopying activities under scrutiny in that case would likely be viewed as systematic and, as such, infringing (ibid., 245); not until 1992, however, was that viewpoint supported in a court of law (American Geophysical Union v. Texaco.

Copyright Act of 1976 (PL94-552, Title 17, US Code): Section 107 on Fair Use

The Copyright Act of 1976 culminated more than 20 years of Congressionally-authorized study and debate. Although the sections that relate to library activity are stated briefly in the enacted law, those sections have engendered volumes of interpretation, and continuing and heated debate in the library and publishing communities.

Section 106 of the Act states that the owner of the copyright has the exclusive right, subject to sections 107-118, to do or authorize the reproduction of the copyrighted work, and to distribute copies to the public by sale or other transfer of ownership, or rental, lease, or lending. Following sections provide for exceptions to this exclusive right, permitting uses far beyond those derivative ones formerly allowed; prior to the 1976 revision of the Copyright Act, "in no instance did fair use mean reproduction of a work to use it for its own sake." (Horowitz and Curtis 1984, 67)

Section 107 articulates limitations on exclusive rights imposed by fair use considerations and permits reproduction in certain circumstances: "Notwithstanding the provisions of Section 106, the fair use of a copyrighted work, including such use by reproduction in copies or photorecords or by any other means specified by that section, for purposes such as criticism, comment, news reporting, teaching (including multiple copies for classroom use), scholarship, or research, is not an infringement of copyright." This general rule of fair use has been restated thusly: "One may make a use of the copyright of a work to the extent that such use does not unduly harm the copyright owner." (Patterson and Lindberg 1991, 200) In effect the law establishes limits on the copyrights by reserving to others the right to make certain reasonable uses without obtaining permission of the holder.

Of course, what constitutes "reasonable" use is arguable. The law's Section 107 cites four factors to assist in determining whether a use is fair, but these factors have been construed judicially as examples. According to Martin (1992, 348), they do not provide a "definitive list or rigid test."

The first of these factors considers "the purpose and character of the use, including whether such use is of a commercial nature or is for nonprofit educational purposes." Fair use as applied to photocopying in a library setting can, from the examples given in Section 107, encompass both productive use (as in the creation of other critical, research, or scholarly works) and intrinsic, or passive use (Reed 1987, 6); and "commercial" use is not always excluded from the fair use claim, while nonprofit educational use is not always found to be "fair." (Martin 1992, 349)

The second factor is the nature of the work to be copied. Fair uses of creative works, such as fiction, and works whose value lies in their brevity of expression, such as poems or songs, are likely to be narrowly defined; uses of works of reference or nonfiction, where content rather than method of expression is paramount—works more of "diligence than originality or inventiveness" (ibid., 349)—are more likely to be construed as fair. The author's intended use is also a consideration under this factor (Reed 1987, 7). Fair use of a textbook or workbook, intended for individual purchase and use, would likely be viewed restrictively.

The third criterion relates to "the amount and substantiality of the material used." Seeking a solely quantitative answer to what amount is permissible reproduction ignores the question of "substantiality": of what importance is the amount copied, however brief, to the whole?

The fourth criterion for fair use is perhaps the one most called into question with document delivery services, at least from the publishers' perspective: the effect of the use upon the potential market or the value of the copyrighted work. Considered to be the most important element of fair use, it is also "the least understood and, as a consequence, the most misapplied of the factors." (Latman 1986, 250) Some publishers contend that any library reproduction intrudes upon the potential market and, therefore, does not constitute fair use or fair dealing. "The existence of `fair-use' clauses in various national copyright laws, which permit the supply of a single article copy to a private user, has had the effect of providing for unlimited use of a journal without payment. This not only erodes the individual article reprint business of publishers, but also led to the publishers' belief that subscriptions were being canceled as material was freely available through photocopying." (Stern and Compier 1990, 79)

Librarians might counter with a different scenario. A likelier alternative to purchase, were single article photocopying not an option, particularly in budgetary hard times, might be to suggest other, on-hand resources, or to direct a patron to a nearby holding library. Still, "the most likely of the four factors to defeat a fair use defense" is this one. (Rinzler 1983, 27)

Copyright Act of 1976: Section 108 on Library Photocopying

Kenneth Crews' (1990) study of university copyright policies illustrates the confusion surrounding the copyright law and library practice. He cites a wide range of library statements on photocopying. Some universities permit single copies only, others cite CONTU guidelines and permit up to five articles, and others assert no copying limits at all. It is clear, though, that libraries are grappling with the issue of copyright, albeit in ways that may appear inconsistent.

Section 108 of the Copyright Act provides the statutory basis for library activities relating to the single-article copying that comprises most document delivery. It addresses both activities within libraries and interlibrary activities. Section 108(f)(4) extends protection to some activities that may not be considered to be fair use as defined above, but does not limit or alter those exemptions provided by section 107, stating that "nothing in this section . . . in any way affects the right of fair use as provided by section 107." And the ALA has emphasized that "rights of fair use granted under Section 107 are independent of and not limited by those rights granted under Section 108." (Library of Congress 1983, 57)

It is not an infringement for a library to reproduce and distribute *one copy* of a copyrighted work if three conditions are met:

- reproduction is made without any purpose of direct or indirect commercial advantage

- the library's collections are either open to the public or available to nonaffiliated researchers in the field, such as through interlibrary loan
- the reproduction includes a notice of copyright.

A single copy of one article or other small excerpt of a copyrighted work may be provided to a user by a library under certain conditions:

- the copy becomes the property of the user
- the library has no reason to believe it is intended to be used for purposes beyond private study or research
- a prescribed warning of copyright is displayed in the library where the request is accepted and on the request form. In instances where the requester does not actually enter the library to place a request, the warning notice must be displayed in other ways. It may be read over the telephone, for example, or faxed to the requester. The copy may be made from the collection of the library where the request was placed or may be provided from the collection of another library.

An entire work or a substantial portion of such a work may be legitimately photocopied under some circumstances:

- an archival reproduction of an unpublished work in the collection of one library may be made for preservation, security or for deposit in another library for research use
- a replacement copy may be made for a damaged, lost, or deteriorating work, for which reasonable investigation indicates an unused replacement cannot be obtained at a fair price
- a copy of an entire copyrighted work may be made if, on the basis of a reasonable investigation, it has been determined that a copy cannot be obtained at a fair price.

Under Section 108, rights extend to isolated and unrelated copying of a single copy of the same materials on separate occasions; they do not extend to related or concerted copying of multiple copies of the same materials on one occasion or over time, or to systematic copying of single or multiple copies. "According to Professor Melville Nimmer, subsection (g)(2) prevents a requester from asking for single copies of different articles from the same issue of a journal on separate occasions, thus eventually getting an entire issue, or substantial part thereof, without purchasing it." (Heller 1986, 30) While libraries can intercede in cases of requests that might be considered related or concerted when they are made at the same time, it is unlikely, in light of both the effort involved and the privacy implications, that a library could or would track any individual's requests over time to determine whether that user is engaged in a systematic, infringing activity.

Interlibrary loan arrangements are not prohibited from coverage under section 108 provided they do not have "as their purpose or effect that the library or archives receiving such copies or photorecords for distribution does so in such aggregate quantities as to substitute for a subscription to or purchase of such work." (108)(g)(2) Some examples of prohibited practices, provided in Senate Report 94-473, include the following:

- coordinated collection development agreements in which a library develops a subject collection and establishes itself as a source for copies, leading other libraries with similar collections to cancel or refrain from purchasing subscriptions
- the reliance upon reproducing photocopies for staff members' use, in lieu of multiple subscriptions to a needed journal title
- the establishment of agreements where branch libraries serve as article distribution sources for periodical titles intentionally not carried by other branches in the system.

ARTICLE SUPPLY THROUGH INTERLIBRARY LOAN

One document delivery concern with copyright among librarians relates to photocopies obtained from other libraries, through traditional ILL arrangements. Such arrangements are allowed under Section 108 of the Copyright Act of 1976 so long as they do not have the effect of substituting for subscription or purchase; indeed, the OCLC Interlibrary Loan Subsystem logged nearly two million photocopied "loans" in 1991, which accounted for 44 percent of all items loaned in that year (Nevins and Lang 1993, 38). The American Library Association Interlibrary Loan form, commonly used in libraries, provides for indication of compliance with the law or the CONTU guidelines. If a request falls under "fair use," for example, as articulated by the Law, or is excluded from copyright by any other stipulation of the Copyright Law, the designation "CCL" (Conforms to Copyright Law) is given; if a request is permissible under the CONTU Guidelines, as one of five or fewer in a year, from a given journal title within the last five years, "CCG" (Conforms to CONTU Guidelines) is indicated.

CONTU Guidelines

Congress established the National Commission on New Technological Uses of Copyrighted Works in 1974 to *consider and provide recommendations* on issues relating to copyright and emerging information storage, processing, retrieval, and reproduction technologies. The CONTU Guidelines on Photocopying Under Interlibrary Loan Arrangements were developed as part of this process, and agreed to by principal library, publisher, and author organizations. They remain today the standard articulation of compliant interlibrary photocopying activity. According to the American Library Association (Library of Congress 1983, 53), the guidelines

are intended "to help to define the point at which a library could proper-
ly be expected to expand its collection into an area of new interest. More
specifically, the guidelines define the parameters of a safe harbor."

The sticking point of Section 108(g)(2) of the Copyright Act is the
phrase "such aggregate quantities." Identifying the point of economic
harm to publishers from library photocopying activity would rest upon
the determination of when the activity is substituting for a subscription or
purchase. CONTU addressed this through the development of specific
guidelines defining compliance for a variety of situations. The Guidelines
for the Proviso of Subsection 108(g)(2) are intended to apply to the most
frequently encountered interlibrary case: the delivery of photocopies of
articles from periodicals published within five years prior to the
request date. They identify a level of borrowing that constitutes
a substitution for a subscription.

In brief, the guidelines define reproducing in "aggregate quantities as
to substitute for a subscription or purchase" as being the provision of
more than five filled requests in a calendar year for articles from any periodical title (not issue) published within five years prior to the request
date. Known as the "rule of five," this guideline does not address the
copying of articles from periodicals published more than five years prior
to the request date, which many librarians assume, in effect, to be "out
of copyright" in relation to photocopying.

Nonetheless, "the guidelines speak only to material which was published during the five-year period immediately prior to the interlibrary
request, but it seems reasonable to believe that if and when the question
should come before the courts, it would be held that they would apply to
older material as well." (Lieb 1983, 137) The limit of the rule of five does
not apply if the requesting library has a current subscription to a title or
has a subscription on order; this instance is viewed as if the copy was
made from the library's own collection, and the copyright restrictions of
section 108 on photocopying within a library would prevail. It also does
not apply if the copy is being requested because the borrowing library's
own copy is not available at the time that the request was made.

Although monitoring compliance is the responsibility of the requesting
library under the CONTU guidelines, librarians remain confused as to what
the responsibility of the supplying library should be. In the lively interlibrary loan electronic discussion forum on BITNET, one participant stated
that her library will provide only one article per issue from within the past
five years, unless the borrowing notes indicate compliance. Another participant on the list had noted, earlier, that "as a lending library we will not
send more than five articles to a single library." Yet, in the words of another correspondent, "There is absolutely NO reason for lending libraries to
worry about what the borrowing libraries are asking for." Another writer

noted, "Keeping track . . . is the requesting library's responsibility, since obviously if they made one request of one institution and one of another, neither of those institutions would know about the infringement."

The requesting library is required to maintain records of all requests made under these guidelines for three years after the end of the calendar year in which the request was made. It has been pointed out that the library performing the actual copying may, in fact, be liable for infringement: "Without the requesting library's representation of compliance with the Act or the guidelines, royalties may be due the copyright owner, and permission to copy may be required. Acting on an incomplete interlibrary loan form clearly puts the burden on the copying institution to comply with the Copyright Act." (Heller 1987, 48)

The National Interlibrary Loan form, modified in 1977, provides for the requesting library to indicate that the request conforms to copyright law (CCL) or CONTU guidelines (CCG). The use of these codes on interlibrary loan requests has also come into question: if the lending library is not responsible for monitoring compliance, why use these codes at all? In a statement that may not find agreement in the publishing industry, Patrick Brumbaugh (1991) wrote: "Many libraries believe that what the law requires is not a paranoid adherence to numerical limits, but rather a record-keeping system that monitors your requests, so that you can avoid repeated violations on a particular title and can keep your collection manager information of patterns that clearly DO mandate purchasing a title." The American Library Association has commented that "the CONTU Guidelines, useful though they may be, are not, in any sense, inviolate rules." (U.S. Library of Congress 1983, 57) Laura N. Gasaway, Director of the Law Library at the University of North Carolina at Chapel Hill and an expert on copyright issues and libraries, even refers to these guidelines as "suggestions," emphasizing their nonstatutory status.

Library Cooperation and Copyright

Senate Report 473 provided examples, given above, of some library photocopying situations pertaining to cooperative arrangements that are barred from exemption by Section 108(g) (Senate 1975). One interpretation that extends these examples has recently come to the fore (see the survey, below): whether all article copy provision arising under resource sharing or cooperative collection development arrangements is not excluded from exemption under the Copyright Act. At a recent workshop on document delivery, one publishing industry spokesperson expressed the opinion that copyright permission and royalties are due on *every* photocopy of copyrighted works provided through established library resource sharing arrangements, rather than exempted by the CONTU "rule of five." And a published statement from the Association of American Publishers (1992) reinforced that opinion in stating that "the Act, legislative history, and case law make quite clear that . . . de facto coordination of copying and purchase activity among customers, and the development of services—whether 'private' or 'public'—for the specific purposes of providing customers with copies, are beyond the limited exceptions to copyright owners' rights." Some in the library field, however, have not accepted this interpretation and have expressed the belief that only after a year as a member of such a consortium could a library amass enough use data to determine whether the requests for a canceled title exceed the "suggestion of five."

Even membership in library networks has raised a red flag vis-à-vis copyright. One writer notes that although "the Register of Copyright believes that library networks may not make use of section 108 and doubts the applicability of Section 107," the American Association of Law Librarians does not believe that membership in networks necessarily precludes libraries from rights under those sections (Talab 1986, 75). Yet another writer from the library community cautions that "groups of libraries which participate in cooperative acquisition programs must make certain that the document delivery component of the program complies with copyright law and guidelines." (Higginbotham 1990, 74) As Robert Wedgeworth of the American Library Association, (Library of Congress 1983, 68-69) commenting on the 1983 Report of the Register of Copyrights, stated: "Indeed, reliance on networking to substitute for a subscription to a periodical is not only illegal; it is inefficient and expensive."

AN INFORMAL SURVEY, A RANGE OF OPINIONS

In an informal survey taken in 1993, participants on the ILL-L discussion group and other library practitioners were asked to respond to the following scenarios. While their responses, presented below in composite format, express neither authoritative legal opinion nor even consistent library practice, they illustrate the range of operational decisions surrounding copyright and document delivery, and the difficulty in applying the law and the guidelines consistently.

Q. A library closely monitors interlibrary loan requests for article copies and pays royalties at regular intervals. If CONTU guidelines are exceeded for a title, must permission for additional copies be obtained and/or royalties paid prior to the filling of the requests or is payment "after the fact" sufficient?

A. One library director, and a member of the legal profession, commented that "the CONTU guidelines are 'guidelines,' they are not the law," and added that strict adherence to the guidelines would require obtaining permission (but not necessarily paying the royalty) prior to making the copy. Some librarians indicated that they would not exceed the guidelines without obtaining permission from the publisher in advance or using a "copyright-cleared" supplier; others stated flatly that they would make the copies and deal with royalties later. One respondent commented that simply by exceeding the guidelines, a library has not necessarily exceeded its rights under sections 107 and 108. The specifics of each instance of photocopying would need to be analyzed to determine fair use or other exemption. And another mentioned that for some publishers, those participating in copyright cleared services (such as the CCC, for example), permission is implicit. As one respondent pointed out, "Seeking permission before copies are made is a slow and very expensive operation and often results in no reply, which leaves the user and the librarian not quite sure what to do next." (Cornish 1993b)

Q. A library user would like copies of six articles from one recent issue of a journal. If this is the only request this year from this title, are royalties due on one article only or all six?

A. The variety of responses here indicates the confusion surrounding this issue. Several respondents believed that royalties would be due only on the sixth article requested, as providing copies of the first five would be permissible under the CONTU guidelines. Another said that an argument could be made, by applying the four fair use factors, that all six are fair use. Another felt that only three articles per issue were acceptable, and thus royalties would be due on at least three articles. (This respondent suggested that the patron purchase a copy of the issue from the publisher.) In two separate opinions, the point was made that this copying would be outside the limits of fair use because of the number of articles being requested from a single issue for a single user; their libraries would pay royalties on five of the six requested articles, with only the first copy permitted under fair use. This latter interpretation is the one with which the authors concur.

Q. If this request was for articles from a 1975 journal issue, would royalties be due?

A. Most respondents felt that royalties were probably not required for pre-1976 articles, more than one even stated that articles older than five years were "fair game." One respondent felt that royalties would be due even on an older issue. Graham Cornish noted that "making these copies in the United Kingdom without the necessary royalty payments would be totally illegal."

Q. Five libraries form a consortium. Resource sharing is one of many stated purposes of the consortium. The consortium does not have an official collection development policy, although lists of expensive titles being considered for purchase or cancellation are often shared. Reciprocal interlending is practiced among the members, with each library maintaining records for royalty payment and collection development decisions. Under these conditions, does the "rule of five" still apply or are royalties owed on every copy provided? What if the consortium were founded primarily for resource sharing purposes with joint decisions being made regarding purchases or cancellations among member libraries?

A. Some respondents believed that photocopying under reciprocal borrowing agreements, without concerted collection development, to be within the guidelines. One respondent mentioned that, while the rule of five applies individually to each member library, the collective copying practices might rise to the level where they became "systematic" and thus not exempt. Others saw no problem with copying under such arrangements, whether or not the consortium practiced shared collection development. An interesting facet of this scenario was addressed by one respondent, who felt that the relationship between the libraries was an important consideration. He argued that libraries with the same director, funding source, and processing facilities should be considered a unit, with the sharing of resources not "interlibrary" but "intra-library." Libraries linked within a state-wide institution, but with separate administrations, would not be viewed as sharing a single collection, and thus would be engaging in interlibrary, not intra-library activity. Other respondents expressed interest in this situation, and uncertainty about the implications. At a recent workshop on copyright law, Laura N. Gasaway pointed out that activity between branch libraries would be considered "inter-library" if each had separate interlibrary loan functions and if each were monitoring use independently under the "suggestion of five," and "intra-library" if the ILL activity and its monitoring were done centrally for the branches as a single library unit.

COPYRIGHT AND LIBRARIES IN COMMERCIAL SETTINGS

While Section 108 of the Copyright Act of 1976 specifically excludes from protection any reproduction made with "purpose of direct or indirect commercial advantage," libraries in corporations or other commercial organizations are not necessarily disqualified from inclusion under that section by their for-profit context. The 1976 House Report states that a library in a profit making organization would not be authorized to (1) use one subscription or copy to supply its employees with multiple copies; (2) use one subscription or copy to systematically provide, on request, copies of individual articles to employees; or (3) use interlibrary loan to obtain photocopies in aggregate quantities to substitute for subscription or purchase (House 1976). The Report does state that "isolated, spontaneous making of single photocopies by a library in a for-profit organization, without any systematic effort to substitute photocopying for subscriptions or purchases, would be covered by section 108, even though the copies are furnished to the employees of the organization for use in their work."

Virginia Boucher (1984, 57) wrote that "it is the library or archives within the institution that must meet the 108(a) criteria, and generally not the institution itself." Additionally, the prohibition on commercial advantage has been viewed as directed not even at the library, but at the photocopying *activity* (Martin 1992, 355). Charles H. Lieb (1983, 132) has pointed out that a clear contradiction exists in two Congressional reports on the Act. The phrase "direct or indirect commercial advantage," according to the 1975 U.S. Senate Committee on the Judiciary report, "is intended to preclude a library in a profit-making organization from providing photocopies of copyrighted materials to employees engaged in furtherance of the organization's commercial enterprise, unless such copying qualifies as a fair use." The 1976 report of the House Committee on the Judiciary, however, found that "the 'advantage' referred to in this subsection must attach to the immediate commercial motivation behind the reproduction or distribution itself, rather than to the ultimate profit-making motivation behind the enterprise in which the library is located."

While Lieb's (ibid., 133) interpretation holds that the intention is "to preclude a library in a commercial or profit-making enterprise from copying copyrighted materials without the consent of the copyright owner," others' readings of the law oppose that view. For example, Heller (1986, 28) opined that "the fact that an attorney's research is done with profit-making motive should not negate a finding of fair use . . . [and] . . . should not be deemed a predominant purpose of commercial gain and therefore, an impermissible use." This application of "fair use" by a profit making corporation was recently tested in a seven-year long class action suit brought on the behalf of scientific journal publishers.

At the center of this case (American Geophysical Union v. Texaco, Inc., 802 F.Supp.1. S. Doc. NY 1992) was the common practice of researchers at Texaco (and at many other corporate research facilities, law offices, and other commercial enterprises) of obtaining, for their own files or use, photocopies of articles of interest from journals held in the library. Texaco, unlike some of the other petroleum companies, did not avail itself of the "annual authorization" licensing arrangements offered by the Copyright Clearance Center (although it was a member of the CCC) that would permit this level of photocopying activity. In July 1992, in the U.S. Court for the Southern District of New York, U.S. District Court Judge Pierre Leval ruled that the photocopying activities of the scientists were outside the protection of Section 107 of the Copyright Act, and were not therefore "fair use." Leval clarified the differences between this case and an earlier test of photocopying [Williams & Wilkins Co. v. United States] in which the activities were found to be fair use. In Williams and Wilkins, the scientists involved were "engaged in pure research activities with no commercial motivation." (Goldberg and Bernstein 1992, 3) Despite Texaco's argument that the copying done by and for its researchers "served to advance science and the social good from research" (Decision in Texaco 1992, 13), Leval held that the research was commercially motivated. In addition, the options for accommodation provided by the CCC licensing services negated any claim that, as in the earlier case, the progress of science would be impeded by the burden of compliance. [In a brief filed by Texaco, however, both the motives and the mechanisms of the CCC were called to task: "Given the CCC's goal of advancing the financial interests of publishers, the record demonstrates, not surprisingly, that it is neither fair to users nor efficient." (Court of Appeals, 2d Cir 1993c, 32)]

This decision affirms that "notwithstanding the omnipresence of the photocopy machine, publishers and other copyright owners must be compensated for the commercially motivated, unauthorized reproduction of their works." (Goldberg and Bernstein 1992, 3) Fearing that the effects of this decision may extend beyond the photocopying activities of the special library, an amicus curiae brief was filed by a group of library organizations (Association of Research Libraries, American Association of Law Libraries, Special Libraries Association, and others), and a separate one by the ALA, in March of 1993. The ARL, et al., brief stated, "The decision of the lower court, if upheld, threatens a longstanding, reasonable and customary practice in for-profit and non-profit institutions alike—the spontaneous photocopying of single copies of published scientific, technical, and other research and scholarly journal articles, notes and comments by researchers, scientists, and scholars for their own research use." (Court of Appeals, 2d Cir 1993b, 3) The brief stated that the trial court committed errors in the application and balancing of the "fair use" factors, and that the court should not have addressed Section 108 at all: ". . . its conclusion that Texaco's scientists made copies for 'commercial advantage,' as that term is used in Section 108, flies in the face of legislative history expressly stating that the type of copying at issue here is not for 'commercial advantage.'" (ibid. 4) The ALA brief concluded this way: "In effect, this Circuit has been asked to hold that copying factual research, which ultimately may benefit a commercial entity, is presumptively not fair use. If that holding becomes the law of this Circuit, how do libraries protect themselves from even potential liability other than barring users from copying information, which may even remotely result in a benefit to a for-profit entity? How do libraries conduct this inquiry without a direct assault on constitutionally and statutorily protected rights to privacy? How do libraries pay for this inquiry?" (Court of Appeals, 2d Cir 1993a, conclusion)

In discussions on the CNI-Copyright Electronic Forum, participants have voiced concerns about the effects that this decision will have upon libraries in both the for-profit and the non-profit sectors. While the case only applies in the district in which it was decided, it is anticipated that appeals may take it to the Supreme Court, where it would become the law of the land (Gasaway 1993, 12). Ann Okerson, director of the ARL Office of Scientific and Academic Publishing, commented, "University attorneys will read this case and caution their librarians to be very careful about how much copying is done. You also have to wonder if this is the doorway to an action where a research library that provides interlibrary loan and document delivery would be affected." (Rogers 1992b, 110)

THE COMMERCIAL INFORMATION BROKER AND THE FEE-BASED SERVICE

The introduction to the CONTU Guidelines (Library of Congress 1977, A3:7) anticipates what is rapidly becoming predominant in the document delivery environment. "The point has been made that the present practice on interlibrary loans and use of photocopies in lieu of loans may be supplemented or even largely replaced by a system in which one or more agencies or institutions, public or private, exist for the specific purpose of providing a central source for photocopies. Of course, these guidelines would not apply to such a situation."

The Association of American Publishers (1992) has clarified its interpretation of the copyright implications of the activities of commercial and fee-based document delivery services. The statement says that "a commercial document delivery service engaging in the copying and redistribution of single and multiple copies of copyrighted articles must secure permission from and (if requested) pay royalties to the copyright holder. The case is not materially different for the newly-emerged, fee-based and technology-enhanced copying and distribution services of libraries. These activities are indistinguishable in purpose and effect from those of commercial document suppliers." Many of the established commercial document delivery services have recognized an obligation to collect and distribute royalty fees, and are rolling royalties into their charges; however, the library that uses commercial providers for documents will find that it is paying, on many occasions, a royalty for a single article copy that should fall under the fair use exclusion.

James S. Heller (1987, 43) has written: "The business of the information broker is to provide documents—either originals or copies—on demand. The company profits by providing the documents and any copying done pursuant to that end would likely be considered to be for a commercial purpose. Such copying would, in all likelihood, be considered unfair." C. H. Lieb (9183, 138) explained that brokers and document suppliers have limited rights under section 107 and none under 108 because "in conducting their business they act as principals for their own account and not as agents for their customers who may or may not have fair use rights." The Register of Copyrights, in the 1983 report *Library Reproduction of Copyrighted Works* (Library of Congress 1983, 120), addresses this issue directly:

> The term "information broker" covers firms with all levels of copyright compliance, ranging from those who obtain permission or pay for all copies they use to those who conduct their business, with no copyright permissions, by using the collections, staffs, and facilities of large academic and other libraries. If a person seeking a photocopy (whether locally or via ILL) works for any information broker, then the purpose of the copying is not "private study, scholarship, or research." If the library has notice of such person's status, then section 108 does not authorize the copying. As commercial copying of the most obvious type, section 107 would likewise be unavailable, meaning that in the absence of explicit permission from the copyright owner, photocopying for "information brokers" is an infringement of copyright.

Information brokers who include document delivery as one part of their package of services feel that their position is less cut and dried than the AAP statement would indicate. As Ruth Dukelow (1992, 120) has suggested: "The brokers could claim that this use is not contrary to section 108(d), because they are selling not the *photocopies*, but their *services* to locate the articles." Dukelow does mention the objection that the copies requested by brokers do not become the property of the requesters (brokers placing the request), as stipulated in Section 108, but get passed along to clients. The Copyright Clearance Center has added yet another consideration to this discussion. A posting on the Interlibrary Loan electronic discussion group in 1993 called attention to the Copyright Clearance Center's Catalog of Publisher Information's statement that "the system of authorizations (to reproduce articles) does NOT include . . . reproduction for resale to the general public." This certainly seems to exclude the large-scale photocopying activities of such brokers, and perhaps even the services of nonprofit, fee-based library services as well, from this licensing arrangement.

What about the cost-recovery document delivery services within libraries, also mentioned by the Association of American Publishers? Both the photocopy provider's purpose and the requester's purpose may come into play in a determination of copyright exemption. If the library, even indirectly, realizes a profit through its fees, the use may be viewed as commercially motivated. James Heller (1986, 29) writes that, while libraries may charge fees to hire staff in order to provide such services, if the library used the revenue from copying "to supplement existing resources, the library may actually be profiting from the copying." Laura

Gasaway (CNI-Copyright . . . Forum) stated that "most of the copying within higher education (absent coursepacks) is fair use. The other exception that I see is if a library is providing a commercial document delivery service to outside users. In this instance, the library would need to pay royalties." While the library providing such services to outsiders probably has no way of knowing if the proposed use is for commercial gain, Heller (1986, 28) expands upon this stipulation: "Under most circumstances, the extent to which a library may provide copies through an in-house, fee-based photocopying service to persons or organizations outside the parent institution will have to be justified under either Section 107 or 108 or the activity will be considered infringing."

TELEFACSIMILE AND COPYRIGHT

Telefacsimile offers a solution to the response time dilemma. It ". . . addresses the one weak link left in interlibrary loan procedure, slow document delivery." (Ensign 1989, 805) The newer generations of fax machines are increasingly affordable; the more rapid transmission speeds—now under 12 seconds per page, in some machines, as opposed to the six minutes per page in the 1960s—make telecommunication charges less forbidding. Features such as large capacity paper trays and automatic dialing let the fax machine work alone, as though it were a staff member on the night shift.

While the copyright law does not as yet address library use of the telefacsimile machine, many librarians assume that the sections of the Copyright Act of 1976 and the CONTU guidelines that apply to fair use as regards library photocopying and interlibrary lending continue to apply to these processes even when expedited by the fax machine. If fax is considered no more than a faster vehicle for interlibrary loan transactions, that assumption may indeed be justified.

Telefacsimile use normally requires that a single sheet original be used for transmission. Thus, the supplying institution must photocopy the requested pages from a book or journal resulting in two copies of that original being made: the photocopy that remains with the sender, and the second copy that appears at the receiver's fax machine. The first copy made is unlikely to be retained for future use, while the copy thus transmitted will normally be delivered to a patron. The question has arisen as to whether the transaction should count as two copies under CONTU guidelines or one. Proposing a parallel between telefacsimile and computer programs that are "copied" in order to be used, David Ensign (ibid., 811) writes: "Assuming that the copy made for transmission by the lending library is no longer needed and is destroyed after transmission, the effect upon the holder of the copyright to the material should be no different than the effect of the reproduction of the computer program. The inference can be made that making this copy is not an infringement." If a supplying library regularly "backs-up" fax delivery with mailed copies, the receiving library should also destroy one copy.

Because the CONTU guidelines place the onus for compliance upon the borrowing library, it seems accurate to assume that it is not the use of telefacsimile by the lending library that will alone determine infringement. In situations where availability of duplicated copies of a publication substitutes for a purchase, the method of obtaining those copies (locally through the use of a photocopier or remotely through the use of fax technology) will be less important than the effects of that activity.

OPTIONS FOR COMPLIANCE

Even with an understanding of the exceptions and a mechanism to track usage, the process of compliance is often cumbersome and slow. Any library that intends to comply with the spirit of the law by tracking usage and paying royalties on a regular basis may discover that this "after the fact" compliance, however commonly accepted, does not adhere to the letter of the law. If libraries need to obtain permission to exceed the "fair use" guidelines from the copyright holder before it provides the photocopy (unless some other arrangement, such as use of the Copyright Clearance Center, is in force), then rapid document delivery might well falter.

One route to copyright compliance in the delivery of documents may be through subscription. A library may enter a current subscription to a journal for which it gets many requests, thus qualifying to exceed the limit of the CONTU guidelines of five article requests per title from the past five years, per year. Copyright considerations aside, the costs involved in filling repeated requests from the same journal are likely to be greater than the purchase of a subscription, and such a purchase may thus also be justified in anticipation of future demand.

A request frequently put to librarians, and one vehemently discussed on the Interlibrary Loan electronic discussion group, is for multiple articles from a "single-theme" issue of a journal. A library is unlikely to choose the continuing commitment of a subscription to a narrow or very specialized publication for which the demand is momentarily high due to some event-related timeliness or an individual's research agenda. The ILL-Listserve discussions showed some disagreement about how much of an issue could be provided, ranging from one article to 49 percent of an issue; while Section 108(d) clearly states that the limit is "no more than one article . . . to a periodical issue," one librarian felt that in some instances, the "aggregate quantities" of Section 108(g)(2) allowed for greater leeway in interpretation for interlibrary loan requests.

Librarians commonly respond to such requests by providing a table of contents to enable the user to prioritize and, perhaps, limit the requests. In some instances, however, where a table of contents has some editorial content, that page would also be counted as one of five allowable copies under CONTU! One route might be to provide the permitted single article and to pay royalties (assuming that the publisher agrees, as a participant in the Copyright Clearance Center, for example) on the additional article copies. One participant on the ILL-Listserve in 1993 suggested that these special, one-topic issues might be cataloged as books rather than checked in as regular issues, or that such issues be loaned, as an exception to the usual noncirculating policy for periodicals.

Another way to provide access to these single-theme issues, rather than request photocopies, is to borrow entire issues from other libraries. In light of libraries' reluctance to part with original journals, particularly when bound in volumes, this idea may not be easy to implement. The Interlibrary Loan Policies Directory (Morris and Morris 1991) does list libraries that lend whole issues. One can also try to purchase an entire volume from a publisher—although many publishers do not retain back issue inventories. Agencies such as USBE, or vendors such as Faxon or UMI, are also possible sources, although holdings may not be complete and fulfillment may not be timely enough.

Several avenues exist for compliance through paying royalties to the copyright holder. Many libraries will fill multiple requests from a single issue when copyright compliance, through the payment of royalties, is indicated on the request form. Commercial document providers may structure a royalty collection mechanism into their fees; but this is not invariably the practice. Eamon T. Fennessy reported that, in the U.S.A., in addition to the 7.6 million photocopies provided through the interlibrary network annually, 3.1 million copies are provided by 260 document suppliers. He believes that only a fraction of photocopies are being paid for by those within the document supply industry (Fennessy 1991, 31).

Royalties may also be paid directly to a publisher. This requires that the library request permission to make the photocopies in advance, and accept the delay of awaiting a response. Such a request to a publisher should include the following information: title, author, and edition of publication; what material within the publication will be used; the number of copies to be made; how the copies will be made; the use to be made of the copies; the manner in which they will be distributed; and whether the copies will be sold. Some who have attempted to use this method report frustrations. "Dealings with publishers have varied widely. For example, professional and trade journals generally respond to permissions requests faster than the major trade publishers. . . . Some publishers refuse to accept requests sent via facsimile machines and others have installed recordings and electronic mailboxes to accept messages, making it almost impossible to contact their permissions department by phone." (Tackett 1992, 33) The time involved in this option may preclude it for most document delivery operations.

A new service offered through the Association of American Publishers PUBNET Permissions electronic mail system allows requests and responses for copyright permissions to be transmitted instantaneously, thus overcoming some delay. That service, however, is designed primarily to serve bookstores and copy centers that sell course packets. The list of publishers is limited to textbook publishers, making it of negligible use to libraries but certainly a possible model for other services. As Eamon Fennessy (1991, 31) points out, neither delays involved in obtaining permission, nor high royalty fees, legitimize noncompliance with the copyright law: "The US law does not say 'don't obey Title 17 if the rightsholder doesn't give you permission right away.' It says 'the rightsholder has exclusive right to reproduce the work.' Nor does the law say obey Title 17 only if the fee charged for the photocopy is acceptable to you the user."

"Out of print" does not equate with "out of copyright." Section 108 of the Copyright Act does provide for the duplication of entire out-of-print works if it has been established that a copy cannot be obtained at a fair price. The librarian would have to take steps such as contacting the publisher and established out-of-print vendors in order to determine that a copy cannot be purchased. For out-of-print works, such as textbooks, a service such as the Copyright Permissions Service, an arm of the National Association of College Stores, will provide a "generic" format copy for ten cents per page plus royalties and shipping (Blumenstyk 1992, A35).

Entering into a license arrangement, through special arrangement with a publisher or an organization such as the Copyright Clearance Center, is another option for compliance when repeatedly or regularly copying from a publication. Licensing was a topic of interest in July of 1993 for the participants in the CNI-Copyright electronic discussion group. One participant noted that a journal publisher had included, with renewal notices, a flyer offering "a new subscription plan for subscribers who need to distribute articles . . . on an organization-wide basis." For double the subscription price of a given journal, unrestricted photocopying is permitted for internal distribution. Electronically capturing articles by scanning or keyboarding for inclusion in an internal database is permitted; and two additional print copies will be sent to the same address as the original subscription (Cook 1993). Another participant on the listserve where this was posted objected that this sort of license "assumes away fair use."

Whether a library may, through contract or license, be required to relinquish the rights accorded by the Copyright Act has also been a topic of debate in the CNI-Copyright discussion group. Section 108(f)(4) of the Copyright Act does state that "nothing in this section . . . in any way affects . . . any contractual obligations assumed at any time by the library or archives when it obtained a copy or photorecord of a work in its collections." This apparently allows for a licensing agreement that precludes the exemptions permitted under the same section of the law. In another example, a newsletter publisher recently mailed a letter to subscribers stating that unauthorized photocopying and distribution of the publication constitutes copyright infringement and is illegal. The letter goes on to define infringement as when "any part of the publication is reproduced to be distributed to anyone, within or outside of the subscriber's organization." It remains unclear whether conditions of subscription in this instance constitute a contract (as in 108(f)(4)) that would abrogate exclusion under Section 108; the question of whether fair use can be summarily excluded is also raised by this approach.

Reproduction Rights Organizations

Reproduction rights organizations, or permission-payment clearinghouses, such as the Copyright Clearance Center in the United States, act as intermediaries in the document delivery process, and provide a mechanism for authorizing copying of published materials for which permission and/or royalty payment is required. These organizations do not provide documents, but rather serve as centralized collection and disbursement agents for the copying of registered documents by registered users. They have no responsibility for enforcement of copyright laws nor for monitoring of violation or infringement.

The foundations of the Copyright Clearance Center were laid in 1977 by the Association of American Publishers and other users in anticipation of the 1978 implementation of the Copyright Act of 1976. In order to facilitate compliance, the suggested structure called for member publishers and authors to establish and register publications that could be copied (in excess of the copying permissible by statute or fair use) and to individually set fees for permission to copy. (Indeed, the very existence of the Copyright Clearance Center was viewed, in the Texaco decision, as a factor in the finding against Texaco.) Membership in the CCC does not prohibit publishers from negotiating licensing agreements independently with users as well.

The Copyright Clearance Center's original service (then called Precoded Permission Service) provided for identifying codes printed in the registered publications. Now incorporated into the Transactional Reporting Service, this program provides permission for copying on an individual document basis. Publishers of scholarly, technical, medical, legal, and trade journals, business magazines, newsletters, books, proceedings, and other publications participate, with over 1.5 million titles from over 8,600 publishers worldwide providing immediate authorization in this way (Alen 1993, n.p.). Upon registration, and payment of a registration fee (as well as an annual charge), each user is assigned an account number and sent a Catalog of Publisher Information (COPI), which lists participating titles and authorization fees. The user provides the Center with a regular report of use, and the Center calculates the royalty fees and bills the user.

More recently the CCC developed the Annual Authorization Service. The common practice of purchasing one or two subscriptions to journals or newsletters and providing multiple photocopies to staff members and departments within for-profit organizations is coming under increasing legal scrutiny, as evidenced by the Texaco ruling. Through the Annual Authorization Service, established in 1983, participating corporations and special libraries need make only a single annual payment for all internal copying and distribution done from registered titles, therefore both minimizing the delay and eliminating the record keeping constraints heretofore imposed by compliance.

Growing interest in networked and electronic access to the full text of selected documents has opened a new area of involvement for the CCC. Several pilot programs coordinated by the CCC are in place that expand and redefine the possibilities for copyright compliance in a number of information delivery scenarios. The CCC now offers "unique databases composed of a limited number of specific titles for a defined and limited group of users." (Copyright Clearance Center 1991) In one project, the CCC has assisted the participating organization in obtaining authorization to provide network access to specific documents over a corporate computer system. The CCC's University Pilot Program in 1990 involved six academic institutions in a study of the types and quantities of materials copied for a variety of purposes. (Crews 1992) Other pilot programs involve the inclusion of copyrighted materials in a company-wide electronic current awareness system. (Wilson 1992, 18) The Copyright Clearance Center is located at 222 Rosewood Drive, Danvers, Massachusetts, 01923, phone 508/750-8400.

International Copyright Considerations

Copyright laws vary from country to country, and consideration of these nationally established laws lies beyond the scope of this book. For an overview of various national copyright treatments as they relate to document delivery, the authors suggest the article by Graham Cornish (1989a, 117-123), "Copyright Law and Document Supply: a Worldwide Review of Development, 1986-1989" in *Interlending and Document Supply*, as well as that journal's ongoing considerations of the topic.

Although information is increasingly readily available through the ever-widening electronic webs, with little more than a nod to its country of origin, copyright protection is not similarly synchronized. "Legal support and constraints are not adequate at present, nor are they harmonised interna-

tionally. Many existing copyright laws still pre-date even the photocopier."
(Plassard and Line 1988, 68) While copyright is defined and established in
national law, it is the international agreements such as the Universal
Copyright Convention and the Berne Convention that provide for cross-
border protection of copyright. The Berne Convention, to which the
United States became a signatory in 1988, is the major multilateral, international copyright agreement establishing minimum standards for signatories.
Both of these agreements are based upon the principle of national treatment, meaning that member nations protect works of other members'
nationals to the same extent as they protect the works of their own nationals. For the librarian grappling with issues related to photocopying of foreign materials, adherence to the concept of "national treatment," and
reliance upon the national reproduction rights mechanisms when required,
will likely prove the best course.

Questions of document delivery and library photocopying are not clearly resolved through these conventions. While copyright owners are guaranteed reproduction rights both through Berne and national laws, exceptions exist for purposes of teaching, private use, and "fair use," under certain conditions; and many national laws impose more restrictive and/or
more specific exemptions. Perhaps because copyright is difficult to monitor and to enforce, and exemptions are not consistently defined, a trend
has been identified in international copyright from control of copying to
payment for use (ibid., 68). "Some countries have revised their copyright
laws to depart from the exclusive (but too often unenforceable) rights
model to a system of legal licenses, authorizing the copying but ensuring
compensation to the copyright owner." (Ginsburg 1992, 185) In many
countries, reproduction rights organizations provide a means to administer
permissions and royalty payments both internally and internationally.

Reproduction rights organizations have become established worldwide
in the last decade, as technological advances increasingly challenged copyright controls. They provide collective administration of copyright that,
according to Joseph Alen (1992a, 74), is the most efficient way to manage
the sheer volume, the different national languages and laws, various royalties in different markets, and different systems of enforcement:
"Reproduction rights organizations are the only answer. They are adaptable
to different economic, political, and legal structures, and they approach the
problem collectively." (Alen 1992a, 74) The RROs, like the U.S. Copyright
Clearance Center, operate within their countries to grant authorization to
users to make copies of copyright protected works.

RROs also operate internationally, through bilateral agreements with
other national RROs; based upon the Berne Convention, they function as
agents for the collecting and disbursing of royalties on behalf of foreign
copyright holders. For example, in 1987, the Foreign Authorization pro-

gram was initiated by the Copyright Clearance Center, which made it possible for U.S. copyright holders to collect royalties due from photocopying of their works in foreign countries. Almost twenty-five percent of the royalties collected by the Copyright Clearance Center are now for foreign rightsholders; almost five million dollars in foreign royalties have been collected by CCC for publishers in the United States (Alen 1992b, 119).

Reproduction rights organizations in different countries have devised mechanisms specific to their situations and to their national laws. For example, in Norway, Kopinor, founded in 1980, represents Norwegian authors' and publishers' organizations, collecting and distributing their royalties, and also represents foreign "rights holders" for copying of their works within Norway. One mechanism used by Kopinor is a per-page fee for student copying (Vokac 1991, 200). In Germany, VGWort uses a machine levy, collecting fees for the manufacture and importation of photocopying machines according to the machines' capabilities and locations (Alen 1992a, 75). In Britain, the Copyright Licensing Agency (CLA), formed in 1982, monitors the interests of copyright holders, institutes legal proceedings to enforce copyright holders' rights, and collects and distributes fees for copying, through licensing arrangements. Its licensees do very little record keeping; the organizations perform regular but infrequent sampling, which enables the CLA to establish fees and determine a basis for their distribution to publishers. The CLA is particularly interested in document delivery services, and sees its licenses as a way for document suppliers to comply with the restrictions of the Copyright Designs and Patents Act of 1988.

The International Federation of Reproduction Rights Organizations is a membership organization for the RROs. Three aims of this organization are to encourage creation of RROs worldwide, facilitate agreements between and on behalf of members, and, through educational programs, increase public awareness of copyright and the mechanisms offered by RROs for compliance (International Federation of Reproduction Rights Organizations 1991). Information on these organizations is available through IFRRO.

Some publishers believe that much royalty-due photocopying involving document delivery between nations is done without compensation to copyright holders. In 1992, the Association of American Publishers surveyed a number of academic libraries in an effort to quantify transnational document delivery activity. Their letter, summarized on the FISC-L electronic discussion group, defined "cross-border document delivery" as "a transaction by which an entity in the United States obtains a copy (other than an original) of a copyrighted work or portion thereof from a source outside the country." The AAP asserted its support for publishers' rights on an international basis, and posed questions to ascertain how much document supply in this country comes from overseas sources, and which overseas and domestic sources of document delivery are used most frequently. As might be expected, the librarians queried greeted the survey with some suspicion, another indication of the widening gulf between librarians and publishers over the copyright issue.

ELECTRONIC DOCUMENT DELIVERY ISSUES

It seems certain that even more complex copyright and information supply issues will attend the development and growth of electronic document delivery. The electronic document delivery service, and the timely and affordable access it promises, grows increasingly among both interlibrary and commercial document supply facilities. Commercial document providers rely upon scanning, storing, and transmission of articles as the preferred method of delivery, with new products and services, such as tables-of-contents indexes and linked delivery of articles, dependent upon these technologies. Libraries are incorporating scanning and networked delivery into their regular document delivery operations; some have launched digitized information pilot projects of their own. The prediction, in a report to the Council on Library Resources (Information Systems Consultants Inc., 1983, 44) a decade ago, that electronic delivery is unlikely to "displace more familiar techniques such as U.S. mail, courier, and truck delivery systems within the foreseeable future," only serves to highlight how far document delivery has evolved in this short time.

The confusion over copyright in the electronic publication, storage, and transmission of information is compounded by several factors central to the nature of the beast. In electronic publishing formats, problems of authentication and of identification abound. The controls offered by "fixing" the text on a page do not apply:

> In electronic form, information and ideas resemble images, which don't vary as much as words or expressions of ideas. Second, the new technology is more difficult to use without making a copy of it . . . into a computer memory or onto a screen. Third, with electronic technology, ideas can be put to entirely new uses, and can be more easily stored, disseminated, and compiled for different purposes. Finally, electronic information can be taken apart and recombined to create new works (Ogburn 1990, 260).

Gary Cleveland (1991a, 126) points out that some dispute even exists about what constitutes a copy in the electronic environment: "The process of delivering a single document electronically can be argued to involve the making of a number of copies: one when the document is transmitted to a remote computer's memory, another when it is displayed on a screen, another when it is saved to disc, and yet another when it is printed on paper." Lenzini and Shaw (1992, 14) make another distinction, between "an electronic representation of a page during a photocopy or telefax operation . . . and the storage of that representation for later re-use in producing another copy." Both the electronic scanning and transmission of documents and the storage of document images are a source of concern to publishers and copyright holders. As Carol A. Risher (1993) of the Association of American Publishers expresses it:

> The 1980 amendment to the copyright act made clear that input into a computer system is making a copy and that it is different from other forms of copying. The copyright act specifically speaks about "photocopying" in libraries (in section 108) and there is a definite difference between photocopying and electronic storage. One must get permission to convert copyrighted material to electronic form whether to do so is to "create a derivative work" (an exclusive right of the copyright holder) or to "make a copy" (another exclusive right of the copyright holder) or to distribute over the network (another exclusive right) or to display on screen (another exclusive right).

Although Risher's comments referred to electronic reserve room issues, the circumstances described could as well apply to the scanning and transmission activities of document delivery.

How does fair use apply in this electronic arena? Some argue that it pertains just as it does in the print context. "Just because something is stored or sent electronically, its copyright status does not change nor does its fair use status. If it is permissible to make a photocopy, it is permissible to make an electronic copy. Publishers don't say this, but the copyright law does." (Gasaway, 1993, 10) Cleveland (1991a, 127) points out, however, that electronic availability of documents may result in a "magnitude of copying . . . that . . . may seriously undermine the potential market for a particular document," which is one characteristic of the fair use test.

Licensing is one approach to managing copyright for electronic documents. A license may, for a single fee, grant limited or unlimited copying permission to a defined group of users, as does the Copyright Clearance Center's authorization service. Pricing under a license may be based upon surveyed usage in the particular environment or a similar one, or may be related to other measures such as the number of user stations that can access the document. For example, a document may be made available over a local area network for a licensing fee that takes into account either total potential users or potential simultaneous users.

In recent years, the "shrinkwrap contract" has emerged as an informal agreement. The term applies to an attempt to place restrictions on the use of a publication, such as one in a CD-ROM format, through implicit agreement upon purchase. "Producers put a notice on software packages indicating that it is 'licensed' and not sold to the individual. The statement generally indicates that opening the package [constitutes] acceptance and agreement to the licensing conditions and restrictions listed on the package." (Nissley and Nelson 1990, viii). A "shrinkwrap contract" may attempt to bind the purchaser of information to certain limitations on the use of that information, which may exceed those stipulated by law. This agreement may affect both the in-house use of a document (paper or electronic) and the supplying of a document to remote users or other libraries. "It is unlikely that such a contract is enforceable with regard to any use of information forbidden in the contract but permitted under the Copyright Act." (Bennett 1993, 89) Gasaway encourages librarians to try to negotiate aspects of licensing contracts, rather than simply accepting all standard conditions as written.

Copyright holders find it nearly impossible to control photocopying of their works; electronic reproduction offers not controls but manageable accounting possibilities. Graham Cornish (1993a, 14) says that "in the end the question surrounding copyright materials is not so much 'How shall we prevent access and use?' as 'How shall we monitor access and use?'" He describes a consortium of interested parties involved with publishing, libraries, software, hardware, networking, or databases from six countries who, under the name of CITED (Copyright in Transmitted Electronic Documents), developed a model for monitoring the copying of copyrighted materials that are stored or processed digitally. The CITED model captures and records usage of information, detecting actions that are not permitted (either generally or by an unlicensed user), reporting on usages, and even suggesting to users how they can legitimately execute requests that have been refused. It is hoped that this model will lead to an "acceptable minimum standard of protection and access." (ibid., 19)

Major benefits of the emerging electronic document delivery environment are the speed and ease of delivery. Those advantages will be undermined, however, unless the efforts to control electronic copying are not resolved to the satisfaction of all. It is imperative that users, publishers, vendors, and librarians soon agree on copyright issues.

Chapter 8

International Document Delivery: Regional and Worldwide Systems

Scholarship, research and professional study transcend linguistic and national boundaries. We live in an age when people of all nations are more interdependent than ever before, and technology has accelerated and intensified this interdependency. The education and research community is global. Students from almost every country can be found in every other country . . . the more we interact with other nations, the more we become interdependent for information sharing (Brown 1989, 74).

For economy and efficiency, each country should provide access to materials published within the country itself and to the resources of its own major libraries, but often that is not enough. Each country has to decide whether to build up its total national resource (the combined holdings of all its libraries) by subject specialization or other means, or to rely on international interlending and document delivery (Line 1989a, 110). Self sufficiency of 70 to 80 percent should be each country's aim. Access costs make levels below 70 percent not cost effective: to develop a collection to meet user needs beyond 80 percent requires investments too large to justify the marginal return they would bring. Maurice Line (1979, 164) demonstrated this equation using BLDSC as an example: "In the UK, two-thirds of the 49,000 journals [1979 figures] currently received by the British Library Lending Division are rarely used; 1,500 journals account for 50% of satisfied demand, 5,500 for 80%, and 9,000 for 90%. . . . twice as many journals are required to meet 75% of demand as are needed to meet 60%, and twice as many to meet 90% as 75%." Line (ibid., 164-5) says that a core collection of approximately 8,000 journals, most of which would likely be already available in the major libraries of each country, would meet 80 percent of demand. Meeting 80 percent of a country's needs for non-journal materials is more problematic and not as easily predicted as journal usage.

IMPERATIVES FOR INTERNATIONAL DOCUMENT DELIVERY

No country, however wealthy or however effectively it plans for national availability, can be totally self-sufficient, so international document supply becomes necessary (Line 1986, 5). Line (1979, 162) says that "efforts to improve bibliographic access must be accompanied by an improvement in document supply, with the ultimate aim of Universal Availability of Publications. . . . UAP aims at ensuring that everyone, everywhere, can obtain access to all documents, whenever and wherever they have been published" and that "extended access to bibliographical references will be merely frustrating unless the documents to which they refer can be obtained."

IDD Requirements

With the rapid development of linked electronic research networks, it seems reasonable to assume that international document delivery may escalate tremendously in the next few years. Line, long a leader in developing international standards for document supply, defined the following requirements for an effective international document supply system (ibid., 163-164):

- the documents must be available
- requests must be transmitted quickly
- documents must be transmitted quickly
- accounting must be simple
- procedures and forms must be simple and standardized
- local (or national) channels should normally be used first
- each country must have a national center or centers, to plan, coordinate, and monitor.

These principles (slightly revised) were adopted by IFLA as international lending guidelines. Although most countries accept the IFLA philosophy, not all comply. Roderick S. Mabomba (1989, 59) sums up the feeling in some Third World Countries: ". . . they sound like preaching 'principles of nutrition' in an environment plagued by drought and famine." He goes on to hint that, while speed may be desirable, ". . . the most important thing is 'whether' the document gets delivered: 'when' or 'how' are often secondary considerations in Third World environments." (ibid., 61)

Even large developed nations, such as the United States, have trouble adhering to the principles. Large libraries here are likely to be more directly involved with international ILL because the Library of Congress, which previously served as a national center for receiving foreign ILL requests, no longer provides this service (Wright 1993). It will be interesting to see if this leads to more foreign libraries joining the major United States bibliographic networks, such as OCLC and RLG, with their large international database collections of library holdings. Early evidence that such is happening occurred when the Consortium of University Research Libraries (CURL) in the United Kingdom formed an alliance with the Research Libraries Group. Many of these libraries, and other major national libraries and research centers also use OCLC, but many have been "nonsuppliers" (i.e., do not accept ILL requests from OCLC) in the past.

IDD Impediments

International document delivery has always posed a number of hurdles. Electronic research networks eliminate some barriers created by distance, difficult terrain, and poor mail or transportation systems by making library holdings accessible electronically, and by providing an avenue for the electronic transmission of some documents. Many countries have developed integrated library networks, but there is still considerable variance. Europe counts at least thirty library networks within the European Community (EC) alone. Even with language and cultural variations, it may prove easier to standardize those few compared to ten times that number in the United States. In Europe, national research networks and national library networks are more likely to have been linked from the beginning, whereas in the United States the two systems developed separately.

All European networks are now linked through the IXI backbone (Cleveland 1991b, 78). This standard linkage is important because regional differences make total cooperation unlikely. According to Maureen Pastine (1992, 126): "It is doubtful if there will ever be one centralized European database of library holdings." Much of the cooperative networking effort is guided by the Commission of European Communities' (CEC) Plan of Action for Libraries in the EC (ibid., 119).

International ILL/DDS projects are increasing as networks reduce communication barriers. In one cooperative project the National Library of Canada, BLDSC, and the Library of Congress tested the use of ISO ILL protocols for the exchange of interlending messages in the international arena (Cleveland 1991b, 101). Libraries in the Netherlands, the United Kingdom, and France are currently involved in the PICA (Project for Integrated Catalogue Automation) RAPDOC Project (DerWers 1994). Around the world libraries are installing Ariel, the software from RLG, which allows electronically scanned documents to be quickly transmitted from libraries thousands of miles apart.

The CEC has funded many international document delivery projects. FIZ 4 was designed to solve the problem of rapid delivery of patent texts, including associated diagrams, chemical formulas, characters, and symbols. Parliamentary reports and other documents of European Community interest were transmitted to users in member countries in the EURODOCDEL project (Gurnsey and Henderson 1984, 90-91). The CEC Libraries Programme is currently sponsoring EDIL (Electronic Document Interchange Between Libraries), running from January 1993 through June 1995. EDIL involves three organizations in France, PICA in the Netherlands, BLDSC in the United Kingdom, TIB in Germany, and one Portuguese library. Pilot experiments are planned for the first six months of 1995 (CEC Libraries Programme 1994, 6).

Many international projects involve the transmission of electronically stored documents. The success of such projects could lead to more international resource sharing agreements and cooperative conservation projects— once copyright legalities are resolved. Whether preservation copies are produced on microform or by digitizing, the processes are expensive and labor-intensive. It is not cost effective if works already duplicated by one library or publisher are recopied by another. Once a master copy produced, further copies can be made quickly and at low cost, and can be made available economically to other libraries for preservation or interlending purposes. Conservation projects may thus provide inexpensive copies which could be loaned overseas without the high costs and risks to materials currently prevalent in international document delivery (Line 1986, 10).

International document delivery is plagued by a lack of standardization, including different ILL request forms, variations in payment mechanisms, customs practices and postal regulations for research materials. Libraries sending mail or fax requests should use IFLA International Loan Request Forms which are recognized by libraries throughout the world. The forms are available from the IFLA Office for International Lending, c/o BLDSC, Boston Spa, Wetherby, West Yorkshire, LS23 7BQ, U.K. *A Guide to Centres of International Lending and Copying* (Barwick 1990) is the best source for determining if special request forms are required by national interlending centers. The guide is essential for any library doing extensive international lending or borrowing. Besides telling which libraries serve as international lending centers, it describes payment mechanisms and procedures adopted by IFLA libraries as acceptable for international ILL/DDS activities (ibid., 1990). Unfortunately, it does not go into customs and postal regulations for the countries covered; nor have past editions covered electronic transfer of requests or photocopies. The *Model Handbook for Interlending and Copying* (Cornish 1988) and the *Interlibrary Loan Practices Handbook* (Boucher 1984, new edition forthcoming) also provide guidelines and procedures for international document supply.

Libraries that borrow heavily from U.S. libraries may also wish to acquire the *Interlibrary Loan Policies Directory* (Morris and Morris 1995, in press) to determine whether the material desired is likely to be loaned. This source is particularly good for policies related to dissertations, theses, microforms, etc., and also provides information on charging policies (although fees cited may not be up-to-date). A similar work exists for Canadian libraries (Smale 1994a, 12). Sending a request which has no chance of being filled is expensive and wastes the time of the patron and all libraries involved. International requests should always come from a library, not an individual, and should always indicate copyright compliance on photocopy requests. Otherwise, the request will almost certainly go unfilled or be delayed.

Documents sent across international borders are often delayed by customs inspections. A customs declaration is not necessary on packages labeled "international loan between libraries," but that does not guarantee that materials will not be delayed while customs staff check procedures! Microfiche may receive particular scrutiny. David Bradbury (1992) told of instances where microfilm came under suspicion as espionage material and was held up until the film could be viewed prior to forwarding to the borrowing library. Will computer disks meet a similar fate?

Libraries often insure books sent abroad, but insured mail must have a declared value, which requires a customs declaration, thus slowing delivery. The receiving library might also have to pay customs brokerage fees, adding to the cost of the transaction. Some countries do not accept insured mail at all and will return the documents to the lending library without the material ever reaching its destination. Each library should acquire a copy of its postal system's regulations for international mailings, such as the *International Mail Manual* for the United States (Postal Service, 1993).

International book loans will probably always be a weak link in the chain of international document delivery. It is expensive to ship books airmail. Many libraries request that the borrowing institution pay shipping fees both ways (or they assess an international lending fee that is high enough to compensate for higher shipping rates). Surface mail shipments can take months and books are often returned damp or moldy from being in a ship's hold for weeks. For this reason, some libraries refuse international loan requests. Books that have been scanned and stored electronically provide a future solution either through electronic transmission or by air mailing disk copies. However, copyright issues, particularly those involving electronic transfer, are problematic, with no uniform international standards. While most countries have laws covering traditional intellectual property rights, they have yet to determine how to protect electronic information.

International reply coupons (available from postal services all over the world) are accepted by many libraries as payment, especially when fees only cover postage. Some international lending centers sell their own prepaid vouchers or require deposit accounts. Other libraries exchange ILL coupons offering a future free loan or copy in return for one provided. Transactions among libraries exchanging their own coupons should be balanced between loans and borrows. Coupons are useless to net lenders, who deserve compensation, especially if they also provide verification or location referrals for other libraries.

Lending and/or shipping charges lead to a "free or fee" controversy over international loans. Until recently no general agreement existed on what charges, if any, should be assessed (a library's standard lending fee, direct costs, copy charges or shipping charges). The International Federation of Library Association (IFLA) has now approved and begun to promote a voucher payment scheme. The voucher's initial cost is approximately $8 (U.S exchange rate), but it can be reused until it wears out. Half-vouchers are also available to be applied to long photocopies (Gould 1994, n.p.). The vouchers will remove the necessity for payment by bank notes in the national currency of the lending library, thus reducing internal processing costs significantly. If universally accepted, the new IFLA voucher payment scheme could replace all supplier-specific voucher and coupon payments. This is probable because unused vouchers can be returned to IFLA for a cash refund, which benefits net lenders. Since the vouchers can be reused, their exchange between libraries with balanced lending:borrowing would keep costs low.

In many cases, local, state, or federal governments subsidize regional resource sharing endeavors, but each library must be able to at least cover the costs for transmitting requests and packaging and returning materials in a manner equal to that used by the lending library. In some instances this may preclude international loans, restricting international document delivery to the supply of materials that do not need to be returned.

Limited budgets may prevent wide-scale international borrowing, but other factors are equally relevant. One might expect countries with poor collections to be heavy users of ILL/DDS, but data from a UNESCO study covering European interlending during 1989/90 indicates otherwise. The countries that generated the highest number of requests per 100,000 population were those with comprehensive library collections as well as well-developed online communications networks, union catalogs, and other facilities necessary to develop a good interlending network. However, a low volume of interlending did not always equate with poor collections. The UNESCO study indicated that many countries at the bottom of the interlending scale had considerable library resources that were not being shared. The countries of Eastern and Southern Europe, in particular, had widely varying interlending patterns. Some countries with high levels of industrial research and development generated large volumes of requests, but in Spain, Italy, and the former countries of Czechoslovakia and Yugoslavia, many industrial and commercial organizations bypassed library and information networks, going directly to the sources that best served their needs (Cornish 1991c, 23-24).

Political and economic sanctions may also limit international resource sharing. Librarians may be trained as keepers of the books, rather than for public service. Gassol de Horowitz (1988, 94) referred to Latin American libraries as "stockpiles and cemeteries of irrelevant materials." In a similar vein, D.H. Borchardt (1979, 149) described a notion that is still prevalent in some developing nations of libraries as "book museum[s] or bibliotemple[s]". Further, "it is this notion of custodianship which prevents . . . the acceptance of the idea of open access libraries and free lending facilities." And, "there seems to persist an overriding fear that lending in general and lending to other libraries in particular is not to be encouraged. . . . It may even be necessary in some countries to introduce enabling legislation so that librarians need not fear prosecution if books lent either to persons or to institutions are lost."

Clement (1989, 6) expressed similar views. "Constitutional and legislative guarantees regarding human rights and access to government and other information relate to important issues like equality of access to information and to traditions of free library service. These in turn affect the charges that can be imposed and the ability of interlending systems to recover costs. . . . Different economic systems, for example, socialist and capitalist, affect the way libraries are planned, governed and funded, as does the involvement of the public and private sectors." Even under such conditions, seeds of change are being sown, often usually beginning within a municipal area, or regional or special library organizations.

Cultural diversity and language variations pose additional problems in international (and even national) cooperative efforts. Arundale (1991, 18) describes one such project hampered by language barriers. "To date, all attempts to plan a European on-line newspaper archive have foundered on the question of language." Individual countries have limited electronic access to newspaper texts when a single organization makes its own publication(s) available, but it is rare for an independent host to bring together a range of different publications. Arundale (ibid.) describes one such project hampered by language barriers: "To date, all attempts to plan a European on-line newspaper archive have foundered on the question of language."

Individual countries have limited electronic access to newspaper text when a single organization makes its own publication(s) available, but it is rare for an independent host to bring together a range of different publications.

More than 60 percent of scientific and technical information is published in English. Indications also point to scientific papers as increasingly technical, using unfamiliar jargon not easily translated. It is a challenge to provide multilingual access to bibliographic records, many of which were not standardized, even in their print origins, anywhere that English is not the national language or in multilingual regions. Language barriers are particularly acute in Third World countries where competency in English as a second language seldom exists.

Countries need to depend less on protocols and manuals developed in English-speaking nations and develop systems in their own languages. Multilingual thesauri and computer aids being developed will help eliminate language barriers and ease time-consuming and costly online searching of free text in a language in which one is not fluent (Dubey 1988, 49-50, 52). Integrated services digital network (ISDN) telecommunications systems will also help by providing a common user interface to digital networks all over the world (Pastine 1992, 124). This problem is being tackled by the Universal Dataflow and Telecommunication Core Program of IFLA, which is working on multiscript and multilingual records and an expert system aid to automatic classification being carried out by OCLC. (ibid., 120, 124)

The growth rate for the Western European online market increased 40 percent per annum from 1982 through 1985. At that time, Great Britain had about 41 percent of the share of online activity, followed by the Federal Republic of Germany (16 percent); and France (15 percent); the remaining EC countries had a combined share of 28 percent. A 30 percent per annum growth rate was predicted for the period 1982-1990 (Schwuchow 1989, 365, 367), which turned out to be pretty much on target, although Germany has outpaced the United Kingdom in the sale of CD-ROM products due to the development of the former East German sector. France and Spain also show an above average market in CD-ROM sales. Frost and Sullivan (1993, 66) projected continued strong growth in the European market for CD-ROMs through 1994/95, with end-user access to CD-ROMs in Europe now primarily in libraries.

Online services throughout the world are dominated by larger organizations. Usage patterns indicate a continuing trend toward "one-stop-shopping," a trend that favors larger hosts, leaving smaller hosts with specialized databases in vulnerable positions in the marketplace. In Europe, however, the role of noncommercial public databases is also significant (Jones and Dowsland 1990, 2).

As the international online market becomes increasingly competitive, relations sometimes become strained between database hosts who wish to protect the privacy of registered users and database producers who need more data on consumer interests to maintain a competitive edge. It is questionable, for example, as to how relevant the information in online databases may be for developing nations. Chemical Abstracts Services, Biosciences Information Service, and NTIS have indicated that users in such countries find relatively little material of interest in their databases, with the exception of those institutions in Latin America and the Middle East where databases for scientific activities have been available for some time (Dubey 1988, 51-52). Mexico and Chile are making progress in establishing gateways on the Internet (Feick 1994, [2]). Document delivery issues become a factor when databases cite documents not easily obtainable in isolated regions (Kinney 1989, 285-286).

CD-ROM technology has helped to ease both language and delivery problems. It allows more time to search without increasing costs. It is also beneficial in areas that are geographically isolated with poor or expensive telecommunications systems (ibid., 283). Problems persist due to lack of computer expertise and difficulties in obtaining parts and supplies, but CD-ROMs at least provide a variety of relevant information—particularly agricultural and medical databases, with full-text retrieval available for many journals, newspapers, and reference tools such as encyclopedias and dictionaries. Donors are often found to finance original equipment purchases, but ongoing costs are still impediments to the widespread use of CD-ROMs.

In the excitement over the possibilities opened up by developing document delivery technology, it is easy to forget that such services are unavailable to much of the world's population. Networked document delivery relies on advanced telecommunications systems, but those systems are relatively primitive in many areas of the world. Telephone lines and switches need upgrading in most developing nations before online communications can be introduced on a broad scale.

In some countries, telephones are rare, if available at all. Eighty percent of the world's telephones are in North America and Europe. Some developing nations may have only one phone per 100 people (Lavery 1988, 43). In many countries, only one online communications node may exist, and that one may be afflicted with numerous technical problems. Transmission may be affected by the slightest rainfall. Equipment necessary for networked communications may be economically impossible to acquire.

Monopoly suppliers may display take-it-or-leave-it attitudes, leading to communications systems that bring more problems than solutions and that are prohibitively expensive to most potential users. A half-hour of online searching via intercontinental connections equals the average monthly income in some developing nations. Local area networks are rarely used in developing nations to extend the use of online databases to remote users. Networking of any type between computer locations is almost nonexistent, mainly because communications facilities, particularly switching devices, are not suitable for transmitting data (Matta and Boutros 1989, 62).

Yogendra P. Dubey (1988, 47-48) has identified three levels of development and performance of information systems and services in Third World countries:

1. Countries in which the information infrastructure is fairly well developed including computerized facilities for information flow in the most critical sectors.
2. Third World countries in which information facilities have been set up in key sectors of the economy, and an operational base for scientific and technological information has been established through universities, documentation centers, etc.
3. Countries in which information activities on science and technology are at a very low level.

Solutions are needed to resolve conflicts between information-poor nations and those with excellent information resources. Tension between advanced nations, which view information as a valuable commodity to be protected, and developing countries, which desire greater access to such information so that they can progress technologically, economically, and socially, is increasing. Science-based technology is a key factor in such changes, but not the only one. Gassol de Horowitz (1988, 17) states that "the problem does not lie in the importation of technology . . . but in the absence of selectivity and, above all, in paying for technology without actually securing control over it, much less over its reproduction and adaptation. In this context, development of science and technology has become primarily a political and social issue, not a technical one."

Libraries and documentation centers are the basic institutions of information in most developing nations. Unfortunately, low demand for service implies that many such institutions are ineffective instruments in the dissemination of information and have failed to create an awareness about the potential value of information. Librarians and information professionals need better training for their role in fostering improved information delivery (Dubey 1988, 48-53) in helping to eradicate illiteracy, which in many developing nations is as high as 70 percent (Gassol de Horowitz 1988, 21). On the other hand, in nations where libraries are effectively promoting information literacy, the move toward full usage of information resources in libraries is raising users' expectations and forcing changes in traditional interlending activities.

Developing countries suffer even more obstacles to the free flow of information than other nations, such as the following:

- a low volume of internal publications and lack of union lists or regional surveys to help identify nearby holdings, forcing dependence on more advanced nations—which in turn depletes the local economy
- few document suppliers or document supply systems for foreign materials, combined with poor local selections, making for collections which fail to meet basic needs
- national policies—including restrictive copyright regulations—that inhibit the acquisition or exchange of foreign materials
- substandard telecommunications and postal services that create geographical barriers to long-distance document distribution due to expensive and inefficient means of transporting documents (foreign shipments may take months to arrive, air mail may be prohibitively expensive, and fax may not be available as a means of receiving articles)
- new technology for electronic dissemination may depend upon expensive hardware and a computer-literate staff
- devalued currencies and inflated costs from publishers and information suppliers decrease buying power (because costs for everything from production to advertising are high in countries that produce few publications, it is devastating when those investments lead to materials loaned instead of sold due to cooperative acquisitions and resource sharing)
- absence of national bibliographic control systems limit access to internal publications
- variations in interlending forms and methods of transmitting requests.

Nations with access to electronic networks can overcome many of those barriers. E-mail communications with peers around the world may provide more publishing opportunities and certainly a wider exchange of ideas to assist those in developing nations. Unfortunately, in cultures where the working class is primarily unskilled and jobs are scarce, automation of any kind may be perceived as a threat to employment opportunities rather than an aid to communications and document delivery. Available equipment, much of older vintage, may not be compatible with the latest technologies. Newer processes, dependent upon more sophisticated equipment, may be too complex for use in nations with a shortage of skilled technicians. The problem is often magnified by "brain drain", when those who leave the country to receive training often do not return, preferring to work in a developed nation (Matta and Boutros 1989, 62-63). The cost of acquiring modern equipment is likely to increase the need for economic assistance from advanced nations. Even when equipment and personnel are available, shortages of electrical power and supplies, such as paper and toner, may persist.

Some steps can be taken to improve access to information even in underdeveloped nations. Countries designing a national library and information system can include interlending and document supply as an integral element from the beginning (Line 1979, 165). Almost every country has a national library or a major library that someday may be so designated. Forwarding a copy of each bibliographic record to a central library of material that is added to every library in the nation, will serve as the foundation for a future nationwide union list (Borchardt 1979, 148). A regional union list is not a prerequisite for cooperative resource sharing, but some means of making users at other libraries in co-operative arrangements aware of each library's holdings must exist. A general survey of the collections will suffice when only a few libraries or small libraries are involved (Rydings 1979, 79). Each country can endorse the IFLA concept of Universal Availability of Publications and become a resource for its regional publications (Cornish 1989b, 251).

Developing nations rarely produce scientific or technical publications and often rely heavily on advanced nations for such items, but each library within a country or region can develop a collection policy to acquire documents in at least one area of specialization and then be responsible for sharing with other libraries in the region (ibid., 253, 258). Such cooperative acquisition programs will only succeed, however, if all participants perceive them as beneficial to their institutions (Rydings 1979, 79). Resource sharing programs are often the first to go during budget cuts, or during administrative or political changes. In developing nations, small scale resource sharing agreements are more likely to succeed than complex programs that are too expensive, too uncertain, and too slow (ibid., 80).

Document delivery can always be improved by acquiring the best equipment within one's budget. Some libraries may have to choose between purchasing library collections and furnishings or packaging supplies, photocopiers, telephones, fax machines, or computers needed for alternative access. Such decisions are difficult, but must be prioritized based upon user needs.

Countries that do not now have an interlending and document supply code are encouraged to develop one, or adopt a code in use elsewhere. An appropriate model must reflect the country's geographical characteristics and population density; rural and urban differences; and transportation and communication systems. What works in a small, densely populated country served by high-speed transportation probably will not work in large, sparsely populated regions (Clement 1989, 5). Countries with limited funding for information systems should have a national information policy in place to enable appropriate distribution of scarce resources and to ensure that basic information services are available upon which to build efficient electronically based networks (Dimitroff 1993, 29).

Clement (1989, 13-14) has categorized factors that affect the selection of a centralized or decentralized interlending and document supply code, and the handbooks previously mentioned by Graham Cornish (1988) and Virginia Boucher (1984, new edition forthcoming) provide model codes which may be adapted to meet a country's needs. Developing an interlending code presupposes that a region does have libraries with collections available for lending that are of reasonable interest to other libraries. Size of the collection is not as important as content and uniqueness.

IDD AROUND THE WORLD

The following summarizes document delivery activity from the international perspective. The information was gleaned primarily from published resources, and embellished through interviews and personal contacts in several countries. With political upheaval taking place in so much of the world, up-to-date reports are difficult to come by and some projects may no longer be active. Even so, the document delivery services described are applicable in countries other than those in which the idea originated.

The text provides a regional overview of document delivery activity, followed by specific projects of more local interest that other nations or regions might emulate.

Africa

Poor bibliographic control is a major problem in the 55 African nations. The situation hinders verification of citations and location of holding libraries. Contributing factors include ineffectual provisions for legal depositories, an inadequate number of trained staff in national libraries, and a lack of basic computer-based facilities (Zell 1984, 36-37). The Pan-African Documentation and Information System (PADIS) was organized by the United Nations Economic Commission for Africa to develop bibliographic services, with regional information centers to serve those engaged in the economic and social development of the African states. In addition to PADIS, the Association for the Advancement of Science and the HealthNet SatelLife sponsor projects that provide or subsidize document delivery to African institutions (Hafkin 1994).

PADIS serves as the regional point for the provision of information required by those engaged in the economic and social development of the African States. With headquarters in Addis Ababa, Ethiopia, PADIS now has 39 national centers and 44 institutional centers throughout the continent. PADIS provides online search services with free document delivery within Africa for any document listed in the PADdev bibliographic database of African development literature. Whenever possible, documents are sent by e-mail either by scanning, encoding a machine-readable copy, or by sending a full-text file. Outside of Africa, documents are provided for a nominal per page copy cost. Documents are sent via United Nations mail pouch, assuring fairly rapid delivery worldwide.

In African nations, transmission of requests is most likely to be by letter, telephone, or telex; or by messenger when nothing else is reliable. It is sometimes easier to communicate by telephone/telex with other continents than with countries within Africa. Because the mails are so unreliable, many libraries are reluctant to loan documents for fear of loss or damage; many will lend using registered parcel post, an expensive option if ILL volume is high. Courier services are more reliable, but quite costly. Loan periods tend to be very short with little consideration given to the possible remoteness of the borrowing library. Photocopying and microcopying is done more willingly by the few libraries that own copiers, but only when stringent copyright clearances are obtained prior to lending (Inoti 1992, 40, 45) and the potential supplier has the supplies and postage to comply. All equipment is in short supply and impossible to maintain unless there are local vendors (who tend to offer only expensive models). Telefax is not widely available. Libraries that can afford it use outside commercial suppliers, such as BLDSC and AGRIS (Mabomba 1989, 59-60).

National library boards, documentation centers, or archives exist in most African countries, but the range of activities varies widely. A number of countries have been effective in bringing books and libraries to rural areas. The Tanzania Library Service provides a model of national library services with its network of regional branch libraries, backed by mobile units, a postal library service, and a school library service. Unfortunately, the Tanzania example is rarely emulated. Many countries remain seriously underdeveloped, or lack governmental and financial support for library expansion.

East Africa (Kenya, Tanzania, and Uganda)

Among East African states, Kenya, Tanzania, and Uganda have established meaningful document delivery services. Only the University Library at Dar es Salaam, Tanzania, provides international loans outside of East Africa, but each country has one university library that provides photocopies to libraries abroad (Barwick 1990, 67-68, 117, 121). There is no international interlending center in Uganda, but the Kenya National Library Services is registered with BLDSC as the borrowing library for Kenya (Inoti, 1992, 46).

Since the 1960s, each country has had a centralized public library service for the entire country modeled after the previously described Tanzanian system. A regional union list for East African publications is shared by the three university libraries. Lack of a comprehensive national union list makes document delivery erratic. Special libraries tend to be small, primarily serving only their own clientele, although public libraries are operated by The British Council and the United States Information Service in Kenya. School libraries tend to have inadequate facilities, are poorly organized, and lack trained staff (ibid., 41-42). Acquisition funds for materials are scarce, so book stock is low and interlending crucial, but little effort is expended to coordinate collection development (Ndgewa 1979, 172).

Unreliable transit and mail services in this huge region of over 679,000 square miles limit access to resources, particularly outside major cities. With the exception of the three university libraries, most libraries do not have formal ILL agreements, but few refuse to lend to other libraries in the region except for restricted materials—periodicals, research reports, East African publications, and theses—that often are in greatest demand. The university libraries' formal ILL arrangements tend to be more limiting than the informal "understandings" that exist among other libraries. Most libraries charge for photocopies. Unfortunately, each country has its own currency, which is not exchangeable in other East African nations, so payment has to be made in an external currency, such as sterling or dollars, a time-consuming process for the small amounts involved (ibid., 173-176). Checks are accepted in Kenya and Uganda as a form of payment from countries outside East Africa (Barwick 1990, 68, 121). Libraries in East Africa are more likely to borrow from abroad (particularly from BLDSC) rather than from neighboring states, because of the many impediments.

North Africa

Algeria is unique in having an effective legal deposit law that has helped that country develop one of the most important national libraries in the Third World. The national library and the large university libraries in Algiers and Constantine compare with similar libraries in developed nations. The Biblioth_que Nationale publishes a national bibliography and operates an exchange system with over 200 foreign libraries and scientific institutions (Taubert and Weidhaas 1984, 61, 65).

Algeria and Morocco are linked to the EARN network. Little has been published regarding networked document delivery activities specific to these or other North African nations, except Egypt, where document delivery activity is more closely linked to that occurring in the Middle and Near East regions.

South Africa

The State Library maintains union lists of periodicals and monographs in South African libraries. The union lists are distributed in microfiche editions and help to identify holdings within the region. The State Library serves as the national interlending center. For many years, little interlending took place outside of South Africa due to political differences (Mabomba 1989, 59)—a situation that has improved now that many nations have lifted political and economic sanctions related to the country's former apartheid policies.

Canada

Interlending in Canada is decentralized, with libraries outside Canada going directly to libraries known to hold an item. The National Library of Canada acts as a clearinghouse for locating holdings for the humanities and social sciences, referring such requests to appropriate Canadian sources when materials cannot be supplied from the National Library's collection. International requests for photocopies of scientific materials should be sent directly to CISTI (Canada Institute for Scientific and Technical Information): CISTI does not loan materials outside Canada (Barwick 1990, 20-21). OCLC is gaining a foothold among Canadian libraries, but the two major bibliographic systems in Canada are DOBIS and ISM (formerly Utlas). DOBIS, the National Library's automated system and online union catalog, will be replaced in 1994 by a new system, AMICUS (Smale 1994b). Canada has played a leading role in the development of standardized ILL protocols for electronic document delivery.

The Canadian Library Association endorsed "National Guidelines for Document Delivery" in June 1994; ASTED, the Canadian French-language library association, is also considering adopting the guidelines. The CLA/ASTED Interlibrary Loan Code is being revised, and will include many of the same changes in the new code for the United States (ibid.). The revised codes of these neighboring countries place more emphasis on resource sharing. A key strategy in Canada is the development of service partnerships, with the ultimate goal of establishing one or two partners within each region that will focus on interlending services (Lunau and Dinberg 1991, 59). Cross-border resource sharing has also existed for some time, but the actual volume of lending between United States and Canadian libraries has been quite low, averaging four percent borrowing from the United States and two percent lending to the United States. Technological improvements, such as Ariel and AVISO, may change that, particularly for copy transactions. While the United States and Canada have long had an open border, customs officials seem to relish tying up book loans between the two countries. Some Canadian libraries have established postal boxes in neighboring United States cities to receive materials and post book loans to United States libraries (ibid., 60).

CAnet is the Canadian TCP/IP backbone of library/research networks. Approximately 82 percent of academic and college libraries have access to this network; OPAC access is the most common form of information resource available. Twelve percent of academic and college libraries use telefacsimile for document delivery, with more libraries indicating plans to do so in the future (Cleveland 1991b, 73). A number of Canadian libraries have acquired Ariel for transmitting documents via the Internet. AVISO, an electronic messaging system widely used in Canada with ENVOY 100, is also used for sending requests over the Internet.

Europe

Many examples of document delivery activities in various European nations have been given throughout this work, but the activities of the Commission of the European Communities (CEC) deserve additional mention. CEC has taken the lead in coordinating library programs for EC nations. The Document Interchange Between Libraries (EDIL) project is underway. Another pilot project, EUROPAGATE, is endeavoring to design and implement an ISO SR/ANSI Z 39.50 gateway so that library users in any EC nation can access any library catalog seamlessly and the incorporate the record into an e-mail request for document delivery (Kelly 1994, 6-8).

End-user Courses in Information Access through Communication Technology (EDUCATE) involves libraries in France, Ireland, Spain, Sweden, and the United Kingdom. It aims to develop a self-paced, multilingual, user education course to be distributed over academic networks and designed to give hands-on experience in a real world setting, providing interactive access to online information systems, library OPACs, electronic news conferences, and electronic documents. educate will include instructions related to file transfer and e-mail functions, as will as navigational tools such as Gophers, WAIS, WWW, ALEX, and MOSAIC. The project began in February 1994 and will run for three years. Initial subject areas cover physics and electrical and electronic engineering (Fjälbrant 1994, 8-9).

Another CEC-sponsored project, the European Initiative in Library and Information in Aerospace (EURILIA), is a three-year project with participation by several EC nations as well as the United States. EURILIA will establish a new service based on a standardized pan-European system for information access, retrieval, image browsing, and document delivery. The document delivery component is based on mail and Group III and Group IV fax. The project is meant to develop and establish the feasibility of a multisite, multicountry document retrieval and delivery mechanism, with a standard user interface. Although the pilot project testing is taking place in aerospace libraries, the developed system should be applicable as well to other areas (O'Flaherty 1994, 9).

General access issues still to be resolved throughout Europe include retrospective conversion of paper records into machine readable formats, large backlogs of uncataloged materials, and diverse ways of dispersing government materials. Obstacles to document supply are high costs related to international telecommunications, taxes, and postage (Pastine 1992, 125-126).

Networking is just getting off the ground in some European nations. Portugal has begun a network, but its use is limited to a few libraries (Cornish 1991c, 23). Spain is trying to establish an interlending center similar to BLDSC, but has made little progress. Networking in Spain is developing piecemeal among libraries. Libraries in the Barcelona region have developed a union catalog on CD-ROM, but other libraries have no union list upon which to rely (ibid., 24). Swiss libraries on SIBIL systems are to be linked with the SWITCH academic network (Dempsey 1990, 13).

Eastern Europe

Alexander Dimchev (1991, 16) states that "libraries in Eastern Europe are behind their Western partners because of financial problems and lack of technological development. In this respect it should not be expected that there will be a fast change in their strategy and methods of work in the field of interlending. . . . A possible way out could be found if governments in the East European countries pay serious attention to finance and equip libraries with new technologies." Networking is becoming a feature of the information infrastructure in Hungary, Spain, Italy, and Yugoslavia; and to a certain extent in the Czech Republic, the Slovak Republic, Poland, and Romania. Greece is more technologically aware, yet has no basic infrastructure for interlending (Cornish 1991c, 23).

Albania. In libraries that provide national and international document supply, collection development is oriented toward specialized areas of research. Items borrowed focus on materials about Albania and the Balkans, and on science, technical, and medical publications (STM). Materials published in the Balkans and Western Europe and sociopolitical-political works are the most frequently loaned, with 60 to 70 percent of the interlending requests met. While little research is carried out in the areas of science and technology, information is needed for applications and adaptations. Albanian libraries are interested in developing relations with Austrian and Italian libraries because postage costs can be met by vouchers converted into Albanian leks rather than hard currency (Mosko 1991, 13). Little development of new document delivery technology in libraries has been undertaken in Albania; no research networks are in place.

Bulgaria. Interlending in Bulgaria is centralized with the National Library as the main functionary. Due to decreased funding, more reliance is placed on interlending. Increasingly, requests are sent to foreign libraries, primarily in Western Europe. Payment for such transactions is a major concern, often depending on contacts with libraries in Western countries and financial support from international organizations. Interlending in the past has depended highly on libraries in the Soviet Union and other former socialist countries. The transition to a market economy in those countries is expected to bring changes in payment mechanisms and tariffs. Dissemination of "grey" literature is very limited. Scientists in Bulgaria have free access to numerous databases worldwide, but libraries do not stock the documents indexed. Dimchev reported on two studies indicating that Bulgarian libraries can only supply 35 to 53 percent of the references cited by online databases, while international norms indicate most library systems should be able to satisfy 80 to 85 percent of such requests. According to Dimchev (1991, 17), "only 7 to 8 percent of the world's periodical titles and about 1 percent of monographic literature are available in Bulgarian libraries."

Document delivery in Bulgaria is hampered by a level of technological development that lags other European libraries. The country's libraries are not automated, only a few have telex, and photocopy machines are in short supply (ibid., 17-18). The Bulgarian National Library closed its doors for two months in the summer of 1990 to bring its deplorable operating conditions to public attention. In November, 1990, many librarians joined a general strike, but by the time the government was brought down, the country was in such dire need of basic necessities that the problems of cultural institutions were of minor concern. These problems have affected international document supply. The total number of transactions has dropped considerably since 1989 because the library does not have the funds to send materials by airmail. While the National Library is still a depository library for all Bulgarian publications, it has neither the space to shelve new books nor the staff to classify them. Nondepository acquisitions depend upon exchange programs and gifts (Paskaleva 1991, 19-20).

Confederation of Independent States (CIS). Information related to document delivery since the dissolution of the former Union of Soviet Socialist Republics (U.S.S.R.) was difficult to find. There are a number of listservs that reflect significant activity and no dearth of information regarding the political turmoil in Eastern Europe and Eurasia. Sadly, while the newly independent nations struggle to establish their new identities, many of the listservs seem embedded in the past, having names that begin with "former" and "ex." Indeed, the Internet resources guide that covers the CIS carries the title *The Guide to On-Line Resources for Researching Central/East Europe and the Former Soviet Union.* (Brown, n.d.) Another CIS database, *CERRO Archives,* was developed by the Central European Research Organization, but here again we find file names such as "Post Soviet Study resources on the Internet", "Ex-USSR Database", and "Russian and East European Studies Home Pages via WWW." (CERRO,n.d.) The Scientific Library of the Moscow Lomonosov State University and the Library Computer Network Company provide a document delivery service pertaining to Russia. For more information send an e-mail message to inf@lib.msu.su. query to ILL-L or FISC-L listservs usually brings a quick response for current information on document providers in the new confederation.

Czech Republic and *Slovak Republic.* Political reforms in the former Czech and Federal Republic (CSFR) since 1989 have brought many changes to libraries that have affected their document delivery capabilities. A democratic government has eliminated censorship and now allows use of all library stock; 960,000 volumes removed twenty years ago from Czech Republic libraries have been returned. Hundreds of new publishing companies and periodicals have been started since the state monopoly for selling and distributing books was dissolved. Libraries can now acquire foreign literature without restriction and establish links with foreign suppliers. Many libraries rely on book donation to supplement inadequate materials budgets. Unfortunately, donations are often outdated. collection development based on gifts suffers from lack of coordination. Libraries are critically short of space, forcing the storage of many volumes elsewhere and seriously affecting access (Richter 1991, 25).

The two national interlending centers are the National Library in Prague for the Czech Republic, and the University Library in Bratislava for the Slovak Republic. Other research and special libraries also participate in international document delivery, particularly in the Slovak Republic where international interlending is decentralized. The Slovak Academy of Sciences and the Slovak Technological Library supplement the collections of the University Library, which served as the de facto National Library of Slovakia until 1954 and is still the only designated depository library for Slovakian publications. The two national interlending centers have developed a national union list of foreign publications available within the former CSFR. Czech and Slovakian users are supplied with information personally, by post, telephone, fax, and telex. A new library act drafted by both republics will include legal regulations governing interlending and international document supply (Jakubekova and Sedlackova 1991, 32-34). Rampant inflation (including the cost of library materials, postage, and rent on library facilities), devalued exchange rates, and budget reductions for international lending services have all hurt international document supply activities. The number of requests received from abroad dropped twelve percent in 1990 (Richter 1991, 26).

Equipment, particularly computers, communications devices, and copiers, was substandard in CSFR libraries but has been improved in the Czech Republic. A few Czech libraries now have online catalogs, some of which are networked (Borovansky 1994). A national online catalog was planned, but shortage of funds, the poor quality of the communications systems, and the lack of a public packet switching network are major obstacles to be overcome before links with other international networks are common. In contrast, the Center for Scientific, Technical and Economic Information, the National Information Centre (NIC) and the Institute of Applied Cybernetics are well equipped and offer access to selected foreign databases as well as some of local origin. NIC has access to over 700 databases in all subject fields, but the high cost for using such systems causes them to be under used. Other libraries depend more on CD-ROM technology than online access (Richter 1991, 26-27).

Hungary. About 15,000 libraries in Hungary have a combined stock of over 100 million volumes, but escalating prices have curtailed acquisitions (Gulacsy-Papay, Karacsony, and Sonnewend 1991, 112). In 1988, for example, natural sciences research libraries subscribed to 1,800 foreign periodicals for 47 million forints; in 1989, they subscribed to 10 percent fewer titles for 61 million forints—10 percent less for 30 percent more! The situation was even worse in agricultural libraries, where subscriptions decreased by 11 percent but costs went up 40 percent (Szabo 1991, 107-108). A citizen's average monthly salary would buy a year's subscription to three or four Hungarian periodicals; a three-month subscription to a foreign daily newspaper; or three or four English or German technical books. Libraries, particularly public and school libraries, therefore play a vital role, and interlending is an important service (Gulacsy-Papay, Karacsony, and Sonnewend 1991, 112). Depository copies are central to the acquisition of Hungarian publications, with sixteen legal deposit copies going to libraries in five regions and several special libraries, as well as to the National Library. Fewer deposit copies are required for audiovisual materials (ibid.).

The Szechenyi National Library (OSZK) maintains a union catalog for books and periodicals. It provides materials location information to other libraries and it serves as the national lending center for foreign requests. To improve document supply, OSZK developed a storage library to hold documents for use in national interlending, for which rules were established by the Minister of Education in Decree No. 19 of 1980, and revised in 1990. Interlending is free, except for photocopy costs. The Decree provides only three to five days for meeting requests, but that service goal is far from standard. Regional libraries have high fill rates for Hungarian documents, particularly in Budapest. Besides OSZK, a few other specialized libraries are directly involved with international interlending, particularly the Hungarian State Agricultural Library and the National Technical Library. In 1990, the National Library received 22,000 requests—90 percent from Hungarian libraries; 10 percent from abroad. About 14,000 to 15,000 requests are sent to foreign libraries (primarily scientific and academic requests) with about 10,000 requests filled—60 percent with copies. German, British, Dutch, Swiss and French libraries fill most of the Hungarian requests, while items loaned go mainly to German, Czech Republic, Slovak Republic, Austrian, Confederation of Independent States, and Bulgarian libraries (ibid., 114).

Hungarian library services provide modern computer databases with online and CD-ROM access. Most libraries are "extremely interested in rapid and reliable document supply, including the most expedient solution which is the international lending system. . . . The international inter-library lending system should preferably be expensive but reliable and quick rather than cheap but unreliable and erratic. . . . We are keen to see the payment system for international lending flexible and not burdened with unnecessary or unwieldy administrative costs." (ibid., 115)

France

Since 1980, France has developed a national cooperative acquisition network CADIST (Centre d'Acquisition et de Diffusion de l'Information Scientifique et Technique) made up of twenty university libraries. Government funds are available to each library to acquire and share materials in specific subject areas. Outside the university systems, interlending responsibility is divided between the Centre de Prêt of the Bibliothèque Nationale, which loans duplicates of monographs in its collections, and INIST Diffusion, which supplies photocopies of periodicals published since 1940. (A deposit account or a VISA card is required for payment.) ILL activity in France increased by 45 percent from 1981 to 1986, and by 5 percent per year since then, peaking in 1991/92 at 1,194,483 transactions. More than 80 percent of the requests were for photocopies. TPEB, a computerized messaging system on the SUNIST host, is used for transmitting requests and replies, but not for locating holdings. As of 1988 there were 200 university libraries and 50 private libraries or documentation centers using the system. The average fill rate was 85 percent with the CADIST filling approximately 26 percent of the transactions (Deschamps 1991, 35-36).

France serves as a model for national promotion of computer literacy. France Telecom originated an electronic telephone directory in 1979, which went public in 1983. Any end-user in possession of a terminal or microcomputer with a modem card, or a Minitel terminal can access the telecommunications network after signing an agreement with the national Ministry of Education (Ménil 1993, 32). Minitel terminals are systematically distributed free to all want them. The Minitels support both videotext and ASCII data communications. Online charges are included in monthly telephone bills (Guyot 1989, 3).

The TRANSDOC experiment by the Centre Francais de Copyright was the first to set up a copyright payment scheme in Europe (Gurnsey and Henderson 1984, 38). The project intended to demonstrate that any existing bibliographic database in ISO format could be linked to an automated full-text delivery system, considerably advancing state-of-the-art online retrieval considerably and overcoming a major obstacle to the widespread use of bibliographic files (ibid., 90).

France participated in Project ION with the PICA Foundation in the Netherlands and LASER in the United Kingdom. This pilot project demonstrated OSI connections between three European interlending systems (Deschamps 1991, 38). The Trés Grande Bibliothèque (the new Bibliothèque Nationale) will be a high technology library where old texts will be digitized, creating an electronic library, with online books, films, videos, and telecommunications (Pastine 1992, 121).

Another major French undertaking was FOUDRE (FOUrniture de Documente sur Réseau Electronique), an electronic document delivery project carried out over eighteen months (ending June 30, 1992) in twelve academic libraries. Its purpose was to provide rapid access to high-quality reproductions of documents by scanning and digitizing upon request (Ménil 1993, 32, 34). Each requested document was stored on an Optical Numerical Disk (OND) and the contents sent to the requesting library by the ISDN (Integrated Services Digital Network) NUMERIS. Each lending library kept a database of all stored documents from which subsequent requests were automatically supplied without human intervention, saving time and labor, and preventing further damage to original documents (Deschamps 1991, 38).

Each document is digitized page by page, with software providing a choice of text, photo, or mixed mode and a choice in strength at a density of 300 dpi. One of the unique features of the indexing system (based on ISO Standard 9115) is that it creates a unique, periodical identifier that is not a sequential accession or order number, but instead is made up of alphanumeric characters comprised of the ISSN, the year of publication, the volume, issue, supplement, and the article's first and last pages. This system makes it easy to home in for instant retrieval of future requests. Transmission of a previously stored document took about seven seconds per A4 page considerably less than retrieving a volume from the shelves, photocopying, and preparing for mailing or faxing. There was a low rate of recalled requests, however, only 2 percent. The elapsed times to perform various operations included one minute for entry of a request; fifteen seconds to transmit ten requests; forty seconds to digitize a page; one minute/page to store the digitized document, and three minutes to transmit ten pages. Quality was superb and the project had a high satisfaction rate of 80 to 90 percent. Although the hardware and software created some problems, the project was proposed for extension as EDIL (Electronic Document Interchange between Libraries), an international cooperative venture with BLDSC, PICA, and TIB Hanover (Ménil 1993, 33-34).

Germany

Most German municipal libraries provide interlending services (usually at a cost of one mark for the request form). The State Library sends books by mail to people living in rural areas. Besides the "farmers post," interlending is usually handled by checking holdings of libraries within one's region and sending a direct request. If the document is not available regionally, the request is sent to one of twelve other regional interlending centers. If the patron indicates that the request should "circulate to all central catalogues," it is passed from one center to another until a holding library is located and the document supplied. Otherwise, no more than three central catalogs are checked. Restricted circulation is the general practice. International sources are used as a last resort (Henschke 1991, 47-48).

Six months before East Germany merged with West Germany, interlending experts in the German Democratic Republic and the Federal Republic drafted procedures for document delivery when the nation reunited. Twenty-two libraries were designated key libraries in the former GDR. These were mainly university and state libraries and a few general research libraries that were well stocked and equipped. They were to assume service functions for other libraries in the new and old federal states. The old Federal Republic had a decentralized system with the workload spread among many libraries. The Special Collections Area Plan had been in force since 1949, providing financial support for five Central Special Libraries and seventeen state and university libraries, as well as centralized acquisitions, cataloging, and interlending of designated subject areas. The former GDR, on the other hand, had a centralized system with the Institute of Interlending and Central Catalogues at the German State Library in East Berlin playing the main role.

The consolidation of such diverse systems has had its difficulties, but actions focus on commonalities rather than differences. The central serials database, the ZDB, functions well as a national union list. Communications between the regions is poor—telephone, fax, and the postal service are all at unreliable stages of development. It is also difficult to exchange data between individual cooperative systems (ibid., 1991, 48-54).

Document delivery demand is highest for medical, technical, and business materials, with most requests coming from the private sector. Forty percent of the demand is for German and English publications (Barwick 1991, 18). Zentralbibliothek der Medizin (ZBM) in Cologne supplies the majority of medical requests, particularly articles indexed on DIMDI. ZBM also offers online search services. In 1990, 335,817 documents were supplied within Germany and 14,444 items abroad, with a fulfillment rate of 93 percent (Gerrard 1993b). UB/TIB (Universitatsbibliothek/-Technischebibliothek), Hanover, which supplies documents in technology, engineering, and related sciences reported in 1991 that ILL transactions were up due to budget constraints, the recession, and the inclusion of libraries from the former GDR. After operating at a deficit in 1992 and expecting worse in 1993, UB/TIB considered canceling subscriptions and charging for book loans (ibid.). Document supply for other scientific materials is provided by the German Information Centre for Energy, Physics and Mathematics at FIZ Karlsruhe, an independent company jointly owned by the Federal Government, the eleven former West

German Landes, and several scientific research institutions. FIZ produces databases relevant to aeronautics, astronautics, astronomy, computer science, energy, mathematics, physics, and space research; and also supplies grey literature in those areas (ibid.).

The Bavarian State Library's Eastern European interlending statistics for ten years (1980-89) showed 8,864 volumes supplied vs. 572 volumes received. The majority of the items supplied went to the former Communist states of Czechoslovakia, Yugoslavia, and the USSR, as well as to Hungary. Materials were also supplied during that time to libraries in Albania, Bulgaria, Finland, Greece, Poland, and Romania. In every case, the Bavarian State Library was a net lender by a considerable margin, with the former USSR being the only country to supply over 100 items (actually 279). Of the Bavarian State Library's international loans, 25.8 percent went to Eastern European libraries, but those nations supplied only 16.9 percent of the international documents received (Pleyer 1991, 67). A lending charge was instituted for non-German materials in 1987 because so many foreign libraries were requesting materials not published in the Federal Republic of Germany. After that, the volume of materials sent abroad dropped approximately 20 percent (ibid. 70).

Italy

The database industry in Italy remains limited, but some well-established online services provide legal and business information to national markets. The European Space Agency Information Retrieval Service (ESA/IRS), in existence since 1969, has expanded to include dial-up access and is available over Euronet. It is one of the largest host organizations in the world, offering a variety of scientific and technical databases (Jones and Dowsland 1990, 16-17).

The Netherlands

The Netherlands is integrating bibliographic networks, such as LIBRIS and PICA (Project for Integrated Catalogue Automation), into the national Dutch research network, SURFnet. National networks have switched to X.25 protocols implementing OSI standards. The bibliographic utility PICA supports shared cataloging, interlibrary loan, and acquisitions from its database of over three billion (3,000,000,000) titles. (Barwick 1990, 24) Remote log-in, file transfer, and e-mail are supported. SURFnet connects with other European and international networks, so PICA's resources are accessible outside the Netherlands (Cleveland 1991b, 85).

The testing of electronic transmission of ILL requests and documents over PICA is underway in the international document delivery projects RAPDOC and EDIL. The Dutch libraries involved consist of the Netherlands Royal Library, which is supplying materials in the humanities and social sciences, the Technical University at Delft, and the Agricultural

University Library at Wageningen (DerWers 1993). The Royal Academy of Sciences Library is also a national lending center for biomedical literature. The automated interlibrary loan system on PICA contains a distribution component that provides automatic load leveling, although libraries with reciprocal commitments can program in their own distribution circuit (Willemsen 1989, 53).

Much of the world's scientific, technical, and medical publishing is concentrated in the Netherlands—a nation struggling with strong inflationary pressures that translates into high production and product costs. Aggressively and competitively serving a worldwide market requires producing sales brochures in multiple languages (including non-roman type), supporting a well-traveled sales force and several international offices, and maintaining editorial, indexing and production staff while trying to stay abreast of the proliferating volume of manuscripts offered for publication (Dijkstra 1993).

Copyright abuse, particularly among large resource sharing institutions and commercial document suppliers, is a major concern among Netherlands publishers. They were among the first to experiment with single article document delivery by scanning documents and storing bit-mapped page images electronically on CD-ROM databases. That effort developed into ADONIS, a product now marketed internationally (Compier 1993). Current document delivery product development revolves around the TULIP project.

Scandinavia

Denmark. Denmark's welfare society has made all educational activities, including access to library resources, free of charge. The libraries of most research and educational institutes come under the aegis of the Ministry of Education; public libraries under the Ministry for Cultural Affairs (Kirkegaard 1985, 95). For many years, public libraries were subsidized by the government—including funds for free interlending and support to the larger county libraries to assist smaller parish libraries and the inhabitants of the parishes. Every subsidized library could "supplement its own collection with loans from bigger libraries, but it must always possess the nonfiction literature which can give the most basic information in different fields . . . [and] . . . access to the best works of the Danish literature and of the foreign literature translated into Danish." (ibid., 98-99) The depressed economy of the 1970s and 1980s, however, brought legislation (the 1983 Library Act) that ended the nearly 100 years of subsidies to libraries (ibid., 96, 105).

Now interlending and document supply is decentralized among a network of public and research libraries. The Royal Library is the "National Library," but the Danish Loan Center at the State Library at Århus is the international lending center (Barwick 1990, 30). The National Technological Library of Denmark (DTB) at the Technical University of Denmark is the national resource for engineering and applied sciences and the national center for the European Space Agency-Information Retrieval Service.

DTB's OPAC is part of an integrated automated library system that allows users to browse, search. and order documents online both from the library and remotely via the university network DENet or Datapak. The OPAC also provides online access to several national and international databases. (Gerrard 1993b) The State Library has also instituted an e-mail document delivery request system which eventually will link holdings to 250 Danish libraries (Barwick 1991, 19).

The State Library is a supplier on the OCLC ILL Subsystem. Besides OCLC, two national online union lists exist: ALBA, which covers material in research libraries, and BASIS for the holdings of public libraries. The two databases initially covered only non-Danish material, but Danish publications have been included since 1980 (Barwick 1990, 30).

Finland. The decentralized Finnish ILL/DDS system is based on a network of research and public libraries. Each library has to be approached separately as no national interlending center exists. Requests from abroad for material published in Finland are likely to be successfully filled if directed to Turku University Library. Requests for Finnish publications published in the Swedish language should be sent to the Library of Åbo Academy (ibid., 36-37).

Helsinki University Library serves as Finland's National Library. It does not Finnish materials, but it provides OPAC access to its library system LIN-NEA (Library Information Network for Academic Libraries) through FUNET, a TCP/IP-based research network. VTLS (Virginia Tech Library System) software has been installed in all Finnish university libraries, the National Repository library, and the central system. All of these libraries are now interconnected via FUNET (Soini 1990, 117).

Norway. Every Norwegian municipality is mandated to have a public library with qualified staff. The Cultural Fund is used to buy 1,000 copies of every Norwegian literary publication for distribution to public libraries throughout the country. This system has the effect of increasing authors' income, reducing book prices, ameliorating economic risks for publishers, and promoting the reading of Norwegian literature (Granheim 1985, 51).

The University Library of Oslo served for many years as the National Library, but other libraries with important special collections now share responsibilities for national resource sharing. Each of these libraries is authorized to collect in its subject field(s) for the national network and receives financial compensation for doing so (ibid., 47-50).

Norwegian libraries have developed a centralized cooperative integrated library system, BIBSYS, which connects all higher educational institutions through UNINETT, the national research network. Access to catalogs and databases is available through remote log-in for subscribed users. Besides the union list of holdings of thirty Norwegian research libraries, other excellent databases are available, including several of interest to music libraries (Cleveland 1991b, 81-84).

Sweden. Almost every Swedish library uses a different online system, but all academic institutions are developing campus networks to connect to SUNET, the Swedish national research network. Access will include Sweden's national bibliographic database LIBRIS, OPACs, and other information resources via remote log-in using IP or domain name addresses to all SUNET users with user names and passwords (Olausson, 1990, 129-132).

United Kingdom

Britain's document supply system depends primarily on a single source—the British Library Document Supply Centre—rather than cooperation among local libraries. Requests not sent directly to BLDSC are usually routed through seven Regional Library Systems. Membership in regional systems is optional and is composed predominantly of public libraries. Subscription fees are based on the scope of services offered (Smith 1993). Regional systems fill less than 20 percent of the ILL requests originating within the regions (Bonk 1990, 232).

BLDSC is the largest library in the world (7,000,000 titles) devoted to document delivery. It handles an average of 14,000 lending requests per day (Gerrard 1993a). The collections cover all subjects and languages and include books, journals, and extensive holdings of technical papers, patents, conference proceedings, musical scores, and dissertations. Seventy-four percent of the requests received relate to science and technology. Of these, sixty-seven percent are for serials, twenty-two percent for monographs, and eleven percent for conferences, theses, music, and official publications (Plaister 1991a, 162-3). The fact that many other major commercial suppliers (RLG's Citadel, UnCover, Information on Demand, etc.) use BLDSC as a resource attests to its excellent collections. Forty-seven percent of the requests come from special libraries; thirty-nine percent from academic libraries; and fourteen percent from public libraries (Bonk 1990, 231). Twenty-four percent of the requests are from foreign libraries (Allen and Alexander 1986, 61). Backup libraries for BLDSC consist of other parts of the British Library system, copyright deposit libraries, and selected scholarly and society libraries, such as the University of London School of Oriental and African Studies Library, the London School of Economics, and the Royal Society of Medicine, each of which are compensated by BLDSC when they provide loans (Bonk 1990, 233).

BLDSC's fill rate averages above 90 percent (Gerrard 1993a): 87.6 percent are filled from BLDSC's own collections, 2.4 percent from back-up libraries in the United Kingdom, and 0.3 percent from locations outside the country. Only 7.7 percent of the requests go unfilled, although referral locations are supplied for 2 percent of the unfilleds (Plaister 1991a, 161). Most requests are processed within two or three days if held at Boston Spa. Foreign requests tend to take longer to process because so many are unverified "last resort" attempts that have a high level of inaccurate citations.

United Kingdom libraries provide the best example of a centralized nationwide document delivery system. A study by Brenda White showed the regional transport scheme to be the most reliable transport mode for 24-hour receipt of book loans. Nonetheless, the United Kingdom's mail service proved to be just as quick and, if the delivery system exceeded 24 hours, sometimes faster (White 1986, 48-49). Mail and the regional transport system are also used regularly for delivery of photocopies, but fax, ARIEL, and other modes of electronic transmission are being used with increasing frequency.

During a 1993 tour of several British libraries, staff members of inter-lending departments were quite positive in describing services received from BLDSC (fast service, high fill rate, and a consensus that although expensive, it was not overpriced for the services rendered). British library administrators, on the other hand, expressed interest in decentralizing document supply. The reasons given for desired change are based on economic constraints, improved access to the holdings of other libraries and online periodical databases, improvements in document delivery technology in many libraries, and implications that BLDSC was overstaffed, under-automated, and not as committed to rapid progress in document delivery as some of the libraries it serves. BLDSC was established before automation emerged in most UK libraries, and its mostly manual procedures were fast, cost effective, and not dependent upon automation on the part of its users (Braid 1991, 101).

Automation is progressing at BLDSC, but the process is extremely expensive and time consuming for a library of its size. BLDSC is set up as a document suppler on OCLC, RLG (Inside Information), CARL (gateway through UnCover), DIALOG DIALORDER, DATASTAR, Orbit (ORBDOC), BLAISE-LINE, DIMDI, and ESA Primordial (Nelson 1993, 28), but the complete catalog is not available on any online database.

The Consortium of University Research Libraries (CURL) set about in 1987 to provide bibliographic access to the OPACs of the seven largest United Kingdom university libraries over JANET, the country's major research network. The United Kingdom Office for Library Networking (UKOLN) was established in 1990 to assist libraries in their networking endeavors and to serve as a liaison on international projects and between systems vendors or utilities and libraries (Bryant and Smith 1993). Over 60 OPACs are now accessible on JANET (Bryant 1992, 68), but there are problems similar to those in the United States with variations in addresses, log-in procedures and command languages (Cleveland 1991b, 74). A list of databases and OPACS accessible over JANET is available online in the file *Network Services Available Over JANET* (Brack 1992). In addition to OPACs, a variety of other databases are available, including some that are full-text.

Many British libraries now have e-mail addresses for interlending (available in the *Directory of E-mail Addresses for Libraries, Services and Staff*). Collectively, the ILL offices share a single e-mail distribution service, LIS-ILL, which is coordinated by the Forum for Interlending (Stone 1991, 134). E-mail ILL requests between libraries may begin to supplant requests being sent to BLDSC; thus, interlending among UK libraries may be on the rise, particularly among academic and public libraries in the London and South Eastern Library Region (LASER). Several CURL libraries installed Ariel software in the spring of 1993, and a test of its viability for photocopy supply was carried out. (Friend, Bovey and Bugden 1993) SUPERJANET is being developed to utilize X.25 protocols. Project ION, EDIL, and RAPDOC are continuing projects intended to expedite international document delivery (Smith 1993).

The move by some libraries towards decentralizing document supply presents a "Catch-22" situation, because were BLDSC to lose a large part of the United Kingdom's interlending volume, it would probably have to raise its rates. Except for deposit copies, BLDSC purchases all of its stock for lending, so a drop in revenue would impact collection building. British libraries would certainly need to commit more staff and equipment to the lending side of ILL than they now do if they began to rely heavily on a reciprocal system. Whether cost savings would result or whether costs would simply be transferred is difficult to predict. One service advantage in reciprocal ILL systems is that usually fewer restraints are placed on who is "eligible" for ILL. This, too, increases volume and could create turnaround delays in understaffed ILL units. A 1989 study on the economics of interlending presented a cost-analysis model that allows UK librarians to estimate the financial impact on local and national sectors that might result from changes in interlending practices (British Library Board 1990).

Because of the heavy dependence on BLDSC for interlending, automated ILL systems must be compatible with ARTTel (Automated Request Transmission by Telephone) system if they are to be widely used. Of the requests sent to BLDSC, 41.5 percent are received via ARTTel. [An additional 4 percent are transmitted electronically from other bibliographic database hosts; 8 percent by Telex; 2.4 percent by fax; and 2.7 percent by telephone for "premium services"; while 41.4 percent are still received through the mail (Nelson 1993, 29)]. AIM, BLCMP (Birmingham Libraries Cooperative Mechanisation Project), EXILE, Libertas from SLS, TINLend, VISCOUNT, and packet switched networks using JANET are among the online systems used within the United Kingdom for transmitting requests electronically (Leeves 1991). The OCLC ILL subsystem is not widely used except by academic libraries doing international interlending (Friend, Bovey and Bugden 1993). CURL members recently formed an alliance with the Research Libraries Group.

Since January, 1991, users at all subscribing U.K. higher educational institutions connected to JANET have had direct document ordering capabilities through the various databases offered by the Institute for Scientific Information. The Combined Higher Education Software Team (CHEST) leased the databases for an annual flat rate fee to allow users unlimited searching and ordering. The database resides on a host at Bath University. Requested documents are delivered by e-mail over JANET directly to the requester's own mailbox within minutes (Academic. . . 1991, 3).

Document delivery in special libraries is exemplified by the services offered by the British Medical Association Library. It provides online searching and document supply from its collections to 90,000 members and 450 medical libraries that hold "Institutional Membership" in the association. In 1992, approximately 50,000 photocopy requests were filled, while approximately 3,000 books and photocopies were acquired for members through ILL. Most requests are received by telephone and mail. The library also contains Great Britain's largest collection of medical films and videos, which are also available for loan. The library is involved in an international interlending project with Polish medical libraries via a link with the Central Medical Library in Warsaw. Fax and e-mail are used to transmit requests and copies (McSean 1993).

Far East

Cataloging in non-roman languages was not automated to any great extent until the 1980s, which is one reason that countries in the Far East and Mideast have lagged in developing automated online catalogs. Although the Research Libraries Group has provided a Chinese, Japanese, and Korean (CJK) cataloging interface for many years, its use has been mainly by United States libraries with East Asian collections. OCLC offered a similar interface in 1987, and more than 200 libraries in Asia and the Pacific Region now use it. Included are libraries in Australia, China, Hong Kong, India, Japan, Korea, Malaysia, New Zealand, Papua New Guinea, Singapore, Taiwan, and Thailand. More than 800,000 unique CJK bibliographic records now reside in the OCLC Online Union Catalog, which has opened up a new area of international document delivery and interlibrary loan activities. The participants consist mainly of large national libraries, academic libraries, and many special libraries: public libraries appear to be represented only by the Regional Council of Public Libraries in Hong Kong (OCLC Users . . . 1994, 15-27).

Medical libraries were among the first in the Far East to begin network-ing and provide document delivery. The Institute of Medical Information (IMI) and the Health Sciences Library of the Chinese Academy of Medical Sciences (CAMS) serve as the Biomedical Information Center of China. Since 1983, MEDLARS offline retrieval and SDI services have been avail-able to China's medical profession. Online services commenced in 1986. While MEDLARS is highly welcomed, only about 10 percent of its inven-tory consists of Asian journals. Some Asian nations (India, Indonesia, and the Philippines) have established national medical databases. Also, the South-East Asian nations jointly organized the South-East Asian Medical Information Center, which set up a serial database called "PERIND," and IMI/CAMS developed the "China Biomedical Literature Data Base" set up according to MEDLARS protocols. IMI/CAMS provides search and retrieval document delivery services to other Asian countries for medical literature (He 1986, 2-4).

Hong Kong

Hong Kong's City Polytechnic Library is developing and using new technology to make the interlending process more efficient. It has devel-oped its own software package for storing ILL records on a microcomput-er (Barwick 1991, 21).

India

The Indian Association of Special Libraries and Information Centres has drafted an ILL code, but it has not yet improved interlending significantly, especially among university libraries (ibid., 17) The country's entire acad-emic and research community is linked via the Education and Research Network (ERNET). Over 300 institutions serving over 2,000 users are using ERNET for e-mail, file transfer, and database access.

Japan

Japan has two major academic and research networks, but only the National Centre for Scientific Information System (NACSIS) is involved in library linking projects. Library systems have been linked to NACSIS since 1987. NACSIS has developed the Science Information Network, a privately operated packet-switching system, which forms a nationwide information system, covering not only science and technology, as its name implies, but social sciences and the humanities, as well. The system links university libraries, computer centers, and national research institutes. Access to library resources is through remote logon, based on N1 (Japan's first academic research network) or directly via the inter-university computer network.

NACSIS contains a large union catalog of monograph and serial records. The network also provides access to a number of commercial databases, pri-marily scientific, technical, and medical, but also to the full-text *Harvard*

Business Review and to a database of doctoral dissertations granted by Japanese universities. The national union catalog maintained by NACSIS is available to catalogers who can access holdings and authority records and do copy cataloging. Local cataloging systems can install an interface software capable of processing Kanji characters (Naito, 1989, 1-3). A few Japanese libraries have stored microform cassettes in a remote, closed stacks environment. When an item is requested, the appropriate cassette is retrieved robotically and whizzed to the patron waiting in a public service area.

Southeast Asia

The main barriers to cooperative library services within Southeast Asia are based on political and colonial heritage, the lack of traditional links at different levels of educational and library development, and the low level of skilled human resources, particularly those related to library automation. The situation is particularly acute in Myanmar (formerly Burma) and the Communist Indochinese states of Vietnam, Cambodia, and Laos. The non-Communist ASEAN (Association of Southeast Asian Nations) countries of Brunei, Indonesia, Malaysia, Philippines, Singapore and Thailand have become increasingly active in cooperative endeavors. ASEAN is concerned primarily with improving economic ties, promoting social and cultural development, and maintaining a balance of power among Southeast Asian nations.

Technological equipment necessary for cooperation exists in the ASEAN countries, although availability varies. Communications facilities are quite sophisticated, linked by satellites and undersea cables. Microcopying, photocopying and computer facilities are increasing in several libraries. One cooperative effort produced a union list of microform materials relating to Southeast Asia; another relates to AGRIS, an agricultural data input project (Tee 1979, 217-219). Libraries were surveyed to determine the types of cooperative projects in which they would be interested. Programs which might include improved access (such as a joint union list and bibliographic databases) or improved interlending and document delivery were preferred over projects such as shared cataloging or cooperative acquisitions. Plans involving remote storage of materials to facilitate document delivery were not highly popular, primarily because most of the libraries surveyed did not have large collections of little-used materials, which could be stored without affecting service to the home institution (ibid., 223).

University libraries are well supported compared to other libraries in the region. Most have good collections and a relatively large number of trained professional staff and access to all types of equipment necessary for networking and resource sharing (ibid., 221). Institutions in Thailand, Indonesia, Malaysia, Singapore, Brunei, the Philippines, and Australia are now linked by AUSEAnet, a metanetwork which allows the exchange of information about microelectronics techniques. The Southeast Asian nations of Malaysia, Indonesia, and Thailand are also linked to research networks (Cleveland 1992b, A55-A57).

Latin America and the Caribbean

According to Gassol de Horowitz (1988, 97), in Latin America "most of the library development that has taken place in recent times has stemmed from what may be termed the Anglo-Saxon tradition; the patterns that have resulted have been propagated, consciously or unconsciously, by Western librarians traveling abroad on consultant missions and through library literature, much of which has been produced in the United States." National collections in Latin America have deteriorated, and even the major universities hold few serials (Bruer, Goffman, and Warren 1981, 1133). Deborah Jakubs (1993, 78) decried the "national preoccupation with the rising costs of science serials [that] has obscured the problems of foreign acquisitions: budgets dwindle at a time when publishing in Latin America is increasing and when more intensive comparative, cross-regional scholarship is underway."

A document delivery network began in 1981 in Latin America and the Caribbean under the auspices of the Regional Network for Information and Documentation. It was coordinated by the Pan American Center for Sanitary Engineering and Environmental Sciences in Lima, Peru. The network's purpose was to facilitate the exchange of information to meet the region's environmental health goals. Each cooperating center was responsible for providing photocopies of documents that it input into an index and tables of contents (from which a union list was developed). International currency exchanges were a major stumbling block in the early years, because each country determined its own charging policy, but consideration was given to using either a coupon system such as that used by AGRINTER (Inter-American System of Agricultural Information) or a revolving fund (Bartone 1982, 255-260).

ILL/DDS activities in the Caribbean face many of the same hurdles as other developing nations encounter. Graham Cornish indicates that some of these go beyond the control of libraries, such as politics, language, national frontiers, legal issues, and currency. Each country has its own political ideologies. There are four main languages (English, French, Spanish and Dutch) with few people fluent in more than one or two. What is legal in one nation may be illegal in another, creating difficulties related to copyright law and trade and tariff agreements. The wide variation of currencies makes it difficult to arrange for payments for document delivery, although the Eastern Caribbean dollar has been established as a common unit of currency for some of the smaller island states. Feelings of independence, national pride, and bureaucratic red tape also interfere in cooperative efforts, with so many nation states existing in a relatively small area

(Cornish 1989b, 249-250). Postal service both within and between the Caribbean nations is slow and unreliable, but telecommunications between the islands is dependable, so telefacsimile is a possibility for improving document delivery. While some of the island nations look toward each other for resource sharing, most rely on links to Europe and the United States, contributing to the lack of standardization for ILL/DDS in the Caribbean. No comprehensive union list exists for the region as a whole, although individual nations do have acquisitions lists for major libraries (ibid., 257, 259).

Mexico

Mexican libraries have trouble finding United States libraries willing to loan books because poor mail service in Mexico often results in losses or delays. The United States/Mexico Interlibrary Loan Project developed a system that uses diplomatic pouches to transport materials. The Benjamin Franklin Library in Laredo, Texas, and the USIA Library in Mexico City serve as referral centers to receive requests and materials from libraries on their respective sides of the border. The Benjamin Franklin Library verifies holdings among participating libraries and forwards requests to holding libraries. Copies may be faxed directly between libraries, but books are always funneled through the two referral centers. Unfortunately, only Mexican libraries that can conveniently pick up material at the USIA Library are able to participate.

Discussions at the Transborder Library Forums indicate a desire to expand Mexican participation in the network by setting up similar programs in other parts of the country, but little progress as been made. According to Bob Seal, requests for books made up 40 percent of the first year's total (Pfander 1991, n.p.). With the majority of requests being photocopies, Ariel and other forms of electronic transmission should improve document delivery to other Mexican libraries, at least for photocopies. Institutions such as the Universidad Nacional Autonama de Mexico, which has 164 libraries now have access to UnCover, and document retrieval via telefacsimile seems to be gaining popularity (Feick 1994, [2]).

Middle East

In 1990, for the first time since the Gulf War, a conference was held to revive and consolidate formal ILL agreements. Representatives attended from fourteen universities and other institutions in the six member states of the Gulf Cooperation Council, namely Bahrain, Kuwait, Oman, Qatar, Saudi Arabia and the United Arab Emirates. Even though the Gulf States have substantial financial resources, no library collection is totally self-sufficient, and ILL is used as a supplementary aid. An ILL Code, approved in 1985, provided for the development of a union catalog, standardized bibliographic procedures, an automated information network to link participating libraries, and rapid communication systems for both requests and document delivery. Geographic proximity, compatible educational concepts, and good financing seemed to predict success, but implementation proved to be difficult in terms of developing the union catalog or a document delivery mechanism. Revisions were made and adopted in 1989, but still ILL proved unsatisfactory. For example, in 1988/89, of 332 requests made, 323 were for articles and nonbook research material directed to only one lender, Kuwait University. Most of the requests were from within Kuwait: all of the transactions were strictly one-way. There were only 7 book requests and they were all supplied within Saudi Arabia. In other words, all of the activity was between two Arab states and even that was very lop-sided (Al Ibrahim 1993, 21-23).

Several obstacles limit the development of document delivery service in the Arab nations—not the least of which is the general attitude toward user service, and an obsession with secrecy and confidentiality, which prohibits all but commercial documents from being used outside one's own library. The Arab League Documentation Centre in Tunis was unable to get a regional agreement approved due to "suspicion and technicalities." (Aman 1989, 85-86) ILL/DDS activities are poorly promoted and few libraries have established separate ILL/DDS units. There are few union lists and union catalogs, so it is difficult to identify holdings in other regions. A union list for Kuwait University was in process but was destroyed during the Iraqi invasion—along with 95 percent of the collection. The collection is being rebuilt with strict bibliographical control for future union list development (Al Ibrahim, 1993, 24).

New technology is readily available in the Gulf States, allowing access to telefacsimile, telex, and photocopying, and to online international databases and electronic networks. The Library at King Abdul Aziz City for Science and Technology in Riyadh has a special department set up primarily to provide document delivery services for requesters as expeditiously as possible.

In countries that do not enjoy the wealth provided by rich petroleum deposits, such library services are not readily available. Even photocopying is difficult in the less developed regions due to lack of paper, toner, and routine maintenance. In terms of links to research networks, libraries in Saudi Arabia and Kuwait have direct access to GULFNET; Israel has access to ILAN, the Israeli Academic Network, a branch of EARN; Cyprus also has access to EARN. Bahrain, Qatar, the United Arab Emirates, and Iraq only have PDNs (public data networks). There are no networks in Oman, Iran, Yemen, Jordan, Lebanon, or Syria (Cleveland 1991b, A-65). The libraries throughout the Middle East receive the benefit of reduced postal rates for printed materials, but, otherwise, the gap between the information "haves" and "have nots" continues to widen.

Most document delivery requests are for science, technology, and medical publications, which are primarily borrowed from BLDSC with fax or mail as the mode of transmission. Electronic document transmission is under consideration in Kuwait, but appears to be a low priority at this time (Al Ibrahim 1993, 24). In Bahrain, local ILL and reciprocal borrowing agreements are found. ILL from United States libraries is available, but BLDSC is used most often for international document supply (Aman 1989, 84).

Egypt

Egypt has improved access somewhat through electronic systems available on the national network ESTINET (Egyptian National Scientific and Technical Information Network). A part of the Academy of Scientific Research and Technology, a government agency, ESTINET provides electronic document delivery services to special library users throughout Egypt. It is based on a distributed network configuration, with the central organizational unit in Cairo and six regional nodes. Network hosts (all in academic institutions) were selected based on geographic location, staff expertise, and willingness to serve the public. There are five information sectors: agriculture; energy; industrial medicine and health care reconstruction; science and technology; and sociology. A national union list of periodicals has been produced that should promote internal resource sharing rather than dependency on international sources, which is both slow and expensive (Dimitroff 1993, 27-28).

While government and foreign funding has improved scientific and technical document delivery through ESTINET, nothing is in place for document delivery of other types of information. The primary means of overcoming deficiencies in Egyptian library collections is through personal networking, and librarians expend considerable energy in developing contacts with other libraries. The inadequate, indigenous information delivery systems force the use of foreign supported facilities located in Cairo, such as the British Counsel Library and the American Center Library, but the practice diverts hard currency out of Egypt (ibid., 25-26).

Oceania/Australasia

Australia

In Australia interlending does not depend upon the national library, but is spread among libraries throughout the nation. The National Library of Australia (NLA) is located inland in Canberra, while the major population centers lie along the coastal fringes of a large continent. This situation makes centralized document delivery almost impossible. A practice begun in Western Australia and that has now spread to other states, is the funding of public libraries by both the municipality and the state or shire. The municipalities provide the facilities and staff, and the regional government does everything else—acquires and catalogs materials, and routes ILL requests. Each library agrees to share materials with other libraries in the region

(White and White 1993). Past studies by Exon (1987) and Taylor (1989) showed that tertiary academic (university) libraries are responsible for the majority of lending in Australia, with the largest portion of requests being for photocopies of scientific journal articles (Allen and Carman-Brown 1991, 33; Runner Ruda 1990, 1). For the most part, Australian publishers and subscription agents offer no document delivery services (Baker 1993).

AARNet (Australian Academic and Research Network) has been the national research network since 1990 and Australian libraries are rapidly connecting to it. A directory is available electronically (search Veronica for "AARNet Resource Guide") that provides addresses, log-in procedures and information resources available on AARNet. Contact Geoff Huston (G.Huston@aarnet.edu.au) for FTP instructions (Huston 1991). Library catalogs form the bulk of resources available on AARNet. Besides Australiania, notable collections include military history (*Mibilist Database*), Canadian studies, health sciences, and Latin American studies. Access to the catalogs is free to registered AARNET and Internet users (Cleveland, 1991b, 86-87).

Throughout the 1980s NLA initiated several activities and policies that have fundamentally affected the total pattern of library services in Australia. The most important of these actions in relation to document delivery services was the development of the Australian Bibliographic Network (ABN) and the creation of a national online union list, the National Bibliographic Database (NBD). Libraries can select from potential lenders on the NBD, then go to whichever library suits them best without any direct involvement by NLA (a direct-to-best-location interlending model). NLA is taking steps to ensure that bibliographic records for all significant collections in Australia are recorded in the NBD to aid access (Fullerton 1991, 102-103, 105). ILLANET is the ILL subsystem on ABN. NLA reports about 60 percent of ILL requests are now received electronically.

One problem encountered is how to handle electronically the vouchers consistently used as payment for nonreciprocal lending or photocopying between Australian libraries (ibid., 110). The vouchers ($6.00A) purchased from the Australian Library and Information Association may be reused, but net lenders cash in extra unused vouchers while net borrowers must purchase new supplies. Some libraries have conducted cost studies to determine whether the standard voucher fee covers the average cost of an ILL transaction and if the cost of the vouchers could go up.

Allen and Carman-Brown (1991, 38) conducted a study to determine the effect of membership in the ABN national network and having holdings in the national union list on the volume of lending among Australian university libraries. Surprisingly, lending activity remained fairly constant over the 13 years studied (1977-1989), but borrowing activity (which was not analyzed) increased 36.5 percent during the same years. Since lending among the university libraries did not increase, one can only suppose that academic libraries now borrow more from nonacademic libraries, utilize commercial document suppliers to fulfill requests, borrow more from international resources, or experience a higher rate of unfilled borrowing requests.

The postal system remains widely used for ILL, but it is being supplanted by electronic transmission of requests. The use of couriers and electronic modes for delivering documents is also on the rise. Runner Ruda (1990, 16) determined the cost to loan a book within the courier network at Curtin University of Technology was $0.53A and $0.82A to supply a photocopy, while outside the courier network, the cost was $13.12A and $8.73A respectively. A trial with ADONIS tested the possibilities of digitized document delivery (Steele 1989, 73). Several Australian academic libraries acquired ARIEL, and its effectiveness for document delivery was tested, with generally favorable results (Kósa and Tucker 1993).

The University of Western Australia will test the cost effectiveness of access to journal literature via document delivery services over ownership during 1994 (Ellis and Rainford 1994, 3-4). A number of Australian libraries helped beta test Ariel for Windows, so the new version should be more adaptable to international standards, such as the capability to scan and send documents on A4-size paper. Even the "fast track" document delivery system once available primarily through NLA has become decentralized as more libraries use fax, Ariel, and other fast methods (Steele 1989, 73).

South Pacific

The South Pacific region has twenty-two island states of different status. Most are poor and newly independent. The population totals 4.9 million people, of which sixty-seven percent reside in Papua New Guinea. The people speak numerous languages (Williams 1991, 111-112) and, although education is a priority, literacy rates vary and access to information remains poor. Transportation in a region that covers 30 million square kilometers (only two percent is land) is difficult to arrange and expensive, but postal services are improving.

Advances in CD-ROM, computer, fax, and satellite technology have contributed to the development of libraries and document delivery in the region. The University of the South Pacific (USP) provides for the document delivery needs of eleven island states: Cook Islands, Fiji, Kiribati, Nauru, Niue, Solomon Islands, Tokelau, Tonga, Tuvalu, Western Samoa, and Vanuatu. The library at USP is able to meet most needs for documents related to the region from its collections (the main Library in Fiji, and a branch library in Western Samoa, plus small collections in a University Centre in every country but one) and national libraries within the region, but it must also use ILL from the British Library, New Zealand, and Australia to supplement local collections.

There are no formal ILL/DDS agreements between the countries of the region, but the informal system in place since 1972 works well. Document delivery as operated by USP on local, regional, and international levels recognizes that access to document delivery services are essential for scattered island states, but services are expensive and the supply of information documents for small island states cannot be free. Libraries and users must be prepared to meet part of the costs of services (Williams 1990, 128). Even so, rising document delivery costs, particularly from foreign sources, are a major concern (Williams 1991, 117).

In 1982, networking efforts began among the small libraries in the region with the establishment of the Pacific Information Centre, which identifies, collects, and records published and unpublished material originating in the region and material about the region published outside. Several bibliographies, indexes, and union lists have been produced. Interlending and document delivery rely on these bibliographic resources, which are distributed to over 3,000 libraries and institutions within and outside the region. In 1989, USP began computerizing its holdings and bibliographies, providing instant access through remote terminals (ibid.). The satisfaction rate on requests from within the region averages 92 percent, while outside the region where-holdings are more difficult to determine, the satisfaction rate is about 81 percent. Of the requests, 74 percent are for photocopies and 26 percent for books (Williams 1991, 116).

Besides the activities at USP, the Pacific Islands Marine Resources Information System (PIMRIS) Library provides a current awareness service of value to the fishing industry. PIMRIS also provides CD-ROM searching of other libraries that have contracted for the service. ILL/DDS for regional libraries in the area of agriculture is through the Library of the School of Agriculture in Alafua, Western Samoa. Requests are received by letter, fax, and telephone. About 75 percent of the requests are filled from within the region. Materials are provided most often via airmail and fax, but regular postal service, couriers, satellite services, and telephone are also used to provide information to remote locations. Satellite communication has been available since 1972 for distance education programs and is available for library use as well; documents therefore can be sent from one country to another in a matter of minutes to those institutions linked by satellite (Williams 1990, 130-131).

Airmail to and from PIMRIS or USP is actually faster (about 10 days) to the farthest locations, such as the Micronesian countries, because good flight connections exist through Honolulu. From New Zealand, Australia, the United Kingdom, and the United States the average is 8 to 14 days. Closer locations average three weeks because of fewer scheduled flights (ibid., 132).

Depending on the status of the requester, ILL/DDS service is provided free or for the cost of copies. Online searching of foreign databases is paid for by the user. Some consideration is being given to providing ILL/DDS services on a cost recovery basis.

While the libraries in the USP network share some of the same problems described in other developing nations—customs and postal regulations and currency exchanges—document delivery overall operates as efficiently as conditions allow, with a high level of cooperation between libraries. Future goals include extending reciprocal services to developing countries in Asia and Africa, seeking additional financing, and attracting trained people to operate the service (ibid., 134-135).

Papua New Guinea (PNG). Stephen Wright's article, "Library Automation in Papua New Guinea," (1991, 37-50) should be required reading for anyone planning to automate a library located in a tropical, developing nation that experiences frequent power shortages, a lack of skilled computer-literate staff, and limited financial resources. The article provides great insight (or hindsight, in the case of PNG) regarding problems and solutions such as battery back-up during power failures for online systems, combating high telecommunications costs of online searching by contracting with libraries in other countries, and the political clout of libraries in helping to bring a package switching network to PNG (PANGPAC), which has further reduced telecommunications rates.

PNG has about 250 libraries and information services, ranging from unmanned school libraries to numerous government and special libraries (minimally staffed, in most cases), up to the National Library and two university libraries, which employ the majority of the nation's professional librarians. Most of the libraries are located in Port Moresby, the nation's capital, with a few facilities throughout the nineteen provinces. The quality of service in the public libraries varies considerably in the provinces, but the National Library Service does not have the authority to require minimum standards. While the overall library development in PNG is low, in a relatively short time span the larger libraries have developed sophisticated systems that provide a high level of service.

Microcomputer use in PNG is increasing rapidly, and the machines are readily available at moderate costs from a range of local marketers, although the level of support and back-up varies considerably. Most schools in the urban areas are using computers for educational and clerical purposes. A few special libraries have developed small bibliographic databases of their holdings. Several libraries have established links with international networks, particularly the FAO's CARIS network for agricultural research, INFOTERRA related to the environment and conservation, and ASTINFO.

The National Library Service has automated the national serials union list (NULOS) and the Papua New Guinea National Bibliography has been produced on CD-ROM. The University of Technology (UOT) at Lae has established a distributed network of microcomputers, while the University of Papua New Guinea (UPNG) opted for an integrated automated library system (ADLIB) run on the university's PRIME minicomputer. Both university libraries have been active in promoting automated library activities. UOT has converted its library catalog to a CD-ROM database that has seamless access through a LAN. UOT utilizes a satellite communications system (PEACESAT) to link Pacific island nations. UPNG's automated library system covers acquisitions, cataloging, periodicals, and circulation.

In 1985, UPNG became the first PNG library with a link to an overseas database, DIALOG, although it took nearly six months to successfully complete the link, due to the lack of expert local advice. Online searching is very popular on the university campuses, but initially linking to DIALOG in California required an STD call through Sydney, Australia, at over $100/hour. UPNG and other interested organizations were instrumental in securing a direct telecommunications link from PNG. At this time, most of the automated activity is directed towards better access; not much attention has been directed towards document delivery although fax and e-mail are now being used for interlibrary loan.

SUMMARY

Ideally, barring international political, ethnic, or economic conflicts, a worldwide document delivery scheme could be developed. Since the world is not perfect, we can only encourage the development of comprehensive national collections for internal interlending and document supply, to be made available to other countries upon request.

Chapter 9

Document Delivery to Branch Libraries and Other Remote Sites

If academic libraries are to utilize computer-based bibliographic networks effectively, they must improve their capacity to deliver materials to local constituencies from their own collections as well as from the collections of institutions in other localities (Dougherty 1978, 25).

Libraries with multiple sites have unique problems in trying to adjust to the economic crunch. Each branch facility needs a core collection to serve the immediate and frequent needs of its specific clientele, but few library systems can afford to duplicate everything held in the main library and unique materials held in other branches. Getting materials to and from various off-site facilities in a library system takes a coordinated effort and a definite commitment to resource sharing, remote access, and prompt document delivery.

The pressure for immediate access and rapid document delivery increases with every mile between sites, the addition of each branch, the growth of clientele at each facility, and the development of new services at each location. Whenever funding cuts loom, remote access, distance education, and branch services may be the first cuts considered, or at the least, expectations are raised that such services become self-supporting. Budgets at the various facilities may be allocated differently. Priorities for service may be unique, and clientele at some sites may have unconventional needs. Library service to multiple sites requires a mini-network within a local system; access and docdel often provide the foundations upon which most such networks are built.

Branches in public library systems may encompass facilities of varying sizes in subdivisions, suburbs, or small communities nearby. Bookmobiles may serve areas without permanent facilities. County and statewide library systems are likely to reflect networks similar to public library systems, except on a larger scale, involving multitype libraries. The distances between sites may be greater, creating more complex transportation and communication problems. Branches in a school library district may include every library in each school in the district; in some communities, school libraries may form part of a local multitype public library system.

In academic library systems, branches may be separate facilities for specific subject areas (law, medicine, or science) or for certain classifications of students (an undergraduate library) but housed on the main campus, or located on branch campuses elsewhere within the same metropolitan area. Some branch campuses may be quite far from the main campus. Continuing education courses may be sponsored at various corporate, government or public sites, with or without on-site library facilities; external students may be involved in long-distance education taught by correspondence, satellite, fiber optics, radio or other means of electronic communication, with or without a library nearby.

Special libraries—particularly corporate, medical, legal, and government facilities—are also likely to have branch offices, all of which probably depend to some extent upon the information center at headquarters. Government libraries in particular may have to serve branches or clientele scattered in remote areas, at sea, or in the air.

The diversity of what constitutes a branch library is immeasurable. One certainty, however, is that clientele at each facility (with or without an actual physical site) will want immediate access to library resources and a document delivery system that is as close to instantaneous as possible.

A MODEL SYSTEM

The document delivery environment at Arizona State University (ASU) provides a microcosm of document delivery programs relevant to many other types of branch libraries, including those in nonacademic environments. Much of this chapter is based upon workshops the authors presented at an annual conference of the North American Serials Interest Group (Walters and Mitchell 1992), from which was conceived the idea for a document delivery handbook. As in those workshops, we have used many of ASU's document delivery programs (past, present, and projected) to form a "composite" case study of services to branch libraries and off-site locations. Examples of document delivery programs in other libraries are incorporated when appropriate to provide broader coverage of service possibilities.

We aim herein to portray the spectrum of document delivery possibilities using a real library as a model, rather than depict a specific library's exact program. While we might have used a fictitious library, the authors have been involved with document delivery services at ASU for so long, it seemed more instructive to build around that with which we were familiar. We hope the same "literary license" allowed authors of historical fiction is afforded us for this purpose.

The university administration describes the entire ASU system as "one university geographically distributed." More than 40,000 students on the main campus in Tempe are served by Hayden Library (the "Main Library," with holdings in the social sciences and humanities). Branch libraries house collections for science and engineering, architecture, music, archives, and law. "Reading rooms" and small reference collections scattered about campus have mostly been closed over the years as they were expensive to maintain and duplicated works readily available within major campus library facilities. The system includes two branch campuses in the Phoenix metropolitan area. Extended education classes are offered within and without Arizona (even abroad) at various sites, many with no permanent library facilities. Courses are also taught by correspondence, satellite, and via public television. For the most part, technical services for branch library facilities are centralized at the Main Library, and all libraries share an online catalog. Interlibrary loan and document delivery services are also centralized. Document delivery between branch libraries occurs mainly between the libraries of ASU Main and ASU West.

ASU West was established in Glendale, a western suburb of Phoenix and about twenty miles from ASU Main in Tempe, an eastern suburb of Phoenix. From the beginning, ASU West's library resources were intended to rely upon, not duplicate, collections in the main campus libraries.

Philosophy of Access

Unlike libraries that initiated resource sharing programs out of necessity, ASU West began its programs out of conviction. From its inception, the library has recognized that resources can be immediate or remote, and that research and teaching needs can be as well-served by efficient, rapid access as by acquisition. The library's goal is an on-site collection sufficient to support the curriculum and basic research needs of students, faculty, and community users.

To augment and enhance in-house resources, the library relies upon the libraries of ASU Main and the greater information marketplace. Access to remote resources strives to be as "seamless" to the user as possible—transparent, uncomplicated, but for a necessary (but not excessive) delay. Through intercampus document delivery services, almost all of the circulating or reproducible items retrievable through the shared OPAC can be in the patron's hands in 48 hours or less.

Document delivery to ASU West includes materials both owned and unowned by the ASU Libraries System. It includes materials loaned or photocopied from collections at ASU Main, materials borrowed from other libraries, and materials obtained through commercial services. As the West campus collections developed (particularly for monographs), ASU Main began to rely heavily on the branch library for recently published monographs. Much of the main campus materials budget is devoted to serial acquisitions, so collection development of monographs at ASU Main has suffered—a consequence that would be felt even more if it were not for the branch collections.

On the other hand, ASU West had decided early not to bind or store long runs of journals, but to rely on the main campus for retrospective materials. The West Campus library does subscribe to journals heavily used on that campus, based on curricular needs and document delivery records of articles requested. So with ASU Main having good retrospective holdings (both monographs and serials) and ASU West having current monographs and serials, the needs for both campuses are easily met through document delivery available to all students and faculty at no fee, except for the cost of photocopies and microcopies.

Contractual Relationships

The primary component of the branch campus document delivery service is based upon a contractual agreement between the ASU Main and ASU West libraries. Although they are part of the same University Libraries System and share many centralized services, funding for the two campuses is allocated on separate budget lines. Document delivery and interlibrary loan staff at ASU Main literally work for West, but on the Main campus. They are hired, trained, supervised, and evaluated by ASU Main staff, but the salary lines are covered by funding derived through a contractual agreement for services. ASU Main likewise pays West for document delivery services from the West Campus. This system is sometimes referred to as "subscribing" to another library. Subscribing libraries do not have to belong to the same library system, but they must agree on the contractual elements.

The main components of an ILL/DDS service contract include:

A. Standards for processing and turnaround.
   Example 1: Eighty percent of document delivery requests are to be supplied within 48 hours of receipt (excluding weekends and holidays). Unfilled docdel requests are to be returned to the requesting library

within three to five work days. (The extra days allows for re-checking for materials temporarily off the shelves.)

Example 2: ILL requests are to be processed within four work days of receipt. Requests for which no potential supplier can be located are returned (or a status report supplied) to the requesting library within five work days.

B. A set fee for service.

Example: Book loans and photocopy requests are supplied at a set fee based on actual cost of operations, subject to annual review. (At ASU there is no profit margin. Copy costs have varied slightly from year to year, sometimes based on the actual number of pages copied or the type of copying, other years based on a flat fee. Unless the governing body requires absolutely precise accounting, a flat fee is more cost effective, saving time on statistics and bookkeeping.)

C. Regular review of the contract.

Example: ILL requests for the branch campus are processed at a set fee, subject to annual review, based on actual costs or national averages, plus any lending fees and copyright royalties. (At ASU, the same fee applies regardless of whether the request is filled, because an unfilled request may involve as much or more work as a filled request. A tiered price structure, based on the amount of work on each request, was considered, but a flat fee, below the national average, was decided upon. The difference in processing easy, average, and problem requests balances out and the extra record keeping of a tiered price structure is eliminated.)

D. Management reports—statistical and financial

Example: Monthly statistics are supplied by each library. (The data provided depend upon the needs of each library, any may differ for the contracting parties.)

The annual review of the contract should cover the following issues:

- Does the contract, as written, create any processing difficulties for staff at either institution?
- Are the current fees sufficient to cover all costs, or more than needed to meet costs?
- Will any program or curricular changes anticipated in the coming year affect the document delivery program?
- Is an internal time/cost study needed to update cost figures?

One contract change over the years affects how to determine service costs. At first, the volume of document delivery requests was quite low and almost exclusively one way (from Main to West), so West simply paid the salary line(s) of staff involved. As volume increased, so did the bureaucracy, with an administrative preference for actual per item costs—involving time/cost studies and more record keeping that, ironically, adds to costs and slows delivery!

Following is a chronology of the technical changes and staffing requirements to show how the present system evolved. At every phase of its development, the system has worked well and could be emulated in other libraries at a similar stage of development.

1984. Requests were received at the branch campus library by reference staff. Initial processing included checking patron's ID, confirming call numbers of titles at ASU Main, and eliminating any requests for titles held at West. Requests for titles held at ASU Main were telephoned to the ASU Main ILL Lending Unit. If the volume of document delivery requests was too high to make telephoning practical, the requests were sent with the courier.

A part-time student assistant in the Lending Unit of ILL did most of the pulling of volumes, photocopying of journal articles, and checking out materials to ASU West. All processes were handled by ILL staff on the unit's own circulation, photo- and microcopy equipment. Books were checked out to ASU West as the "patron." The request form was put into book loans along with a date due slip, or attached to photocopy requests. No paper work was retained by the Main Campus on ASU West document delivery requests.

Interlibrary loan requests for both campuses were processed together. (There was no separate budget line for ILL staff, only for "docdel" staff.) Book loans were received, a copy of the request enclosed, and the book banded with the date due (shortened by a few days to allow for the time of book movement between campuses). The Borrowing Unit interfiled (by date due) a copy of West ILL loan requests with records of ILL loans to ASU Main patrons. Photocopy requests were supplied with a copy of the request attached. A copy of the request was filed at Main in either the "closed" or the "copyright" file.

Materials were delivered once daily (week days) by a campus mail van. A 48-hour turnaround was the norm. Staff at West recirculated book loans (i.e., cleared West as the patron and entered the actual patron's ID onto the Circulation record); the automated circulation system would thus generate any overdue or billing statements directly to the patron involved. Reports on any unfilled requests were sent by the campus mail courier along with materials going to West.

1985. As volume grew, telephone requests became impractical. Requests were printed and carried by campus mail. ASU West purchased a van and established a second daily delivery; campus mail still made one daily delivery. A 24-hour turnaround was frequently achieved with requests received on the morning run delivered on the afternoon run. Nothing else changed.

1986. Requests were transmitted electronically via an IBM XT using PC-File, an inexpensive software package. West staff keyed in the information from the patron's written request forms into a "docdel" file, using a keyboard template. Crosstalk communications software allowed dialing in to the file, downloading, and printing the requests each day. PC-File tracked requests and provided management and statistical reports. (ASU West eventually moved away from this format for transmitting requests, but another small off-site library used it for several years.)

1987. Branch patrons could submit the printouts from electronic sources as requests. A rubber stamp provided the form for patron data (name, status, telephone number, need-before date, id number). At ASU Main, a half-time library clerk and a half-time student assistant in the Lending Unit were now on West budget lines.

1988. ASU West moved from temporary quarters to a new campus, and Fletcher Library opened. Document delivery functions resided in Circulation and at the Information Desk. The West Library was converted from IBM to a Macintosh environment, and a new document delivery database was created using a Double Helix program from which requests were downloaded twice daily. As before, West telephoned ASU Main ILL when ready to transmit; ASU Main staff then dialed in to begin the process. The PC in the ILL office still determined the format of printout; problems or changes in transmitted fields or formats had to be addressed by West staff but on site at Main (20 miles away). At ASU Main, the equivalent of one full-time employee was funded by and dedicated to West document delivery. Volume was still almost exclusively from ASU Main to West, but several lean budget years were in store for the main campus that seriously affected monographic collection development. ASU West, however, with a new library on a developing campus, had funds to develop a core collection.

1989/1990. The document delivery process was split into three units at Fletcher Library. Verification and patron assistance were still done at the Information Desk and receiving and circulation were still handled by Circulation; but data entry, database maintenance, and the *fulfillment of Tempe requests* were assigned to another unit. ASU Main had begun to use the collections (primarily monographs) of West to meet the needs of library users on the Main Campus. The volume from West to Main was still low and West absorbed it with regular staffing (much as Main had absorbed West's requests in earlier years).

ASU Main had also begun document delivery to two additional off-site locations (a smaller branch campus in the heart of Phoenix, and a private business college that shared a few joint programs with the ASU College of Business). An on-campus document delivery service had also begun on the Main Campus. The document delivery program for the business school was identical to that of ASU West, except the volume was lower. The Down Town Center, however, was a much smaller off-site facility, with courses that placed less emphasis on library resources. Only occasional requests for document delivery were expected, so a simple system was devised, using e-mail to forward requests to ASU Main, and a local commercial courier for deliveries, as needed.

Oddly enough, with document delivery growing as a cost-effective means of stretching the libraries' materials budgets, we ran into a delivery problem: Campus Mail Services decided that library materials were *not* mail. The libraries were forced to contract with commercial couriers or hire library staff for all document delivery between campus libraries.

Although different couriers are employed for the various delivery routes, internal procedures are exactly the same for each courier. Materials are pulled, and checked out or photocopied, and placed unwrapped in the appropriately labeled bin. Bins are taken up to the Mail Room prior to the scheduled pickup. Any bins dropped off by earlier couriers are retrieved and brought to ILL/DDS for processing. A bin is shipped everyday, even on those rare days when nothing is being sent to a delivery site. The courier delivers a bin for re-use for the next day's deliveries from that site. Labels on the bins are reversible, so upon receipt, the label is turned over and the bin is ready for return deliveries.

A number of statewide and regional courier systems began under government-funded programs several years back. Many library systems purchased a delivery van, hired a driver, and took on delivery operations under the system's auspices. The high costs for vehicle maintenance and fuel, and insurance coverage soon killed many such document delivery programs. A number of models, however, have used inexpensive local commercial couriers with great success—for example, statewide systems in Connecticut, Ohio, Pennsylvania and Illinois.

Any library system seeking to provide courier service between large numbers of library branches should look to the transport scheme used by the British Library Document Supply Centre (Boston Spa) and the regional library systems in the United Kingdom. They use a combination of commercial couriers and rail services to transport millions of items each year to over 4,000 libraries in every region. Loans from Boston Spa are sent by overnight train to London, where they are picked up, sorted by area, and sent out on vans to area libraries, providing fast and economical deliveries (Gerrard 1993a; Smith 1993b). Such a transport system is possible only when branch libraries are located in regions linked by inexpensive rapid transit.

Although pale in comparison to the volume of materials being transported in the United Kingdom, the volume between ASU West and Main increased to the point that reorganization was necessary to maintain efficiency. At ASU Main a separate Document Delivery Unit was formed to handle docdel to branch campuses and the on-campus document delivery service, and to reduce the work load of the ILL Lending Unit. ILL became officially "Interlibrary Loan and Document Delivery Services" (ILL/DDS).

Staff levels and procedures did not change, but the DocDel Unit concentrated on serving the ASU library community with commitments for 48-hour turnaround and the ILL Lending Unit concentrated on the four-day turnaround protocols of the OCLC ILL subsystem and mail requests from other libraries.

Requests from "for-profit" libraries and all rush requests from non-ASU libraries were turned over to FIRST, a separate fee-based document supplier within the University Libraries systems. FIRST was established in 1987 to serve nonuniversity library clientele, relieving other public service staff. While the document delivery procedures and operations of FIRST were similar to those of the ILL/DDS unit, each served a totally separate clientele. They did share, however, the characteristic that both were established to be self-supporting. For the DocDel Unit, the contractual arrangements for document delivery to the branch campuses, particularly ASU West, and the revenue generated by the on-campus document delivery service, Library Express, were the main sources of funding. As volume grew for the DocDel Unit, a unit head was hired to take over most of the supervisory functions previously handled by the department head as well as the processing of ILL requests from the branch campuses. This change reduced the volume of requests handled by the ILL Borrowing Unit, which now could concentrate on improving turnaround for the Main Campus.

1991. Software, hardware and the structure of the document delivery system at West were once more changed to accommodate increasing volume. The Filemaker Pro database provided better turnaround statistics, networking capability (the data can be viewed anywhere in the library on a LAN), easier access, and passwording that permitted different levels of access. A fax modem was installed to transmit requests twice daily to ASU Main's ILL/DDS unit (giving West control over the transmitted format and fields in the request). The receiving function was reassigned so that the same staff members who did the data entry would also process received items. The lending operation (fulfillment of ASU Main requests) was moved to Circulation. This arrangement paralleled procedures at ASU Main. Because turnaround was consistently within 48 hours, West ceased notifying patrons upon receipt of materials, notifying them instead only when delivery was delayed.

1992. By this time the volume of requests for monographs was almost equal between the two campuses. Photocopies were still primarily one way (from Main to West) because the branch campus libraries do not maintain large backfiles of journals. In a few cases where the Main campus does not own a title or has canceled one that West owns, backfiles may be held at West, or could be transferred to ASU Main if storage was a problem at the branch site.

Some document delivery programs at institutions such as the University of South Africa (Unisa) maintain multiple copy collections for lending to off-site libraries and students in distance education programs. For popular courses, thousands of copies of some works are purchased (Shillinglaw 1992, 145). This method of document supply relies on a keen sense for predicting user needs, not to mention a huge budget. Many multi-branch libraries also have a large volume of books sitting unused after initial popularity wanes.

Due to years of limited funding for library resources, all of the Arizona State University Libraries are primarily single copy libraries. Reserve collections and a strictly enforced recall system provide equal access to library resources for all students on the various campuses. By the '90s, the volume of document delivery requests from ASU West averaged around 13,000 annually, with a fill rate of over 88 percent. The most common reason for an unfilled is that something is temporarily off the shelf, not unusual on a campus where collections are heavily used by over 40,000 students, faculty, and the surrounding metropolitan area.

Reaching and maintaining this high level of service has not always been easy. Remote access and document delivery were planned as an integral part of service to branch campus libraries, and service to West from Main was well planned and executed. A weak link in the chain turned out to be ASU Main's dependence on the branch campus for monographs. No one predicted that likelihood! Like many other multi-campus and multi-library institutions, ASU was not immune to rivalry and hard feelings as the "child" sometimes seemed better off than the "parent" institution.

Sharon West documented similar problems as the University of Alaska began providing library service to rural students through distance education programs. She states: "As much as a library may believe that its off-campus students and faculty need information resources, the on-campus political environment must be factored into the building of such a program. . . . Since the library was already suffering from inadequate funding to support the on-campus programs, the reallocation of funds to a new program was not initially well received. Some library faculty also shared the concern of the teaching faculty about providing a higher level of service to off-campus users than was being given to on-campus users." (West 1992, 559-560)

Similar complaints were being heard on the Main Campus of Arizona State University. "Why are the new books at West? If we can give West 48-hour turnaround, why can't they do the same for us?" Requests from ASU Main for materials at ASU West began sporadically and were not handled systematically. Each branch submitted requests in whatever format best suited their work flow. The staffers at West were forced to manipulate various formats—Campus Mail, telephone, e-mail, or fax—making it difficult to develop routine handling methods and delaying turnaround. Service to

ASU Main library clientele suffered from the lack of coordination, but few people at ASU Main were aware of (or willing to admit) that volume had exceeded spontaneity. The wake-up call came the year that ASU Main borrowed more books from West than West did from Main! A Total Quality Management Team was formed from Main and West to work out the problems and better coordinate services on both campuses.

1993. Since the total quality assessment, all requests from Main for books at West now go through Circulation at the ASU Main Library. A copy of the OPAC record is faxed to ASU West after checking to be sure there are no prior holds for the work. Books received from West are held for pickup in Circulation. Photocopy requests and materials are still processed by ILL/DDS.

The previous reorganization in ILL/DDS refocused energies to meet contractual agreements with the West Campus, while relieving the rest of the department from trying to balance the rapid delivery to branch libraries with priorities for the main campus. While ILL/DDS activity was still concentrated at ASU Main, some equipment (e.g., photocopiers, fax, and Ariel workstations) was installed in branch facilities where materials were frequently requested; thus, staffers could do more work on-site. This adjustment improved turnaround and caused less wear and tear on collections, and materials were back on shelves and available to in-house users much faster.

At ASU West by this time, staffing levels were being increased in the three units providing document delivery functions. Data entry and materials receiving now formed part of the duties of four staff members. A Library Specialist was employed full time to monitor the document delivery process, train staff, troubleshoot problems, and produce statistical reports from the document delivery database. This is the only full-time staff member assigned to document delivery at ASU West; all other staff members involved in docdel have other work assignments within their departments. ASU West staff also began to make more use of commercial document suppliers, such as UnCover, so that materials not held within any ASU library might be ordered directly from West, bypassing the ILL process.

As ASU West's monograph collections developed, the volume of requests from West plateaued and the staff at Main was reduced accordingly (although the staff simply shifted to the Library Express side of the unit where on-campus document delivery was booming). At ASU Main, staff supported by the contract with West now consist of a Library Supervisor (75%), a Library Assistant Senior (50%), and a Library Aide II (student assistant) (50%).

1994. Arizona State University recently received "Research One" status, a rare accomplishment for a university that is neither a land grant institution nor associated with a medical school. It is even more amazing to achieve such status with minimum duplication of library collections within a multicampus system. Relevant to the university's high ranking among research institutions are its rapid document delivery between branch libraries and liberal use of commercial document providers known for fast service, combined with interlibrary loan activity between other libraries equally committed to rapid turnaround.

While we believe that ASU provides a model for document delivery between branch libraries, the preceding chronology makes it obvious that current procedures were not developed overnight. A high level of service requires regular monitoring of the work flow. Practices are continuously adjusted to meet changing needs and to incorporate developing technology.

Work Flow for Branch Library Document Delivery at ASU

Request Initiation. At West, except for items found cited in another publication, patrons are likely to find most citations in electronic databases, either on CD-ROM or on an online OPAC database. Patrons are expected to search West's serials holdings to determine if a title is held there. If not at West, but available from ASU Main, the patron provides relevant information (call number, location, format) and submits a docdel request form. Requests can be submitted on printed forms, by telephone, fax, and e-mail, through liaison librarians (subject specialists). A self-service station near the Information Desk provides access to the serials list, as well as a stock of all necessary forms and supplies, eliminating for experienced document delivery users the need for mediation .

Request Review. Staff members review request forms for completeness and to identify any special handling needs. The "need before" date is frequently omitted but, when provided, it may allow ASU Main staff time to re-check for items temporarily "not on shelf." Urgent requests are more likely to be faxed separately, rather than included in the batch process. Patron data provides a quantifiable measure that is used to assess services and collection development needs.

As part of the review process, staff members check the bibliographic citation for omissions and obvious errors. They also confirm call numbers and holdings information. Some controversy exists over whether patrons should receive this much "hand-holding." Many libraries just instruct patrons on how to do their own reviews. As the volume of requests at the branch campuses escalates, such checking becomes increasingly labor intensive. (ASU Main uses the do-it-yourself approach—with proper instruction.) After the information has been verified, the form is placed in an outgoing tray checked hourly by the data entry personnel.

Branch Processing of Request. Data from request forms is entered into a Filemaker Pro database. Each operator entering data has a unique password permitting only the appropriate level of access to particular layouts.

The program assigns a unique number and date stamps the entry. Patron data is easily copied on multiple requests. The form includes check boxes for common "unfilled" replies, which standardizes responses from ASU Main. Unfilled reports serve as a "quality alert" for either library.

Transmission of Requests to Main. After they are entered, requests are batch transmitted by telefacsimile twice daily to the Document Delivery Unit at ASU Main. Faxing has the advantages of 1) not needing to be monitored as was the case when we used Crosstalk software, 2) being initiated by the branch campus, independent of ILL/DDS at Main, and 3) making the workstation immediately available for other tasks. Since the database is networked on a local area network (LAN) at Fletcher Library, entry and receipt can take place from several stations simultaneously, and public service staff members have uninterrupted "view only" access to determine the status of a patron's request.

At ASU, each branch library may use a different means of transmitting requests to ASU Main and the requests may come at different times of the day. To a certain extent, this can not be totally coordinated due to different equipment and staff capabilities at various sites, but more effort is being made to coordinate the receipt of requests.

Processing Requests at Main. Upon receipt at ASU Main, the requests (three per faxed page) are counted immediately, matching the count to the numbers on the cover sheet. If a page was skipped or something cut off, a quick phone call has the missing part on it's way by fax. The pages are cut in thirds, sorted by call number and locations, and pulling and copying processes begin immediately. The bulk of the requests are normally in either the Main Library where the ILL/DDS office is located or at the Science & Engineering Library, a few blocks away. ILL/DDS staff do all of the pulling and copying/faxing/Arieling, or circulating of materials in these two major facilities. When materials from smaller branches are needed, an e-mail message can be sent to that branch, where staffers there pull the item and e-mail to ASU Main that it is ready for pick up.

If branch staffers have time they may photocopy the material or fax it to ILL/DDS; otherwise volumes are held for ILL/DDS staff to do the copying. Copying of items in special collections, such as archives or rare books, is done by trained staff from the particular unit. That is about the only instance where docdel staff lose control over processing and turnaround time.

Noncirculating materials may be requested through interlibrary loan if patrons are unable to travel between the campuses, although hourly bus service between the two campuses has helped to eliminate that problem. Indeed, the bus service could be used to provide document deliveries between the campuses if the courier vehicle were temporarily out of service.

A copy of the request is placed with each item ready for delivery. Materials are placed in a bin for pick-up by the West courier twice daily. There is no wrapping or packaging, except for tying multiple volume sets together. Reports on unfilled requests and any materials received via interlibrary loan are sent at the same time. Statistics are kept to a minimum at ASU Main—a volume count and the reasons for unfilleds—using Lotus 1-2-3.

Delivery Mechanisms. At this time most of the actual delivery processes between ASU's branch campuses are physical rather than electronic. Fax is used for rush photocopies. Consideration was given to using fax or Ariel regularly; however, with two courier runs daily, the contractual requirement for 48-hour turnaround is being met and frequently surpassed. Electronic applications are reserved for transmitting requests, tracking, record keeping, and producing statistical and management report, thereby freeing fax and Ariel workstations for other long-distance, ILL/DDS document delivery activities.

Receiving Process at Branch Library. Materials are delivered to West by courier twice daily (excluding weekends). Upon arrival, items are counted and batch processed as "received" into the Filemaker Pro database. The original record is retrieved via the request number. The record is checked against the request slip and the item in hand. If all matches, the data record is updated to "filled," prompt, the date automatically stamped on the record, and the item moved to a book truck for delivery to Circulation.

Notification. Whether patrons are notified varies among ASU branch libraries. Because most patrons expect delivery within 48 hours, notification is not always necessary. Some branches deliver materials requested by faculty members directly to their offices, when no charges are involved. If an item has not been picked up or delivered within ten days, the patron is notified by telephone (or mail if telephone contact was not possible) that any book loans not picked up within twelve days will be returned to the lending library. Photocopies not picked up within 30 days are discarded. Patrons are notified by telephone, e-mail, or postcard that a request was unfilled, or advised that more information is needed before the request can be processed further.

Statistical Reports. The Filemaker Pro database manipulates data for a variety of statistical reports about users and the materials requested. Monthly reports are generated reflecting both document delivery and ILL activity for ASU West patrons. A spreadsheet shows each month's activity and year-to-date cumulation, plus comparisons with the previous year. Patron status reports and call number reports are also produced for management purposes and for collection development.

Another report provides data on document delivery turnaround. The report's main purpose is to monitor internal processing at the supplying branch. The software was amended to eliminate nonworkdays when it was discovered that most requests taking more than 48 hours to be filled were primarily those logged in at West after the last transmission on Friday but not sent to ASU Main until Monday. Thus, 48 hours had passed before ASU Main had even received the requests.

DOCUMENT DELIVERY IN OTHER LOCALES

The ACRL *Guidelines for Extended Campus Library Services* (ALA Task Force 1990) offers recommended levels of service and possible solutions for planning document delivery services to branch libraries and other off-site locations. One library system involved in an extensive distance education program is the University of South Africa (Unisa). Over 54,000 FTE students are enrolled in that program. Requests for library materials are received by postcard, by a telephone answering service, by fax, and through Beltel videotext. An automated handling system accepts the large volume of requests and automatically checks holdings, loan, and shelf status. If materials are unavailable, requests are automatically put on hold and a response generated to inform the requester (Shillinglaw 1992, 145-146).

The ability to pay for services varies among the students, and some do not use document delivery due to the high cost of returning book loans. Also, many assigned texts are not well-suited to South African applications. In such cases, documents supplied can create disinterest in library resources. To counteract this, the library system is paying more attention to supplying portions of appropriate text rather than whole volumes. Shillinglaw expresses great hope in potential options made possible by electronic publishing and electronic text document delivery. Publishers may be asked to modify texts originally written with European and North American scholars in mind. Text extracts may be combined into suitable course packets and even individualized for a particular student (ibid., 147-150).

Distance education programs at the University of Alaska faced similar problems. The institution had to overcome the natural hesitancy students in rural areas had about approaching a large academic library. A toll-free phone number was provided with voice mail available, but few people left messages. The voice mailbox was replaced by an answering machine, which was more widely accepted. Faculty involvement, staff visits and publicity were keys to introducing the service. The program's original goal was to provide 48-hour turnaround from receipt of request to mail out, with all material being sent from the Fairbanks campus by first-class priority mail. Faxing later replaced mail for journal articles.

Meeting the 48-hour supplier turnaround became more difficult as volume steadily increased. The biggest problem was inadequate reference interviews, which required returned telephone calls prior to filling requests. Although most students had access to a computer and modem and the University of Alaska's online catalog, few students were using the online catalog to verify references before calling in a request. In 1991, library faculty began teaching a distance education course on information-seeking skills to help students become more comfortable with telecommunications access to an electronic library (West 1992, 555-559).

A number of library systems operate bookmobiles employing modern technology to serve off-site clientele. The Westminster (Colorado) Public Library was the first to use packet radio technology to provide a wireless link between the bookmobile and the library's mainframe computer, but at least forty other libraries have followed its lead. "Tone bursts" convert computer data from a sending radio modem at the main library, relaying it to an antenna, and transmitting it over an assigned FM frequency to the receiving antenna and radio modem on the bookmobile's terminal, where the data is reconverted for normal use. Unfortunately, this does not work well in hilly terrain or over long distances without repeater stations and extra antennas to relay signals. Start-up costs vary widely, depending on the terrain and distance to be covered, but average $15,000 to $30,000. Additional costs, however, are minimal once the equipment is in place.

Electronic Systems Technology in Kennewick, Washington, and Dataradio in Atlanta, Georgia, are two companies involved in this technology. The systems provide somewhat slow response and have a high rate of downtime due to weather conditions and electrical interference, but the advantages seem to outweigh disadvantages and are being considered by many more libraries. A registry of online bookmobiles is maintained to facilitate networking and problem resolution (Alloway 1992, 43-44).

Other bookmobiles, such as Hennepin County (Minnesota) Library, went online using cellular telephone technology. Cellular phones use a special modem on the host computer and mobile terminals. Delays, unreliability, and high monthly bills (around $500) are disadvantages reported, but initial start-up is quite simple and inexpensive. It may be worth implementing cellular telephone technology on a trial basis to evaluate an online bookmobile potential. A cellular-based fax is used by the Chesapeake (Virginia) Public Library System in its bookmobile to provide almost instant document delivery between the home library and the mobile unit (ibid., 45).

Bookmobiles also use CD-ROM technology to provide full-text access to encyclopedias and other reference resources and access to the OPAC. Catherine Alloway states that "to students and other users in isolated areas, reference service could prove to be more important than the lending collection." (ibid.) Laserdiscs and other optical storage technologies also have potential for document delivery to remote users in bookmobiles.

Gonzaga University employs CD-ROM technology to provide remote access to a branch library, where a distance education program is in place for Native Americans. Similar to the program at ASU West, online information access, retrieval, and delivery strategies were planned from the outset to support the overall library needs of students. On-site print resources are limited to a few required and highly recommended texts for courses. Support from other libraries in the locality is minimal, so virtually all materials are supplied from the main library in Spokane, Washington.

As a member of the Western Library Network (WLN) since 1977, Gonzaga University Library has almost complete holdings accessible to the Canim Band over WLN's LaserCat CD-ROM database and search software. Access to the holdings of other WLN member libraries broadens the scope of resource materials available. Utilities for browsing and compiling bibliographies are also provided (Burr 1988, 39).

Canim students search various online databases. PFS First Choice software automatically saves the online session as a text file, which is then reviewed offline by library staff and the student in order to evaluate database selection and search strategy, and to narrow the choice of desired documents. An edited version of the original text file, containing citations and request information, is uploaded to an e-mail utility and transmitted to the main library. Fax is also used for sending requests (ibid.). When requests cannot be filled from Gonzaga University's main library, ILL resources or commercial document suppliers, such as UMI's Article Clearinghouse, are tapped. Book materials are shipped by commercial parcel delivery service, with a delivery turnaround of 5 to 7 days; journal articles are faxed.

Overall turnaround from submission of request to receipt of materials averages eleven days for books and four days for periodical articles. The program has a high level of user satisfaction, although in the beginning not enough attention was given to developing general microcomputing skills needed to successfully cope with hardware and software problems, such as reinstalling software after a hard disk failure (ibid., 40-41).

Other library networks rely on electronically stored collections to supply materials to network or branch facilities, even directly to the end user. Electronically stored reserve collections are being made accessible over local area networks (LANs) to students in residence halls or in their homes, and in other off-site, distance learning environments. Copyright restrictions and licensing agreements limit the potential of these innovative remote access methods, but virtual libraries are beginning to be developed. The Ohio State University Libraries and CICNet, for example, have established an archival collection of electronic journals that is growing steadily, both in the number of titles available and in the volume (6 to 8 per minute) of remote access transactions (Wiemers and Hankins 1994, 8).

OhioLINK provides another example from the state of Ohio of an electronic document delivery service at its best. Rather than expanding library space on its many academic campuses, the university's library system links the campuses electronically and resource sharing is an integral part of the information technology infrastructure. The holdings of all state-assisted college and university libraries are accessible in an online union list, and a statewide delivery system is in place. Patron-initiated check-outs, 48-hour delivery using traditional delivery mechanisms as well as Ariel, CICNet, fax, and commercial full-text services are all features of this innovative system (Kohl 1993, 42-43).

CENTRALIZED INTERLIBRARY LOAN

Interlibrary loan services should be readily available to library users in branch libraries and distance education programs. The ILL/DDS unit at the main library is usually best equipped and staffed to handle interlibrary loan for all libraries in a system. Once more, we will use Arizona State University as a model.

ILL service for all libraries within the ASU system is centralized at ASU Main. Requests are accepted at all off-site locations—in person and electronically, by fax or e-mail. Fax machines were installed in several of the branch libraries to provide faster transmittal of requests to the central ILL/DDS office. Phone requests are discouraged, although that policy varies at the different locations. An e-mail request form provides a convenient way to transmit requests to a centralized ILL unit from branch libraries or from patrons at remote sites.

E-Mail Document Request Forms. Initially, ASU patrons who wished to submit ILL or document delivery requests by e-mail were simply provided directions regarding use of a standard format. That is sufficient when e-mail volume is low. A higher e-mail volume warrants a template request form that is automatically accessible on all e-mail systems available to library users. One coordinated e-mail system that works for everyone is highly recommended. A system that can capture a citation from any online database, automatically transferring the citation onto an e-mail ILL request form is particularly useful.

At ASU, the ILL e-mail template and accompanying help screens had to work with four different sets of commands, as well as being used for slightly different purposes on each campus! Help screens serve a dual role—they help patrons use the e-mail form and they provide an easily up-dated electronic version of ILL/DDS policies and procedures. The help screens also define the patrons' responsibilities in submitting electronic document delivery requests regarding copyright compliance, paying for charges, and returning materials by the date due. Each electronic request, just as with a signed paper transaction, represents senders' electronic signatures and signifies their willingness to comply with all policies and procedures.

Branch Processing of ILL Requests. In whatever manner requests are submitted—electronically or not—the library staffers check for completeness and do some preliminary verification to confirm the accuracy of the citation. How much is done depends upon the resources available at each branch, which varies from the bare minimum in some extended education sites to branches that house excellent reference collections. Any item located within the ASU Libraries system enters the previously described document delivery routine. If unavailable anywhere within the system, the request goes to ASU Main as an interlibrary loan transaction. Over 2,000 ILL requests were processed from ASU branch libraries during the 1993/94 fiscal year.

The ASU Libraries offer a variety of commercial delivery services online through the OPAC. Patrons in remote branch locations use these services frequently for speedy access to documents not owned. If a user does not wish to pay for commercial delivery, however, requests can be submitted to ILL for mediated document delivery. A nine-month user behavior study is underway to determine how faculty members use nonmediated commercial document providers. Library accounts were set up with UnCover, three commercial suppliers available via OCLC's FIRSTSEARCH databases, and with BLDSC for document delivery from Inside Information via RLG's CitaDel document delivery service. Patron-directed services to a multitype library network is also planned using UMI's ProQuest data imaging document delivery system set up on a LAN.

ON-SITE DOCUMENT DELIVERY

A successful document delivery service operating among a group of branch libraries almost surely results in the decision to take delivery one step further—directly to the end user. Many libraries have instituted document delivery services within the organization to add to users' convenience. Such services are most often found on large academic campuses with multiple branch libraries, but a number of special libraries also provide in-house document delivery. In today's electronic age, document delivery complements remote access to library holdings. Not everyone is ready to accept the virtual library concept, but almost everyone at one time or another needs something from a library when it is not possible to physically retrieve the document for oneself. Internal document delivery services are usually designed to meet the needs of primary clientele. Some provide the service free as an integral part of regular library service. Others see document delivery as a convenience, and operate on a cost-recovery basis.

Some libraries may also have fee-based services designed to serve patrons from outside the local environment and may offer services other than document delivery. Such services are the not-for-profit version of the commercial information provider or information brokerage services. Guidelines for establishing a full-service, fee-based information agency are available in other publications. Sue Rugge and Alfred Glossbrenner guide the establishment of a commercial information brokerage service in *The Information Broker's Handbook* (Windcrest/McGraw Hill, 1992). *The FISCAL Primer* is useful for libraries that serve the general community. The packet of information is available from Lee Ann George at Gelman Library Information Service, George Washington University, 2130 H Street N.W., Room B07, Washington, DC 20052. *Fee-based Services in ARL Libraries*, an ARL SPEC Kit prepared by Helen Josephine (1989), is another helpful resource.

The rest of this chapter concentrates on on-site document delivery to a library's primary clientele. On-site docdel services and those offered by other fee-based or commercial information providers differ mainly in that the former's focus is on delivering materials from a library's own collections. Some libraries incorporate the delivery of items acquired through fast-track, access-versus-acquisitions programs into an on-site document delivery program. (The techniques for acquiring documents from outside sources are covered in the chapters on commercial document suppliers and resource sharing networks.) Once received and processed (usually by ILL), these materials may be turned over to the delivery unit for actual delivery to the requester. Library Express at Arizona State University is typical of such operations and serves here as a case study, with procedures from other institutions used to demonstrate additional services.

Direct Delivery at ASU—Library Express

Determining how to convert nonusers into library patrons is difficult at best. The most astute public relations program will fail to get some people into a library. Even at academic institutions, some students pride themselves on getting through college without ever using the library. But when users won't come into the library, the library can be brought to them with a document delivery service. Operating procedures are similar whether document delivery is between off-site branches or within home facilities.

Library Express was conceived primarily as a service to faculty members too busy with research and instruction to visit the library every time one needed a document. At ASU, document delivery to off-site locations is free to students and faculty because it is provided out of necessity and a deliberate service commitment based on access rather than acquisition. The on-campus service is provided for convenience rather than necessity and, as such, Library Express began as a self-supporting, fee-based service.

Patrons wishing to open a Library Express account fill out an application authorizing the library to check out materials to the user's ID. The application confirms the responsibilities of the patron: complying with copyright guidelines, returning or renewing library materials on time, and paying for services used. Upon opening an account, the user requests material as needed. Staffers then pull requested books and check them out to the patron, or photocopy journal articles. Everything is then delivered to a designated delivery point convenient to the user. The service is strictly document delivery—no research services, not even computer literature searches, are conducted for patrons. The procedure eliminates the time-consuming and often frustrating aspects of locating desired materials in the stacks and then standing in line at Circulation; and the inconvenience of getting to the library.

Many libraries, such as at the University of Wyoming, provide free on-site document delivery, but the ASU Libraries did not have the budget to do so. The University of Wyoming Libraries successfully negotiated to receive one percent of indirect cost funds allocated by the University's Research and Graduate Studies Office (Baldwin and Dickey 1986, 705-706). Since Library Express at ASU originated as a convenience service, rather than a necessity, we felt no compunction about charging. Basic costs include $1.00 for every volume pulled and $0.50 for each five-page increment of photocopies. Extra fees are assessed for looking up call numbers or for billing users.

ASU operates the service on a cost-recovery basis, rather than for profit. Our rates reflect the lowest possible level required to cover costs. The service, which began in January 1989, served only ASU employees who had an office within a designated delivery route, but a "Pull & Pick-Up Service" was later added for students and employees with offices beyond the delivery route.

Library Express started small and expanded as volume dictated. It evidences that new services can be instituted even without megabucks in the budget. Some seed money must be available for any new program, but often that money can be found if one looks hard enough. So, what went into starting a new on-campus document delivery service? Library Express planning included:

- determining the need for service
- planning the service
- financing the operation
- laying the foundation
- opening for business
- reassessing needs and adjusting to changes
- enjoying success, or learning from mistakes.

Determining the Need for Service

Much of the Library Express planning followed a strategy used earlier at the University of Wyoming, beginning with a "User Survey." Whereas the University of Wyoming's survey was directed specifically at document delivery service, ASU's aim was quite broad, encompassing all library services. [A copy of the University of Wyoming survey is included in an article by David A. Baldwin and Paul D. Dickey (1986, 707)].

In Wyoming the idea for an on-campus document delivery service had already been conceived, and the survey was intended to confirm interest. At ASU, the idea for the service was conceived because of recommendations made by users for services they wanted from the library. At that time, the idea was "pie in the sky," but eventually, document delivery was seen as complimenting other library services, particularly dial-in access to the online catalog. The "User Survey" at ASU turned out to be an accurate instrument for predicting interest in document delivery, but it was less accurate in predicting faculty's willingness to pay!

A third approach to user surveys is illustrated by the University of Nebraska-Lincoln, which queried faculty members after it had a document delivery service up and running. The purpose of that survey was to determine how many respondents knew about and were using the service (University of Nebraska-Lincoln Libraries Document Delivery Committee— hereafter cited as UNL—1988, 2).

Some libraries have met with resistance when proposing an on-campus document delivery service. This seems to be particularly true when university funding is sought. Richard Dougherty (1978, 29) cites typical negative reactions at the University of California, Berkeley and at the University of Colorado, such as, "'Poor use of library funds,' 'Students can be used to fetch materials,' 'Useless waste of resources,' 'Financially impossible,' 'Faculty are not that busy,' 'Library can go too far in providing services,' 'Wish to retain browsing.'" Even in light of the initial skepticism, UC— Berkeley went on to establish a highly successful on-campus document delivery service which soon converted negative feelings into positive acceptance.

Planning the Service

In July 1986, ASU's on-campus document delivery project became part of the library's five-year strategic plan. A task force performed a literature search, studied similar services on other campuses, and came up with a cost and staffing proposal. A trial service to three departments with large numbers of faculty who were frequent library users was proposed with full implementation by January, 1988. At that point, lack of funding bogged things down, although the recommendation remained part of the library's five-year strategic plan. The original idea visualized an ideal "if-everything-were-new-and-perfect" system with top level staffing.

Financing the Operation

Planning a new service is analogous to designing a new home, wherein the first plan offers a "dream house" with everything in it. Then budget estimates send one back to the drawing board for a final design that is practical without entirely compromising the ideal. Thus, we pared down the original on-campus document delivery plan and looked for alternate solutions. Staffing was tackled first. The ASU Libraries already had a successful document delivery service between the main campus libraries and ASU West. On-campus document delivery involved many of the same job duties, so the staffing proposal was altered to match other document delivery staffing, and the new service was fit into the existing department structure. This required that the supervisor of the Lending Unit and the department head assume many of the responsibilities for getting the service up and running. Student assistants were eliminated from the start-up plan, as they could be employed as needed when the service began paying for itself.

In a large institution, finding surplus office furnishings and equipment is not difficult, and a telephone line already in the ILL Department was dedicated to the new service. The Library Development Officer located donors to cover costs for items still to be acquired, such as a telephone answering machine and the printing of brochures.

Selecting an appropriate delivery vehicle is a key factor. A local hospital librarian used to ride a three-wheeled bike to make deliveries on the hospital grounds, but ASU needed something that could be made secure and that was more suitable for covering greater distances. The University of Nebraska—Lincoln libraries began by buying two backpacks and bus tokens for its delivery staff (UNL 1988, 2)! A few libraries use internal mail systems for document delivery, but more libraries are likely to start on-site document delivery because internal mail systems are unsatisfactory for delivering library materials. The University of Wyoming library shares a van with its Audio Visual Services (Baldwin and Dickey, 706).

At ASU, we considered sharing an electric Cushman cart with the library's mail room staff. By a stroke of luck, a new Cushman had recently been ordered for the mail room, and Library Express was given the old one! Even without this good fortune, one vehicle could have been shared, although a strict delivery schedule would have been mandated for each unit using the Cushman. The old Cushman was modified slightly so that materials would be protected from the weather and from theft. The work was done on campus by physical plant maintenance employees for a nominal fee. ASU abandoned the proposed trial service and went for broke, serving all departments within the delivery route. With these reductions to the original plan, total start-up costs were kept below $5,000.

A decision was also made to have Library Express staff use photocopiers in public areas rather than the equipment in ILL/DDS. There were several reasons behind that decision: 1) the library did not want to jeopardize relations with the commercial copy service, and patrons would normally make these copies themselves using public machines; 2) it did not interfere with regular use of the ILL department's photocopy equipment; 3) meeting tight delivery deadlines was easier with access to any public copier. Nonetheless, as the service grew, copy volume began to tie up public copiers: the next contract negotiations for copy equipment incorporated Library Express' needs.

Laying the Foundation

Much behind-the-scenes work must be done before beginning any new program. Procedures and policies must be established, including determining who will be served and deciding whether the service will be free or how much any fees will be. Job descriptions must be written and staff hired. Brochures and forms must be designed and printed. Publicity to generate business must be prepared. Questions to be answered include the following: Will the number of requests be limited to insure fast turnaround to all clients? How will requests for items not in the library be handled—returned to patron or submitted to ILL? How much preliminary work is the patron expected to do—supply call numbers, verify citations?

Each program has its own special needs. Before Library Express could open for business, we had to get a permit to drive a vehicle on campus, modify the Cushman vehicle to meet Department of Public Safety requirements, establish financial accounts, get an e-mail address for Library Express, arrange for priority photocopying with the Library's copy service when requests exceeded what a half-time employee could handle, and arrange with another department to share bookkeeping services, if the half-time employee could not keep up with everything. A logo (not to mention a name) had to be designed and approved before forms and brochures could be printed or the Cushman painted. The red tape seemed endless.

After the Library Assistant was hired, six months passed before service actually began, but the new staffer was actively involved from the beginning setting up bookkeeping and statistical systems. [At ASU, Lotus 1-2-3 was used for statistical records; the University of Wyoming used PFS:File and PFS:Report (ibid., 705)]. The new staff member also worked with the Department of Public Safety to develop a safe delivery route on a campus milling with 40,000 students, designed the modifications to the Cushman, made preliminary contacts with various departments on campus, and posted or delivered advertising to every part of the campus, all while learning the library collections and routines for document delivery.

Opening for Business

After the preliminary planning, the opening was almost anti-climactic. The main activity at first was advertising the new service. Promotional activities might vary in other libraries, but the following provides a sample of what was done at ASU. Library Representatives in each department within the delivery route were contacted and the new service explained. We attended department meetings to address larger groups of faculty. Library subject specialists promoted the service during liaison activities. Articles were printed in two employee publications. The school newspaper was avoided because students were not originally eligible for service. Announcements were made on the electronic mail system at ASU reminding users that they could use e-mail to submit requests. Permission was obtained to post fliers and place brochures in strategic places all over campus. Brochures were placed in orientation packets for new faculty and were also sent to faculty recipients of research grants—many of whom chose to use Library Express rather than hiring and training students to retrieve library materials. The logos painted on the Cushman turned out to be effective advertising, generating interest among students who wanted to know how they could use the service. We are still coming up with new ideas for promoting the service, but word-of-mouth has proved quite effective!

Reassessing Needs and Adjusting to Changes

Unlike the free service at the University of Wyoming, which generated almost 3,000 requests in the first four months of its pilot stage (ibid.), and the free service at UC—Berkeley, Library Express at ASU did not succeed overnight. Fewer than 300 requests were generated in the first quarter, although each month brought steady growth. Speed and quality of service were the main factors in garnering new clients. Forty-eight-hour turnaround was guaranteed, but most requests were delivered within twenty-four hours, and more often than not, within four hours!

Starting out slowly under a cost-recovery basis was not all bad. It allowed time to work out minor problems, much as the canceled pilot project would have allowed. The experience with BAKER at UC—Berkeley suggests that free services can be so popular that the rapid increase in demand may jeopardize performance. BAKER processed 7,050 requests in its first year and over 16,000 the second. After the second year, the system imposed service fees and a limit on the number of transactions any individual could submit in a day. Richard Dougherty suggests that libraries wishing to avoid fees introduce a document delivery program limiting the number of annual transactions for each department or individual; charges would be incurred if the annual limit were surpassed (Dougherty 1978, 30). A daily limit might also help to spread the work flow evenly throughout the year.

The ASU plan envisioned that the same delivery route would be driven each day, picking up requests and delivering materials on a regular schedule. Business proved too sporadic, however, so clients now call or e-mail Library Express if they have requests or materials to be picked up at an office or dormitory. (There is no charge for picking up requests or materials previously delivered, but disabled patrons are among the few that request pickup.) Not having to stick to a specific delivery schedule and route allowed us to expand the service to ASU students and to employees whose offices were outside the delivery route.

During the fast-paced summer session, a "Pull and Pick-Up Service" was tried, providing services almost identical to regular Library Express services except that materials were held at ILL/DDS for pick-up. This service became an instant success, quickly surpassing the number of requests received through the regular Library Express service.

Because ILL users complain about filling out request forms, we expected that most requests would come in by telephone or e-mail. Telephone requests are limited to no more than three at a time and must be in the English language to avoid transcription errors. Without this limit, the part-time staff would spend all day transcribing phone requests rather than making deliveries. [Problems with incomplete citations received by telephone and the heavy toll on staff time are frequent complaints from other on-site document delivery services (UNL 1988, 5).] While phone and e-mail requests are growing in popularity, Library Express still receives a high percentage of requests on printed forms that patrons have filled out and dropped off personally at the department or one of the branch libraries. Lists of materials can be supplied with one cover request form—a welcome change from ILL, which requires a separate form for each request.

A surprising discovery was that the service did not reduce personal library use. Even though deliveries are made for no extra fee, the majority of Library Express clients choose to pick up materials rather than have them delivered. This isn't because clients have to be in their office or dorm when the delivery is made, because each one designates a staffed delivery point, such as an office secretary or a residence hall reception desk. The service seems to allow patrons more productive use of their time in the library for research and study, with Library Express doing the leg work of actually getting needed documents.

Another surprise is that departments predicted to be heavy users of document delivery services are not the main clientele. Most studies indicate that researchers of scientific, technical, and medical materials frequently have large research grants and a demand for rapid delivery of current materials; consequently, they are the most interested in on-site document delivery. Researchers in those departments do use Library Express, particularly those whose offices are some distance from a library, but most Library Express clients are from departments with small budgets that have difficulty employing student assistants on a regular basis. One explanation may be that the ASU Libraries offer several current awareness/patron-directed document delivery services through the OPAC. Researchers in the better endowed departments may simply be bypassing ILL/DDS, going directly to commercial suppliers—even when the document is readily available within the libraries—while those on tighter budgets rely on mediated subsidized docdel through the library.

Local weather conditions can greatly affect on-site document delivery. In central Arizona, business soars in the summer as desert temperatures climb. In other climates, winter storms and rainy seasons are equally likely to boost business. Promoting the service when the weather is at its worst usually proves effective! While adverse climatic conditions may be good for business, documents delivered physically rather than electronically must be conveyed in a vehicle designed to protect both staff and materials.

Another innovation at ASU involves subcontracting with other document delivery and clearinghouse services on campus. Rather than having staff doing basically the same tasks (pulling books and photocopying articles), such services subcontract with Library Express to retrieve or copy library materials, thus reducing staff requirements and competition for copy machines and access to journals between units.

Most such services in an academic setting operate under continuing or extended education grants or on a cost recovery basis, and they have minimum staffing. By subcontracting with Library Express for document delivery service, their own personnel can better serve clients by being available in their offices rather than running between library facilities.

Some fee-based document delivery systems in libraries serve clients outside the home institution, assessing fees that are usually competitive with other local information providers. While many start out primarily to serve the general public using the library's collections, that source of business may not be sufficient for services operating strictly on a cost recovery basis. Most such services subcontract with similar services in other libraries to expand resources. Many also serve as a library "runner" for commercial document suppliers. By doing so, even though offered by a nonprofit organization, the fee-based services are subject to most of the same rules and regulations that pertain to commercial information suppliers. On-site services, such as Library Express, that serve only their own institution from their own collections, are less likely to be affected by economic recessions or the market-driven competition of commercial document suppliers. In adding the "Pull & Pickup Service" and by subcontracting, Library Express was well established as a university service, processing more than 6,000 requests in fiscal year 1989/90, and recovering almost all operating costs in its first full year of operation.

ILL also began referring requests to Library Express for materials found in-house. It was better to provide the document quickly than to send out batch notices alerting patrons to return to the library to retrieve items. ILL had long felt that the notification letters were a poor use of staff members' time and a disservice to patrons who, for one reason or another, had failed to locate desired materials in the library collections. Such letters call attention to a patron's inability to use the library, discouraging a return trip. Most patrons are delighted to get the actual document quickly even for a fee. A few patrons used ILL as a "call number" service rather than hiring a student assistant to look up call numbers. Those requests delayed processing legitimate ILL requests and were better suited to Library Express.

Service was expanded to include delivery to residence halls, but the "Pull & Pickup Service" remains the program's most popular aspect. Library Express has continued to grow, processing over 8,600 requests in fiscal year 1993/94 with a fill rate of 96 percent. Patrons who have documents delivered to their offices or residence halls have deposit accounts. (No minimum balance is required.) Pull & Pickup patrons either have deposit accounts or pay upon picking up. Some departments have accounts for their faculty and several people have accounts through research grants. Initially, monthly billing (with a slight additional service charge) was available, but that option was discontinued because few people requested it.

Enjoying Success, or Learning from Mistakes

Library Express was designed strictly as a cost-recovery service. During the first year, most income was plowed right back into the service. Soon funds were available to hire a student assistant, so the service no longer had to be operated only part-time. Office equipment was purchased as needed—often acquiring multiple use devices that benefited the whole department, not just Library Express. So even though Library Express was more or less pieced together using equipment on hand, it was soon giving back to the organization more than was borrowed in the beginning.

The operation is a "lean machine" that has grown in accordance with demand for service. The staff was expanded as business grew. Part-time employees are preferred in the Document Delivery Unit because of the tedious and repetitive nature of the work and because schedules can be juggled to accommodate vacations or other absences. AT ASU, 50 percent of FTE employees receive regular benefits, a policy that attracts people looking for part-time employment.

Although Library Express operates primarily on a cost recovery basis, it has become less fee oriented. Free document delivery services were first offered to disabled library users. Disabled Student Resources (DSR) provided additional Library Aides (student assistants) to help Library Express employees pull and copy materials for physically challenged library patrons. Previously, if DSR aides were busy assisting one patron, other disabled users had to wait, even though all they might need was help in reaching a book on the shelf. With the document delivery service, disabled students can call ahead or send an e-mail request and the material should be waiting by the time they get to the library.

Materials are often delivered to the DSR offices, which is convenient for those physically handicapped students who are transported around campus by DSR staff members. Disabled patrons who have access to fax equipment or e-mail can receive documents directly. Faxing became so popular that Library Express began an inexpensive fax service for all faculty and students as another extension of on-campus document delivery.

Besides free services to the disabled, Library Express began to provide limited free services to library staffers needing materials from branches other than the library in which they worked. As the library began to place a higher priority on rapid access, items received through ILL were delivered free to faculty. With so many free services, total cost recovery became more difficult. The library opted to cover one half-time salary line, putting less onus on Library Express to be totally self supporting, while still providing the university community with cost-effective document delivery services.

Other Direct Delivery Programs

Rather than establish an on-site delivery service, some libraries subsidize user-directed document delivery. Each department (or each eligible participant) may be given a specific allocation to acquire documents from commercial current awareness services accessible via the library's OPAC or from accounts held with other online vendors. Some libraries establish accounts with services such as UnCover or buy search blocks from services such as OCLC's FIRSTSEARCH or RLG's CitaDel, but ordering and receipt of materials is left to the end-user.

Most of those services are fairly new and libraries are experimenting with how to control patron-directed, library-subsidized document delivery. Colorado Sate University, for example, allows all library users to order articles from journals not owned by CSU directly from UnCover. An established price limit in the program blocks any document request that exceeds the limit, but the user can e-mail the request to ILL (Wessling 1994).

At Arizona State University, a select group of faculty are involved in a subsidized patron-directed document delivery program. Accounts are set up with various document providers and participants have passwords that enable them to use the accounts in any way they choose. The library tracks how this group uses commercial services as compared to its regular use of the library collections (both in-house and via the Library Express delivery service). Information gathered from the pilot project will be used to determine what mix of delivery services—patron directed, library mediated, or combination of the two—will best serve clientele.

A library can enhance its in-house document delivery program by providing ready access to patron-directed current awareness services (CAS). At the Science & Engineering Library, University of Minnesota-Twin Cities, Minneapolis, faculty profiles specify research interests. Weekly searches of recent publications are transmitted electronically to faculty using a combination of mainframe computing, the campus network, and e-mail. The service was initially tested as a pilot project from August, 1991 to February, 1992 by faculty from the Department of Electrical Engineering. Using Telnet and FTP, microcomputer text files are uploaded to a VAX, with a high-speed (averaging 40KB-64KB/second) connection to the campus Internet. The files are then sent to participants using the VAX/VMS Mail utility and a command program created for the project called Automail. John Butler (1993, 115-118) states: "In contrast to standard e-mail distribution lists, which send one message to many individuals, Automail sends en masse numerous unique messages or files to numerous unique addresses. . . . without operator intervention."

OCLC also provides an e-mail current awareness service, ContentsAlert, based on its ContentsFirst database. Libraries can subscribe to the tables of contents (TOC) service for $14 per title per year. The library can redistribute the TOC to an internal distribution list, or OCLC will e-mail the TOC directly to individual subscribers for $4 per title per year. There is a one-time set-up fee of $50 to establish an account (OCLC 1993, 1). Abstracts are included for the articles, if available. A partial list of holding libraries is also supplied. The user can either request the document from the home library or initiate an interlibrary loan. Eventually the current awareness service will be linked to OCLC's document delivery service (Using 1993, 3-5).

UnCover has started a similar service, UnCover Reveal, where the tables of contents for designated titles are e-mailed directly to end users. UnCover provides the service free, recouping costs whenever documents are ordered, which users can accomplish using the e-mail reply function.

SUMMARY

Future on-site delivery will undoubtedly involve more electronic document ordering and electronic retrieval of documents. These many new services widen the dimensions for on-site document delivery, diminishing the need for staffed document delivery units when commercial services provide fast, efficient service at reasonable prices.

Chapter Ten

Future Directions for Document Delivery Systems

*Librarianship has become not the art of build-
ing collections but the knack of making connections (Wright 1989, 118).*

What does the future hold? Boss and McQueen, in their consultative
report *Document Delivery in the United States* in 1983, found little incentive for change. They identified "major problems in turnaround time within the organizations that supply materials to others" that were unlikely to
"be affected by an improved document delivery system." (Information
Systems Consultants 1983, 4)

Such laissez-faire times are gone forever. At the time of their report, the
library market for commercial suppliers consisted mainly of special
libraries, a market too small to induce document suppliers to enhance services. ILL units then were usually located in an obscure spot in the library
and, for the volume of business received, understaffed. ILL service was
rarely promoted. Even so, most ILL units sought to improve document
delivery between libraries and continued to pursue alternative methods
that might reduce turnaround time.

What change ten years can bring! Fueled by limited buying power, personnel shortages, increased expectations from library users, and improved
technology, "access" is a major part of library missions. The stakes are
high; the players are many; the choices are confusing; the times are exciting. One thing hasn't changed: interlibrary loan remains synonymous with
cooperation and resource sharing, but document delivery, cooperative
efforts, and resource sharing are no longer limited to libraries. Mary
Jackson (1990, 100) summarized the progress of the '80s this way: "During

267

the past ten years, interlibrary loan librarians have revolutionized the trans-mission of ILL requests, witnessed the explosion in the quantity of ILL requests, monitored copyright issues, changed basic library services as the focus has changed from ownership to access, participated in cooperative ventures, standardized the format and content of the ILL request, and have done all of this while handling increased volume of activity in a more time-ly manner."

Oddly enough, because the wheels of change turned so slowly, Boss and McQueen's predictions for the five-year period 1984-1988, are more relevant for the last half of the decade of the nineties. Quoting their study (Information Systems Consultants 1983, 53), those predictions are:

1. Increased electronic transmission of requests via bibliographic utility ILL subsystems, shared circulation control systems and electronic mail sys-tems, including those of the commercial database services.
2. A gradual increase in the number and percentage of documents provid-ed by commercial document services.
3. Increased use of surface courier services by the commercial document services.
4. Heightened awareness by those using electronic . . . delivery of docu-ments.

Their prediction for the "next five to ten years" (the decade of the '80s) is more relevant in the '90s:

> The bulk of the needs of most information users will con-tinue to be met from printed sources—the heart of most libraries' collections. The challenge for libraries will be to augment these traditional resources with information stored in electronic form. When a sufficient body of elec-tronic information has been collected, attention will turn to the transfer of it among libraries and between docu-ment services and libraries (ibid., 58).

Most of Boss and McQueen's predictions are coming to fruition, albeit more slowly than they predicted. Electronic storage and transmission of documents, though still evolving, improves daily. New developments and products are introduced regularly. There is a global trend towards linking the electronic retrieval of bibliographic references with electronic full-text retrieval—much of the activity generated directly by patrons.

Commercial suppliers have got the message that a whole new market for single article document delivery exists among library users. A recent study of serial cancellations in five large midwestern ARL libraries indicates that libraries are canceling large numbers of unique serial titles and core science journals (Chrzastowski and Schmidt, 1993, 101). That trend will lead to more dependence on commercial document delivery, with commercial suppliers competitively vying for the library trade. Allen and Alexander (1986, 8) accurately predicted that "a service which can deliver information rapidly will be used by increased numbers of users if the price is considered reasonable for the perceived value of the service and information."

Publishers generally ignored the single article market until services such as UnCover demonstrated the potential. But publishers, wishing to maintain control over their own products, have hesitated about granting distribution rights to other commercial suppliers that store documents electronically. Since most publishers now store manuscripts in electronic format, they found it relatively easy to develop their own electronic document delivery systems. Thus, commercial document services are multiplying like rabbits.

Every library service with a bibliographic database seems to be marketing a document delivery system, but ILL/DDS units that batch searching, ordering, and payment processes find it difficult to work with so many product-specific systems. Just as libraries use subscription agents to lower overhead costs, ILL departments need to avoid paying hundreds of small invoices for one or two items from numerous suppliers each year.

A six-month document delivery project in 1992 at Arizona State University filled 1,591 photocopy requests representing 1,101 different journals (ASU Libraries Journal Document Delivery Task Force 1992a, 2; Walters 1993a, 4). Such activity was not atypical for ASU, nor would it be for most libraries with diverse clientele. Those figures make it easy to project the potential for large sales, but also demonstrate the difficulties libraries face if they must deal with multiple suppliers for so many requests. As publishers realize the advantages of having articles widely available, there should be fewer "exclusive distributors." As Arundale (1991, 19) puts it, "Customers have less reason to shop around for the right mix of titles when each vendor offers a wide range of products, and they will base their choice of vendor on the ease of use and flexibility of their products."

The future of interlibrary loan and document delivery services will almost certainly involve both libraries and commercial document suppliers. A think tank session during the "Preferred Futures for Libraries" workshop in 1992 proposed the formulation of a new type of consortium for resource sharing that "would be planned with input from publishers and would provide information and documents through many different methods—fax, commercial sources, online file transfer protocol (FTP), etc. Collaborative buying/leasing programs would release funds for the development of the network and the database structure." (Dougherty and Hughes 1993, 17)

This proposal, and other far-reaching suggestions, did not gain unanimous acceptance by the mixed group of scholars, librarians, publishers, information technicians, and administrators participating. We can safely speculate, however, that even without a resource sharing consortium, libraries will augment holdings by offering users a variety of document delivery products and services, consisting of many, all, or more than the samples cited. "The products and services offered may be integrated into library operations or may exist on a stand-alone basis, depending on the library's needs, budget and technical capabilities." (Bluh 1993, 51)

As budgets suffer and resource sharing among libraries becomes even more a necessity, future sharing is less likely to be based on geography. The electronic transmission of documents allows libraries globally to develop reciprocal agreements whenever their collections make an alliance feasible.

The time involved with scanning and the cost of upgrading equipment to keep up with advances in technology will probably mean fewer libraries can afford to serve as net lenders without compensation. Net borrowers have to expect to pay either for access or pay for collections.

On the other hand, as libraries begin to use electronic storage as a means of preserving and accessing their own collections, resource sharing agreements will eliminate duplicate purchases of expensive monographs. Jim Hoover, director of the Columbia University Law Library, cites an example: "Harvard might keep electronic copies of all the Nuremberg trial information, eliminating the need for Columbia to collect any of it." (Bulkeley 1993, 8) While shared resource programs between Harvard and Columbia are not unusual, the libraries involved in future cooperative ventures could be on separate continents and still meet each other's needs equally well via electronic access and document delivery.

Past ILL codes limited nonregional borrowing to "last resort" procedures and also limited what should and should not be requested (Boucher 1984, 137, 139). The revised U.S. Code (approved at the 1994 ALA Midwinter Meeting) removes most restrictions and endorses the concept of access, recognizing ILL as "an integral element of collection development for all libraries, not an ancillary option." (ALA RASD/MOPSS/ILL Committee 1994, 477) Except when local libraries unite for the economic advantages of sharing site licenses, bibliographic utilities, or rapid delivery, institutions with similar needs rather than local multitype libraries are more likely to establish document delivery networks.

While ILL codes are changing to reflect today's needs, institutional attitudes also need to change. Granted that library collections are an institutional asset, in the future access to information resources will be as vital as ownership. Many small libraries will have access to as wide a range of scholarly resources as larger ones. Library collections in academic institutions will probably diminish greatly in importance in drawing or keeping the best faculties. "What will matter in the competition for students and faculty will be the ease of access to information and the extent to which the institution will pay the bills." (Lewis 1988, 296-298) Wright

(1989, 118) summarizes: "The library in its best and future sense, may be more a nexus of information sources than a room full of books."

Publishers acknowledge the budget constraints that libraries face; many even admit to a role in creating the problem. Even so, most believe (with some justification) that resource sharing often exceeds fair use. (Dougherty and Hughes 1993, 19) Publishers cite situations where frequently used library resources are shared without royalties being paid. Such actions will drive up royalty fees and lead to more stringent copyright guidelines—and to the possibility of legal action.

Resource sharing among libraries may be a thorn in the side of publishers, but too much is published for any library to acquire everything needed. Resource sharing agreements should, however, incorporate "fair use/fair dealing" doctrines, with royalties paid directly to publishers, copyright clearinghouses, or commercial suppliers, or through multisite licensing of commercial databases. Without willing copyright compliance, libraries will face tougher restrictions that may force them to limit services to primary constituencies only, regardless of whether access is through other libraries or commercial suppliers (Cargill 1992, 85). We may see more blurring between commercial and library document suppliers, as it is difficult to differentiate between a "lending fee" paid to a nonreciprocal library and a "delivery fee" paid to a commercial supplier. So a bit of healthy rivalry will remain between libraries and commercial suppliers—just enough to keep each on their toes!

If libraries continue to cancel serials without consultation with one another, many will end up with similar collections of core titles (Chrzastowski and Schmidt 1993, 94, 101), very likely matching those most frequently available from commercial suppliers. Demand for unique titles will create more wear and tear on such publications, unless libraries take extra preservation measures. Protecting them by limiting use to primary clientele does not bode well for future ILL lending. More cooperative collection development is encouraged between libraries with reciprocal resource sharing agreements, with consultation on expensive acquisitions and serial cancellations.

When materials budgets are eaten up by continuing subscriptions, the number of books purchased must decrease. ILL requests for recently published monographs will become increasingly common. There should be some effort to maintain national archival collections so that public access remains to all titles—monographs and serials—published within a country. Some libraries are now scanning and electronically storing bit-mapped page images of unique titles in their collections. Legal questions remain to be clarified as to whether such images can be transmitted to users outside a library's own system.

Participants at the "Preferred Library Futures" workshop expressed concern for the preservation of electronically published documents. "There is currently no authoritative body assigned the responsibility for `copies of record' for electronically published information, and no procedures have been established to preserve such copies in perpetuity as part of the scholarly record." To solve the problem a member organization, the "International Collections Points," was proposed. Members would be "service agencies for the rights holders of information housed there. These agencies would be international, identified by discipline, but with redundancy built in to protect against potential gaps." (Dougherty and Hughes 1993, 18)

The future is also likely to bring a blurring between document retrieval services on national electronic networks. Originally for-profit activities were not allowed on research networks, but consumer demand for electronic convenience leads libraries to more reliance "upon information management systems that link and interrelate various library databases and other types of information services" (Cargill 1992, 83), such as document delivery. This need for convenience has forced a loosening of the governance of networks originated for research, education, and government. Libraries share responsibility for commercializing the networks by mounting site-licensed databases on specific networks to provide access to multiple information resources, and by allowing the hosts of information vendors to become database servers on networks. Both methods enhance vendors' services and increase sales (Cleveland 1991b, 113).

To cite one example, four commercial databases from the Institute for Scientific Information are available for unlimited fixed-price searching by registered end-users at academic institutions throughout the United Kingdom, with document delivery available from the University of Bath where the host is mounted (Academic 1991, 3). The number of document delivery requests to ISI has greatly increased through availability over JANET than might otherwise have occurred. While this system is convenient to end-users, some observers question whether such commercial use is contrary to restrictions on network use.

In 1992, the number of commercially registered networks on the Internet surpassed the number of academic networks, which indicates a growing tolerance for charged Internet services (Deutsch 1993, 48). Because library resources form a major portion of the developing information services on the networks, libraries must become more involved in the governance of international research networks, particularly as the networks become more commercialized while issues related to intellectual property are still unresolved for electronic publications. Another issue that will require legislative action relates to subsidized access to the networks. Libraries must work closely with government officials at national and state levels to be sure that as much interlibrary communication as possible is conducted over research and educational telecommunication networks at reasonable rates.

Legislation to promote educational subsidies to reduce commercial telephone rates should also be encouraged (Kountz 1992, 48). It is one thing that library users can select between "fee or free" options, but librarians must actively work to protect public access and dissemination of information. Telecommunication charges and access fees could give new meaning to the phrases "information poor" and "information rich." Equal access for all is a primary goal, but one that libraries can probably achieve only with federal subsidies. The old campaign slogan of "a chicken in every pot" could be rephrased "a computer in every home." France alone has developed policies to make electronic libraries a nationwide reality.

A greater number of scholarly, refereed journals will be published electronically in the future, almost all of which will be accessible on the national research networks. Many of these journals will emanate from academia and nonprofit societies. Some proponents of electronic publishing have proposed that faculty and members of scholarly societies should publish electronically exclusively on research networks (Boyce 1993, 272-273). Trade publishers have been slow to embrace electronic publishing, except for media materials such as audio and video cassettes and computer software. Some publishers considered e-journals to be a "blatant technology push with little or no market demand" (Gurnsey and Henderson 1984, 64), but trade publishers are likely to increase productivity because editorial, postage, and printing costs are much less than with print journals.

Past hesitation was often due to unfamiliarity with marketing and pricing techniques. Much electronic publishing is "on-demand," a concept with different charging and cash flow patterns than traditional publishing's reliance on prepaid subscriptions and high sales to subsidize lesser used products. Most publications are produced with prepaid monies. If the industry develops a fee-for-service base, seldom used items are likely to be high priced or unavailable for electronic distribution except through ILL.

To limit risks, electronic publishing often parallels print publication—offering a familiar product in and electronic format. This works well for bibliographic and reference resources, but the future promises more diversity with unique electronic products. The database industry has begun to develop hybrids that may link data capable of manipulation—financial data from DowVision, for example (Gerdy 1991, 109-111)—with full text and/or bibliographic information. This trend actually began in the early '80s with files such as EMIS (Electronic Materials Information Services) from INSPEC (Gurnsey and Henderson 1984, 38), but is accelerating as computer hardware prices continue to decline and the market expands beyond commercial use into the home market.

Future success for electronically published products depends upon realization of their potential for added value compared to traditional print products. Advantages include speed of publication; output on different media, such as screen and floppy disks; low or high quality printouts; and on-demand publishing services. Speed of publication also provides an opportunity to establish prior rights (ibid., 1984, 66). The portability of computers may also speed evolution toward a paperless society. "Laptops" are as portable as a briefcase and "tablet" or "palmtop" computers, such as the Franklin Digital Book or the Sony Data Discman (Kountz 1992, 42-44) make it as easy to tote a "computer book" as any large paperback book. Kurzweil (1993, 54-55) describes this mechanism for delivering the book of the future: "Resolution, color, contrast ratio, and lack of flicker will all match high-quality paper documents. These truly personal computers will be able to send and receive virtual books instantly through wireless communication." He points out the advantage of a wireless system: "Since any technology tends to be obsolete by the time it is in full production, installing, not to mention updating, a wireless system will be considerably less disruptive." (ibid) Such a format would appeal to a generation reared on TV and video games, but it is also convenient for readers who lack the physical capacity to turn pages, to hold open tightly bound books, or to read from books with small type. Jokes about not wanting to curl up in bed with a computer will soon be passé.

Electronically stored library collections are particularly beneficial to those users with physical or learning disabilities. The visually impaired can employ voice-activated computers to search OPACs. Retrieved text can be vocalized, transcribed into braille, or enlarged on the screen to improve readability. Hearing impaired users can retrieve information electronically faster than by trying to communicate with library staff who may not be able to sign or read lips. Wheelchair-bound library users do not have to contend with the rigors of getting to the library or of reaching shelved items. Computers aid learning by increasing students' comprehension with realistic problem solving simulations and by allowing students to learn at their own pace (Lewis 1988, 292, citing Cyert 1987, 13-21).

Technological developments will impact the capabilities and costs of networking databases. Microcomputers continue to add power at less cost. As CD-ROMs become more compact, fewer discs will be needed per database. Lippert (1992, 136) predicts that as Z39.50 protocols are adopted, search interfaces and databases will be interchangeable. Information seekers in the fast-approaching next century will have less need to visit a library. New library designs will almost certainly give some consideration to the "virtual library" concept. It is unlikely that no books will be found in the libraries of the future, but electronically stored materials eliminate the need for extensive stacks areas, cut down on maintenance, and improve security, while allowing access to many users at once. John Kountz (1992, 47) believes that factors such as "not buying more library space, could be the driving forces in determining 'migration' rates to new delivery systems as we approach the year 2000. Hard savings should begin to be realized as electronic information delivery becomes operational. . . . Annual costs will be about "three-fifths those of traditional library service."

The electronic library provides access around the clock every day—ideal for reserve rooms and ready reference collections. Even new materials can be stored and accessed by multiple users (with permission from the copyright holder)—a good investment for libraries that must buy several copies of current best-sellers, only to have them lie idle on the shelves after the initial rush subsides. This technique is being used by more libraries for reserve room documents, which can now be accessed by students from their residences without worrying about whether another student might be using the material. Multimedia technology allows even more creative development of reserve materials. According to Jennifer Cargill (1992, 83-84), libraries of the future "will depend more upon the use of computers to access and deliver information—rather than persevere in the notion that only librarians can efficiently instruct patrons in how to access data and only librarians can deliver correct information."

At the Chicago-Kent School of Law at Illinois Institute of Technology, the pages of new books are ripped out, scanned, and then discarded. Columbia University's Law Library plans to scan 10,000 deteriorating old books annually, storing the contents on a $1.5-million supercomputer called the Connection Machine (a move designed to provide enough shelf space for all of the new copyrighted material the library will acquire) at a much lower cost than the now-canceled plan for a $20 million building addition. Columbia also intends to put other noncopyrighted material on the computer (Bulkeley 1993, 8).

One can't help but hope that some thought is being given to the possible rarity of the works being destroyed, but possibly all books of today are the incunabula of tomorrow. Libraries have always faced this dilemma—the imperative to encourage use of collections pitted against the need to preserve them. The electronic medium can solve this conflict (Lewis 1988, 296).

We should point out that bound books can be scanned without damaging the original. Better copies can be obtained by scanning new books with wide margins and clean pages, rather than waiting until a book deteriorates beyond repair. Although some libraries do choose not to retain an archival copy, others may prefer to scan a document upon arrival, store the original and circulate the electronic version. The legality of doing this without copyright clearance is still unresolved.

As libraries shift from acquisitions to access, they will need more integration and cooperation between staff, and possibly restructuring of positions. With constrained budgets, libraries face greater service expectations from patrons and a broader variety of information resources. Preston Treiber, president of EBS Book Service, laments that "the 'just in time' vs. 'just in case' mentality so prevalent in the current academic library environment tears at the very heart of the mission of libraries. With decreased academic library materials budgets and more material to purchase, many academic library bibliographers, collection development officers, and acquisitions librarians may be losing the game to faculty who want the library to buy what they need when they need it." (Strauch and Miller 1993, 138)

Nonetheless, many librarians find this to be an exciting time of broadening horizons. The new library environment offers a holistic approach to librarianship, with no units isolated from the process. Bluh (1993, 51) states:

> Librarians from a number of areas will work closely together to select the best products, plan for and implement services, develop training and instructional programs for staff and patrons, and consult with legal experts to draft satisfactory contractual agreements, ones that best reflect the requirements of the organization.

The future will also bring more blending of duties between librarians and information technologists. Some institutions are likely to merge libraries, computing centers, and media centers, as all share in the storage, retrieval, and distribution of information. At the least, they will form strong communications bonds between these entities. Librarians and information technologists also share administrative problems such as questions of copyright, ownership, access to information, fair use and royalty payments, standards, and site licensing (Martin 1992, 77).

One duty will almost certainly become standard in libraries of the future: converting existing books into electronic form. This task will likely be even more daunting than the cataloging retroconversion projects of the '60s and '70s, and the barcoding projects related to automating circulation in the '70s and '80s. Future virtual libraries depend entirely on records converted to machine readable format and on the reliability of online catalogs. While most large academic and public library collections have been converted, all libraries should be encouraged to complete retrospective conversion of card catalogs. Unique collections should have the highest priority for conversion (Mitchell and Saunders 1991, 8). With demand for rapid access growing daily, this is not the time to cut technical services staff. Union lists must be kept up to date, and the contents of full-text databases must be cataloged for access.

While the library user may be calling the shots for just-in-time purchases, librarians will be monitoring statistical data. They will gather information on what was or wasn't acquired in order to reveal the type of material being requested, the number of requests for a title or an article, and demographics of the user population. "Based on information of that kind, a balance can be achieved between the acquisition of and access to information." (Bluh 1993, 51) Libraries have to prepare for the future realistically. Expanding access will not solve every materials budget problem, and instantaneous access is still a dream for most libraries. It very likely will be realized within the next five years or so, but not without a "hard row to hoe."

Interlibrary loan, as we have said many times throughout this work, is no longer just interaction between libraries. As we strive towards virtual libraries that allow user-directed document delivery, Bluh warns that "we must find ways to mitigate the lack of humanity that may creep into our operations." (ibid.) We must devise ways to meet users' needs, and to determine when needs are not being met. We need to develop new skills for serving our publics and establish stronger ties between libraries, publishers, and commercial document suppliers.

Document delivery will play a major role in the future mission of libraries. Goals and objectives supporting that mission will involve the entire library, not just the interlibrary loan and document delivery services. Reciprocal agreements and resource sharing arrangements should be strategically planned to protect and supplement collections for future generations of library users. For resource sharing to be more than symbolic, libraries and their parent institutions will have to revise the traditional competitive environment. Comparative rankings may have to be based more on quality of service rather than quantitative measures. Rather than emphasizing the number of volumes, libraries should be rated according to successful transactions. They should focus on services provided, not on the size of the budget.

Michael Gorman (1991, 4), writing on the future of libraries, stated: "As individuals or corporate entities, we must face the biological fact that we grow and change or we die." Further (ibid., 9), "Change should be welcomed and recognized as a sign of health. The . . . libraries that we serve are of great importance to culture and education. They are worth striving for, and from that striving comes creativity and innovation. In the process of changing and growing we must be prepared, on occasion, to fail and to not become downhearted by failure."

Those professionals involved in interlending and document delivery know that they are in the midst of rapid changes. S. Michael Malinconico (1992, 36) states: "We can deny that significant changes are taking place, we can passively let change happen to us, or we can anticipate the nature and shape of the impending changes and exploit them to our advantage."

It is irrelevant whether one believes the library of the future will be one of virtual or physical reality; what is significant is the capacity to enhance traditional practices with developing document delivery processes. Rapid access and instant document delivery will not come without stress, but the potential for offering every scholar access to worldwide information resources makes the struggle worthwhile.

———————————————————

References Cited

Abbott, Anthony. 1992. Electronic Publishing and Document Delivery: A Case Study of Commercial Information Services on the Internet. In *Networking, Telecommunications, and the Networked Information Revolution: ASIS 1992 Mid-year Meeting.* Silver Spring, MD: American Society for Information Science.

Academic Network Starts Fixed-price Searching. 1991. *Information World Review* 55 (January): 3.

Adkins, Susan L. 1993. CD-ROM: A Review of the 1992 Literature. *Computers in Libraries* 13, no. 8 (September): 20-52.

Alen, Joseph S. 1992a. Collective Administration of Literary Works. *Publishing Research Quarterly* (Summer): 73-78.

_____. 1992b. Collective Licensing as a Practical Solution. *The Bookmark* 50 (Winter): 118-120.

_____. 1993. *Fifteen Years of Copyright Solutions.* Danvers, MA: Copyright Clearance Center.

Alexander, Adrian W. 1993. ACRL Journal Costs in Libraries Discussion Group: A Report. In ALA Annual Conference 1992. *Library Acquisitions: Practice and Theory* 17, no. 1 (Spring): 89-90.

Allen, David B., and Johanna Alexander. 1986. "A Document Delivery Alternative for the CSB Library: Using Commercial Suppliers to Supplement Conventional Interlibrary Loan." ERIC Document, ED 275330. California State College, Bakersfield, CA. Rockville, MD: Educational Resources Information Center.

Allen, G. G., and L. Carman-Brown. 1991. Australian University Libraries Interlibrary Lending 1977-89. *Journal of Interlibrary Loan and Information Supply* 2, no. 1:33-51.

Alloway, Catherine Suyak. 1992. On the Road with Online: The Online Bookmobile. *Wilson Library Bulletin* 66, no. 9 (May): 43-45, 140.

Aman, Mohammed M. 1989. Document Delivery and Interlibrary Lending in the Arab countries. *Interlending and Document Supply* 17, no. 3 (July): 84-88.

American Geophysical Union v Texaco, 802 F.Supp.1, S. Doc. NY. 1992.

American Library Association.   Reference and Adult Services Division. Management and Operation of Public Services Section. Interlibrary Loan Committee.   1994.   "National Interlibrary Loan Code for the United States 1993." In "New National ILL Code on Ill-1 [electronic bulletin board]. [Cited 4 May 1994]. Available from listserv @ UVMVM.BITNET. Later published in *RQ* 33, no. 4 (Summer): 477-479.

American Library Association.   Task Force to Review the Guidelines for Extended Campus Library Services.   1990.   ACRL Guidelines for Extended Campus Library Services.   *College and Research Libraries News* 51, no. 4 (April): 353-355.

Angle, Joanne G.   1982.   Networking and Interlibrary Cooperation.   In *Developing Consumer Health Information Services*, edited by Alan M. Rees, 235-254.   New York: R. R. Bowker.

Ardis, Susan B., and Karen S. Croneis.   1987.   Document Delivery, Cost Containment, and Serial Ownership.   *College and Research Library News* 48, no. 10 (November): 624-627.

Arizona State University.   University Libraries.   Interlibrary Loan and Document Delivery Services.   [1993.]   *Statistics*, 1992/93.   [Tempe, AZ] Typescript.

_____.   Journal Document Delivery Task Force.   [1992a.]   Journal Document Delivery Project: Report, January-June, 1992.   [Tempe, AZ.] Draft Typescript.

_____.   [1992b.]   Report and Recommendations.   [Tempe, AZ.] Typescript.

Arms, Caroline R.   1989.   Libraries and Electronic Information: The Technological Context.   *EDUCOM Review* 24, no. 3 (Fall): 34-43.

_____.   1990.   A New Information Infrastructure.   *Online* 14, no.5 (September): 15-22.

Arundale, Justin.   1991.   Newspapers on CD-ROM: the European Perspective.   *National Online Meeting Proceedings* 12:17-21.

Ashton, Paul.   1994.   Telephone conversation with co-author, Sheila Walters, 11 January.

Associated Press.   1994.   "First-class Tips for Faster Mail."   *Tribune Newspapers* (31 July): A7 (Tempe, AZ edition).

Association of American Publishers. 1992. Statement of the Association of American Publishers (AAP) on Commercial and Fee-Based Document Delivery. Washington, DC. Photocopy.

Association of Independent Information Professionals. 1994. *Membership Directory*. Annual.

Atkinson, Hugh C. 1984. It's So Easy to Use, Why Don't I Own It? *Library Journal* 109 (1 June): 1102-1103.

Bailey, Charles W., Jr. 1992. Network-based Electronic Serials. *Information Technology and Libraries* 11, no. 1 (March): 29-35.

Baker, Colin. 1993. Informal conversations on several occasions, 22-25 March, with co-author, Sheila Walters, Southampton, UK, regarding document delivery in Australasia.

Baker, Shirley K., and Mary E. Jackson. 1993. *Maximizing Access, Minimizing Cost: A First Step Toward the Information Access Future*, prepared for the ARL Committee on Access to Information Resources November 1992 (Revised February 1993). Washington, DC: Association of Research Libraries. Photocopy.

Baldwin, David A., and Paul D. Dickey. 1986. Document Delivery Service. *College and Research Library News* 11 (December): 704- 707.

Ballard, Thomas H. 1991. Resource Sharing Expectations in Public Libraries. *Advances in Library Resource Sharing* 2:12-24.

Barden, Philip. 1990. ADONIS—The British Library Experience. *Interlending and Document Supply* 18, no. 3 (July): 88-91.

_____. 1992. Making Networks Work. *Nature* 359, no. 6394 (1 October): 435.

_____ [Barden, Phil]. 1993. Electronic Publishing: Fears, Myths, and Scenarios. *Learned Publishing* 6, no. 3 (April): 17-eoa.

Barnes, John. 1992. Solving the Physical Access Dilemma. *CD-ROM Librarian* 7, no. 3 (March): 19-20.

Barr, Catherine, ed. 1993. *The Bowker Annual of Library and Book Trade Information* 38:577-606. New Providence, NJ: R. R. Bowker.

Barron, Billy, and Marie-Christine Mahe. 1993. "Accessing On-line Bibliographic Databases." In Libraries.Intro [electronic document] available from gopher.cis.yale.edu. 69 lines.

Bartone, Carl R. 1982. Planning Regional Document-Delivery Services for the Water Decade: The Latin American and Caribbean Region. *UNESCO Journal of Information Science, Librarianship and Archives Administration* 4, no. 4 (October/December): 253-62.

Barwick, Margaret M. 1991. Interlending and Document Supply: A Review of Recent Literature, 19. *Interlending and Document Supply* 19, no. 1 (January): 17-23.

_____, comp. 1990. *A Guide to Centres of International Lending and Copying.* 4th ed. Boston Spa, Eng.: IFLA Office for International Lending, The British Library Document Supply Centre.

Basch, Reva. 1992. Online Deals. In *Online/CD-ROM '92 Conference Proceedings*, Chicago, 14-19. Wilton, CT: Eight Bit Books.

Battin, Patricia. 1988. Endangered Species and Libraries: Collision, Collusion or Collaboration. In *Proceedings of the Tenth Annual Meeting of the Society for Scholarly Publishing*, 1-8. Washington, DC: Society for Scholarly Publishing.

Bennett, Scott. 1993. Copyright and Innovation in Electronic Publishing: A Commentary. *The Journal of Academic Librarianship* 19, no. 2 (May): 87-91.

Berry, John. 1992. CD-ROM: The Medium of the Moment. *Library Journal* 117, no. 2 (1 February): 45-47.

_____. 1993. K. Wayne's World: OCLC Confronts the Future. *Library Journal* 118, no. 9 (15 May): 28-31.

Billings, Harold. 1993. Supping with the Devil: New Library Alliances in the Information Age. *Wilson Library Bulletin* 68, no. 2 (October): 33-37.

Bjørner, Susan N. 1990. Full-Text Document Delivery Online—It Makes Sense. *Online* 14, no. 5 (September): 109-112.

Blair, Joan. 1992. The Library in the Information Revolution. *Library Administration and Management* 6, no. 2 (Spring): 71-76.

Bluh, Pamela. 1993. Document Delivery 2000: Will It Change the Nature of Librarianship? *Wilson Library Bulletin* 67 (February): 49-51, 112.

Blumenstyk, Goldie. 1992. Business and Philanthropy Notes. *Chronicle of Higher Education* 39, no. 4 (16 September): A35.

Bonk, Sharon. 1990. Interlibrary Loan and Document Delivery in the UK. *RQ* 30, no. 2 (Winter): 230-240.

Borchardt, D. H. 1979. Three Aspects of Library Resources Sharing: Notes for the Guidance of Librarians in Developing Countries on Union Catalogues, Interlibrary Loans and Cooperative Storage. In *Resource Sharing of Libraries in Developing Countries: Proceedings of the 1977 IFLA/UNESCO Pre-session Seminar for Librarians from Developing Countries*, edited by H. D. L. Vervliet, 144-151. Munich; New York: K. G. Saur.

Borovansky, Vladimir T. 1994. Notes to co-author, Sheila Walters, July.

Boucher, Virginia. 1984. *Interlibrary Loan Practices Handbook*. Chicago, IL: American Library Association. [2d ed. forthcoming 1995?.]

Boyce, Bert R. 1993. Meeting the Serials Cost Problem: A Supply-side Proposal. *American Libraries* 24, no. 3 (March): 272-273.

Boyer, Janice S., and John Reidelbach. 1990. Document Delivery Pilot Project at UNO. *Nebraska Library Association Quarterly* 21, no. 1 (Spring): 7.

Brack, Verity. 1992. Mailing Lists, Electronic Newsletters and Journals. In *Network Services Available Over Janet*, section 6 [electronic document]. Sheffield, Eng.: University of Sheffield, July. Available from ukoln.bath.ac.uk; Identifier no. BH1D, AP-ZZ3. [992 lines.]

Bradbury, David. 1992. "A Worldwide View of Information: Universal Availability of Publications and International Interlibrary Loan." A presentation at the American Library Association Annual Conference, 28 June, San Francisco, CA.

Braid, Andrew. 1991. The Role of LINC in the Automation of Interlibrary Loan and Document Supply in the United Kingdom. *Interlending and Document Supply* 19, no. 3 (January): 101-104.

Brandt, D. Scott. 1992. Campus-wide Computing. *Academic and Library Computing* 9, no. 10 (November/December): 17-20.

British Library Board. 1990. "Modelling the Economics of Interlibrary Lending: Report, November 1989," prepared by Coopers & Lybrand for The British Library. Boston Spa, UK: British Library Document Supply Centre.

Brown, David J. 1982. Electronic Document Delivery Systems. In *Multi-media Communications*, edited by May Katzen, 68-87. Westport, CT: Greenwood.

Brown, Jennifer C. n.d. "The Guide to On-line Resources for Researching Central/East Europe and the Former Soviet Union." [electronic document.] Available from lib-gopher.uchicago.edu. 121 lines.

Brown, Rowland C. W. 1989. The Role of OCLC in Interlending. In *Interlending and Document Supply: Proceedings of the First International Conference held in London, November 1988,* edited by Graham P. Cornish and Alison Gallico, 74-79. Southampton, UK: IFLA Office for International Lending.

Brownson, Charles. [1993.] Chart and data related to comparison of inventory of three commercial suppliers. Conversations with co-author, Sheila Walters, 1 September and 10 September, Arizona State University Libraries, Tempe, AZ.

Bruer, John T, William Goffman, and Kenneth S. Warren. 1981. Selective Medical Libraries and Library Networks for Developing Countries. *Journal of Tropical Medicine and Hygiene* 30, no. 5:1133-1140.

Brumbaugh, Patrick. 1991. "Two Myths about Copyright Compliance." In ILL-L: Interlibrary Loan Discussion Group [electronic bulletin board]. [Cited 11 December.] Available from listserv ILL-L@UVMVM.BITNET.

Bryant, Philip. 1992. Bibliographic Access in the United Kingdom: Some Current Factors. *The Reference Librarian* 35:57-70.

Bryant, Philip, and John Smith. 1993. Interview by co-author, Sheila Walters, 22 March 1993. Centre for Bibliographic Management and the UK Office of Library Networking, The Library, University of Bath, Bath, UK.

Buchanan, Lori E., Anne May Berwind, and Don Carlin. 1989. Optical Disk-based Periodical Indexes for Undergraduates. *College and Research Libraries News.* (January): 10-14.

Bucknall, Carolyn. 1985. Conjuring in the Academic Library: The Illusion of Access. In *Access to Scholarly Information: Issues and Strategies,* edited by Sul H. Lee, 59-70. Ann Arbor, MI: Pierian Press.

Budd, John. 1986. Interlibrary Loan Service: A Study of Turnaround Time. *RQ* 26 (Fall): 75-80.

Bulkeley, William M. 1993. Libraries Shift from Books to Computers. *Wall Street Journal* Technology Section (8 February): 8.

Burr, Robert L. 1988. The Electronic Branch Library: Using CD-ROM and Online Services to Support Off-Campus Instructional Programs. *National Online Meeting Proceedings* 9:37-41.

The Burwell Directory of Information Brokers. 1994. 11th ed. Houston, TX: Burwell Enterprises. [formerly *Directory of Fee-Based Information Services,* compiled by Helen P. Burwell and Carolyn N. Hill, 1989].

Butler, John T. 1993. A Current Awareness Service Using Microcomputer Databases and Electronic Mail. *College and Research Libraries* 54, no. 2 (March): 115-123.

Cameron, Jamie. 1993. The Changing Scene in Journal Publishing. *Publishers Weekly* 240, no. 22 (31 May): 23-24.

Campbell, Robert. 1992. Document Delivery and the Journal Publisher. *Scholarly Publishing* 23, no. 4 (July): 213-221.

Campbell, Robert M., and Barrie T. Stern. 1990. ADONIS—The Next Stage? *Health Libraries Review* 7:20-21.

Cargill, Jennifer. 1992. The Electronic Reference Desk: Reference Service in an Electronic World. *Library Administration and Management* (Spring): 82-85.

Carrigan, Dennis P. 1992. Research Libraries' Evolving Response to the "Serials Crisis." *Scholarly Publishing* 23, no. 3 (April): 138-151.

Casorso, Tracy. 1991. The North Carolina State University Libraries and the National Agricultural Library Joint Project on Transmission of Digitized Text: Improving Access to Agricultural Information. *Reference Services Review* 19 (Spring): 15-22.

————. 1992. "NCSU Transmission Project." In ILL-L: Interlibrary Loan Discussion Group [electronic bulletin board]. [Cited 6 April.] Available from listserv ILL-L@UVMUM.BITNET.

CEC Libraries Programme. 1994. *IATUL News* 3, no. 2:4-9.

Center for Research Libraries. 1992. FY Fee Structure and Price List (July 1, 1991 - June 30, 1992). [Chicago, IL]. Typescript, Rev. 916, no. 4.4:1-3.

"CERRO Archives." n.d. [electronic document.] Available from lib.gopher.uchicago.edu. 243 lines.

Champany, Barry W., and Sharon M. Hotz, eds. 1982. *Document Retrieval: Sources and Services.* 2d ed. San Francisco: The Information Store. (See Erwin 1987 for latest edition.)

Chrzastowski, Tina E., and Karen A. Schmidt. 1992. Analyzing Serial Cancellations. *ARL: A Bimonthly Newsletter of Research Libraries Issues and Actions* 165 (23 November): 2-3.

———. 1993. Surveying the Damage: Academic Library Serial Cancellations 1987-88 through 1989-90. *College and Research Libraries* 54, no. 2 (March): 93-102.

Clement, Hope E. A. 1989. National and International Models for Interlending. In *Interlending and Document Supply: Proceedings of the First International Conference held in London, November 1988*, edited by Graham P. Cornish and Alison Gallico, 5-15. Southampton, UK: IFLA Office for International Lending.

Cleveland, Gary. 1991a. *Electronic Document Delivery: Converging Standards and Technologies.* ERIC Document ED 350003. UDT Series on Data Communication Technologies and Standards for Libraries Report, 2. IFLA International Office for Universal Dataflow and Telecommunications, Ottawa, Can. Rockville, MD: Educational Resources Information Center. Microfiche.

———. 1991b. "Research Networks and Libraries: Applications and Issues for a Global Information Network." UDT Series on Data Communication, Technologies and Standards for Libraries Report, 1. Ottawa, Canada.: IFLA International Office for Universal Dataflow and Telecommunications.

Cline, Gloria S. 1987. The High Price of Interlibrary Loan Service. *RQ* 27, no. 1 (Fall): 80-86.

CNI-COPYRIGHT AND INTELLECTUAL PROPERTY FORUM [electronic bulletin board]. 1991-1994. Selected comments and general opinions regarding copyright issues. [Selected from citations appearing 1991-1994.] Available from listserv CNI-COPYRIGHT@CNI.ORG.

Coffman, Steve and Pat Wiedensohler, eds. 1993. *The FISCAL Directory of Fee-Based Research and Document Supply Services.* 4th ed. N.p.: County of Los Angeles Public Library; American Library Association.

Colaianni, Lois Ann. 1990. DOCLINE: The National Library of Medicine Experience. In *Riding the Electronic Wave—Document Delivery: Proceedings of the Library of Congress Network Advisory Committee Meeting, November 29-December 1, 1989*, 33-40. Network Planning Paper, 20. Washington, DC: Library of Congress, Network Development and MARC Standards Office.

Colbert, Antoinette Walton. 1979. Document Delivery: Payment Methods for Documents Can Make a Difference in Delivery Times. *Online* 7, no. 4 (July): 78-79.

Coleman, Jim. 1994. Letter to "Current Ariel User," 10 June.

Collins, Tim. n.d. The Future Role of Publishers and Subscription Agencies in Current Awareness Services and Delivery. In *Library Automation and Access to Information*. 14-18. N.p.: EBSCO Subscription Services.

Compier, Henk. 1993. Interview by co-author, Sheila Walters, 25 March, Amsterdam, NE.

Cook, Eleanor. 1993. "A Licensing Question." In CNI-Copyright and Intellectual Property Forum, [electronic bulletin board]. [Cited 21 July.] Available from CNI-COPYRIGHT@CNI.ORG.

Cornish, Graham P. 1988. *Model Handbook for Interlending and Copying*. Boston Spa, UK: IFLA Office for International Lending; Unesco.

_____. 1989a. Copyright Law and Document Supply: A Worldwide Review of Development, 1986-89. *Interlending and Document Supply* 17, no. 4 (October): 117-123.

_____. 1989b. Interlending in the Caribbean. *International Library Review* 21, no. 2 (April): 249-261.

_____. 1991a. Document Supply in the New Information Environment. *Journal of Librarianship and Information Science* 23, no. 3 (September): 125-134.

_____. 1991b. The Impact of Networking on International Interlibrary Loan and Document Supply. *Libri* 41, no. 4 (October/December): 272-286.

_____. 1991c. Interlibrary Loan and Document Supply as an Indicator of Social and Cultural Infrastructures. *Bulletin of the American Society for Information Science* 17, no. 3 (February/March): 23-24.

_____. 1993a. Copyright Management of Document Supply in an Electronic Age: the CITED~ Solution. *Interlending and Document Supply* 21, no. 2 (April): 13-20.

_____. 1993b. Letter to co-author, Eleanor Mitchell, 10 February.

Costers, Look. 1992. "RAPDOC: The Pica Rapdoc Project: From Interlibrary Loan to Electronic Document Delivery." Paper presented at the Second European Serials Conference, Noordwijkerhout, The Netherlands, 9-11 September. Whitney, U.K.: UK Serials Group.

Crews, Kenneth Donald. 1990. "Copyright Policies at American Research Universities: Balancing Information Needs and Legal Limits." Ph.D. diss., University of California at Los Angeles. Microfiche.

_____. 1992. "Copyright Law, Libraries, and Universities: Overview, Recent Developments, and Future Issues." Presented to Association of Research Libraries, October.

Cummings, Anthony M., Marcia L. Witte, William G. Bowen, Laura O. Lazarus, and Richard H. Ekman. 1992. *University Libraries and Scholarly Communication: A Study Prepared for the Andrew W. Mellon Foundation.* Washington, DC: Association of Research Libraries.

Currie, Jean, and Jan Kennedy Olsen. 1985. *Document Delivery: A Study of Different Sources: A Report to the Council on Library Resources.* ERIC Document, ED 262786. Ithaca, NY: Cornell University Libraries. Microfiche.

Daval, Nicola, and Patricia Brennan, comps. 1994. *ARL Statistics 1992-93: A Compilation of Statistics from the One Hundred and Nineteen Members of the Association of Research Libraries.* Washington, DC: Association of Research Libraries.

Davis, Frank L., Robert B. Greenblatt, Gail Waverchak, and John B. Coffey. 1992. The Use of Independent Information Brokers for Document Delivery Service in Hospital Libraries. *Bulletin of the Medical Library Association* 80, no. 2 (April): 185-187.

Davis, Trisha L. 1993. Acquisition of CD-ROM Databases for Local Area Networks. *The Journal of Academic Librarianship* 19, no. 2:68-71.

Decision in Texaco Copyright Suit Rejects 'Fair Use' Argument. 1992. *Database Searcher* 8, no. 7 (September): 13.

Dearie, Tammy Nickelson and Virginia Steel. 1992. *Interlibrary Loan Trends: Making Access a Reality,* edited by C. Brigid Welch. SPEC Kit, 184. Washington, DC: Association of Research Libraries, Office of Management Services.

Deekle, Peter. 1990. Document Delivery Comes of Age in Pennsylvania. *Wilson Library Bulletin* 65, no. 2 (October): 31-33.

_____. 1993. Letter to co-author, Eleanor Mitchell, 30 April.

DeForrest, Janet. 1993. Internet Resources: A Sample. *Library Horizons: A Newsletter of the University of Alabama Libraries*, 7, no. 1 (Fall): 3-4.

DeJohn, William. 1989. Resource Sharing through Networking. In *Effective Access to Information: Today's Challenge, Tomorrow's Opportunity*, edited by Alphonse F. Trezza, 79-90. Boston, MA: G. K. Hall.

Dempsey, Lorcan. 1990. Bibliographic Access: Patterns and Developments. In *Bibliographic Access in Europe: First International Conference: The Proceedings of a Conference . . . held at the University of Bath 14-17 September 1989*, edited by Lorcan Dempsey, 1-29. Aldershot, UK: Gower Publishing Company.

Dern, Daniel. 1994. *The Internet Guide for New Users*. New York: McGraw Hill.

DerWers, Titia van. 1993. Interview by co-author, Sheila Walters, 26 March, RABIN [National Library of the Netherlands], The Hague, NE. The Head of ILL also demonstrated interlending via PICA RAPDOC system.

Deschamps, Christine. 1991. Interlending Between Academic Libraries in France—A Review. *Interlending and Document Supply* 19, no. 2 (April): 35-38.

Deutsch, Peter. 1993. Peter's Soapbox. *Internet World* (September/October): 48-49.

DIALOG Information Services. 1991. *DIALOG Full-text Sources: Subject List*. Palo Alto, CA.

_____. 1992. *DIALOG Full-text Sources Alpha List*. Palo Alto, CA.

Dickson, Stephen P., and Virginia Boucher. 1989. A Methodology for Determining Costs of Interlibrary Lending. In *Research Access Through New Technology*, edited by Mary Jackson, 137-161. AMS Studies in Library and Information Science, 1. New York: AMS.

Dijkstra, Jan Willem. 1993. Interview by co-author Sheila Walters, 26 March, Elsevier Science, Amsterdam, NE.

Dimchev, Alexander. 1991. The Demand for Interlending in Bulgaria. In *East-West Information Transfer: Papers from the meeting on Interlending and Document Supply between Eastern and Western Europe, held at Gosen, February 1991*, edited by Graham P. Cornish and Monika Segbert, 16-18. N.p.: IFLA Office for International Lending.

Dimitroff, Alexandra. 1993. Information Access in a Developing Country: Special Libraries in Egypt. *Special Libraries* 84, no. 1 (Winter): 25-29.

Ditzler, Carol J., Veronica Lefebvre, and Barbara G. Thompson. 1990. Agricultural Document Delivery: Strategies for the Future. *Library Trends* 38, no. 3 (Winter): 377-96.

Dougherty, Richard M. 1978. Campus Document Delivery Systems to Serve Academic Libraries. *Journal of Library Automation* 11, no. 1 (March): 24-31.

_____. 1991. Needed: User-responsive Research Libraries. *Library Journal* 116, no. 1 (1 January): 59-62.

Dougherty, Richard M., and Carol Hughes. 1991. *Preferred Futures for Libraries: A Summary of Six Workshops with University Provosts and Library Directors.* Mountain View, CA: The Research Libraries Group.

_____. 1993. *Preferred Library Futures, 2: Charting the Paths.* Mountain View, CA: The Research Libraries Group.

Dow, Ronald F., Karen Hunter, and G. Gregory Lozier. 1991. Commentaries on Serials publishing. *College and Research Libraries* 52, no. 6 (November): 521-527.

Dubey, Yogendra P. 1988. Information Poverty: A Third World Perspective. In *Unequal Access to Information Resources: Problems and Needs of the World's Information Poor: Proceedings of the Congress For Librarians February 17, 1986 . . .,* edited by Jovian P. Lang, 47-54. Ann Arbor, MI: Pierian Press.

Dukelow, Ruth H. 1992. *Library Copyright Guide* Washington, DC: Copyright Information Services.

Dunn, Dana. 1993. University of Cincinnati. *Computers in Libraries* 13, no. 1 (January): 14-15.

Dyer, Peter Swinnerton. 1992. A System of Electronic Journals for the United Kingdom. *Serials* 5, no. 3 (November):33-35. Summarized by Andrew White, [1993]. Photocopy.

Edmonds, Diana. 1991. Regional Library Cooperation: Regional Roller Coaster: The Accelerating Progress of the Regional Library Systems. In *Handbook of Library Cooperation,* edited by Alan E. MacDougall and Ray Prytherch, 251-263. Aldershot, Eng.: Gower Publishing Company.

Ellis, Arthur, and Anna Rainford. 1994. The Virtual Library at the University of Western Australia. *IATUL News* 3, no. 1:3-4.

Engle, Mary E. 1991. The Electronic Paths to Resource Sharing: Widening Opportunities Through the Internet. *Reference Services Review* 19, no. 4 (Winter): 7-12.

Ensign, David. 1989. Copyright Considerations for Telefacsimile Transmission of Documents in Interlibrary Loan Transactions. *Law Library Journal* 81, no. 4 (Fall): 805-812.

Epstein, Hank. 1989. Technological Trends in Information Services to the Twenty-first Century. In *Trends in Urban Library Management: Proceedings of the Urban Library Management Institute, . . . October 1988 at the University of Wisconsin at Milwaukee,* edited by Mohammad M. Aman and Donald J. Sager, 61-66. Metuchen, NJ: Scarecrow Press.

Erwin, Tracy, ed. 1987. *Document Retrieval: Sources and Services.* 4th ed. San Francisco: Information Store.

Everett, David. 1993a. Full-text Online Databases and Document Delivery in an Academic Library: Too Little, Too Late? *Online* (March): 22-25.

———. 1993b. Full-Text Online Databases as a Document Delivery System: The Unfulfilled Promise. *Journal of Interlibrary Loan and Information Supply* 3, no. 3:17-25.

Exon, F. C. A. 1987. *Survey of Australian Inter-library Lending.* Perth: The Library, Curtin University of Technology.

Feick, Tina. 1994. UnCover in Latin America. *UnCover Update* (Winter): [2].

Feldman, Gayle. 1992. Professional Publishing Goes Electronic. *Publishers Weekly* 139, no. 22 (11 May): 31-33.

Fennessy, Eamon T. 1991. Photocopying: Rights and Responsibilities. *Interlending and Document Supply: Proceedings of the Second International Conference held in London, November 1990,* edited by Alison Gallico, 30-33. Southampton, UK: IFLA Office for International Lending.

Ferrall, Eleanor. n.d. Formal Information Networks: Library Systems, Bibliographic Utilities and Databanks. Tempe, AZ: Arizona State University. Photocopy.

Field of Dreams. 1989. Produced by Lawrence Gordon and Charles Gordon and directed by Phil Alden Robinson. 1 hr., 46 min. Universal City Studios. Videocassette.

Finnigan, Georgia. 1992. Document Delivery Gets Personal. *Online* 16 (May): 106-108.

Fjälbrant, Nancy. 1994. EDUCATE - End-user Courses in Information Access through Communications Technology. *IATUL News* 3, no. 2:8-9.

Friend, Fred J., David Bovey, and Leslie Bugden. 1993. Interviewed by co-author, Sheila Walters, 19 March, University College London, UK, including demonstration of Libertas ILL system.

Frost & Sullivan Report on the European Market for CD-ROMs. 1993. *Online* 17, no. 2(March): 66.

Fullerton, Jan. 1991. Document Supply and the National Library of Australia. *Australian Academic and Research Libraries* 22, no. 4 (December): 103-110.

Gadbois, Lisa Marie. 1994. Telephone conversation with co-author, Sheila Walters, 2 February.

Garrett, John R. 1991. Text to Screen Revisited: Copyright in the Electronic Age. *Online* 15, no. 2 (March): 22-24.

Gasaway, Laura. 1993. Copyright Issues in Electronic Information and Document Delivery in Special Libraries. *At Your Service* no. 26 (September): 10-12.

Gassol de Horowitz, Rosario. 1988. *Librarianship: A Third World Perspective.* Contributions in Librarianship and Information Science, 59. New York: Greenwood Press.

George, Lee Anne. n.d. "The FISCAL Primer". Washington, DC: Gelman Library Information Service, George Washington University. Packet.

Gerdy, Greg. 1991. DowVision—First Customer Reactions. *National Online Meeting Proceedings* 12:109-112.

Gerrard, Kris. 1993a. Interview by co-author Sheila Walters during tour of British Library Document Supply Centre, 18 March, Boston Spa, U.K, including brief discussions with Stella Pilling, Betty Green, Graham Cornish, Frank Wray, and others.

_____. [1993b.] [Notes and packet of material prepared by Ms. Garrard on European document suppliers.] Boston Spa, UK: British Library Document Supply Centre.

Gherman, Paul M. 1991. Setting Budgets for Libraries in Electronic Era. *The Chronicle of Higher Eduction* (14 August): A36.

Gillikin, David P. 1990. Document Delivery From Full-Text Online Files: A Pilot Project. *Online* 14 (May): 27-32.

Ginsberg, Jane C. 1992. Reproduction of Protected Works for University Research or Teaching. *Journal of the Copyright Society of the U.S.A.* 39, no. 3 (Spring): 181-221.

Goldberg, David, and Robert J. Bernstein. 1992. "Texaco" Decision. *New York Law Journal* 208, no. 64 (30 September): 3, 6-eoa.

Gorman, Michael. 1991. The Academic Library in the Year 2001: Dream or Nightmare or Something in Between? *The Journal of Academic Librarianship*, 17, no. 1 (March): 4-9.

Gould, Sara. 1994. "IFLA Voucher Scheme." In ILL-L: Interlibrary Loan Discussion Group [electronic bulletin board]. [7 July]. Available from list-serv @ UVMVM.BITNET.

Granheim, Else. 1985. Special Problems of Libraries Serving a Linguistic Minority: The Norwegian Experience. In *International Librarianship Today and Tomorrow . . .*, compiled by Joseph W. and Mary S. Price, 43-52. New York: K. G. Saur.

Gulacsy-Papay, Erika, Rosza Karacsony, and Peter Sonnewend. 1991. Inter-Lending Needs from the Hungarian Viewpoint. In *East-West Information Transfer: Papers from the Meeting on Interlending and Document Supply between Eastern and Western Europe, held at Gosen, February 1991*, edited by Graham P. Cornish and Monika Segbert, 112-115. N.p.: IFLA Office for International Lending.

Gurnsey, John, and Helen Henderson. 1984. *Electronic Publishing Trends in the United States, Europe, and Japan: An Update of Electronic Document Delivery*, 3. Electronic Document Delivery, 7. Oxford, Eng; Medford, NJ: Learned Information.

Guyot, Brigitte. 1989. Minitel in France. *Outlook on Research Libraries*, 11, no. 11:3-5.

Hacken, Richard D. 1988. Tomorrow's Research Library: Vigor or Rigor Mortis? *College and Research Libraries* 49, no. 6 (November): 485-493.

Hafkin, Nancy J. 1994. "PADIS Update." E-mail note to co-author Sheila Walters, 5 September. [Message id: <7b104460@p101.f1.n751.z5.gnfi-do.fidonet.org>

Hahn, Harley, and Rick Stout. 1994. *The Internet Complete Reference.* 2d ed. Berkeley, CA: Osborne McGraw-Hill.

Halsey, Kathleen F. 1988. "An Evaluation of Document Delivery Service to Interlibrary Loan: A Commercial Firm and A Traditional Library Source." ERIC Document, ED 302261. M.S. thesis, Cardinal Stritch College. Rockville, MD: Educational Resources Information Center. Microfiche.

Hawks, Carol Pitts. 1992. The Integrated Library System of the 1990's: The OhioLINK Experience. *Library Resources and Technical Services* 36, no. 1 (January): 61-77.

Hawks, Carol Pitts, and Adrian W. Alexander. 1993. An Interview with Richard R. Rowe, President and CEO, The Faxon Company. *Library Acquisitions: Practice and Theory* 16, no. 2:93-102.

Hawley, Lorin M. 1992. Faster and Cheaper Document Delivery with Online Searching Skills. *Online* 16, no. 4 (July): 45-48.

_____. 1993. Document Delivery: From Citation to Document Using Online Verification. *Online* 17, no. 3 (May): 70-73.

He, Daxun. 1986. On the Problem of Document Delivery in the Field of Science and Technology in Asia. In *IFLA General Conference (52nd: Tokyo: 1986) Pre-Session Seminar on Special Libraries and Their Role in National Development Papers. . .* , fiche 3-8:1-5. ERIC Document, ED 280464. The Hague, NE: IFLA. Microfiche.

Heller, James. 1986. Copyright and Fee-based Copying Services. *College and Research Libraries* 47 (January): 28-35.

_____. 1987. Fee-based Document Delivery: Permissible Activities under United States Copyright Law. In *Fee-based Services: Issues and Answers: Second Conference on Fee-Based Research in College and University Libraries: Proceedings . . . Ann Arbor, Michigan, May 10-12, 1987,* compiled by Anne K. Beaubien, 41-51. Ann Arbor, MI: Michigan Information Transfer Source (MITS), University of Michigan.

Henry, Marcia Klinger, Linda Keenan, and Michael Regan. 1991. *Search Sheets for OPACs on the Internet: A Selective Guide to U.S. OPACs Utilizing VT100 Emulation.* Westport, CT: Meckler Corporation.

Henschke, E. 1991. A Loud Lament for Federalism: The Decentralised Interlending System in East and West Germany. In *East-West Information Transfer: Papers from the Meeting on Interlending and Document Supply between Eastern and Western Europe, held at Gosen, February 1991*, edited by Graham P. Cornish and Monika Segbert, 47-54. N.p.: IFLA Office for International Lending.

Higginbotham, Barbara Buckner. 1990. Telefacsimile: The Issues and the Answers. *Journal of Interlibrary Loan and Information Supply* 1, no. 1:67-86.

Higginbotham, Barbara Buckner, and Sally Bowdoin. 1993. *Access versus Assets.* Frontiers of Access to Library Materials, no. 1. Chicago, IL: American Library Association.

Hill, Susan E. 1989. Growth of Communication Systems Among Biomedical Libraries. In *Research Access Through New Technology*, edited by Mary E. Jackson, 47-65. AMS Studies in Library and Information Science, 1. New York: AMS Press.

Hoadley, Irene B. 1993. Access vs. Ownerhip: Myth or Reality? *Library Acquisitions: Practice and Theory* 17, no. 2: 191-195.

Horowitz, Irving Louis, and Mary E. Curtis. 1984. Fair Use Versus Fair Return: Copyright Legislation and its Consequences. *Journal of the American Society for Information Science* 35, no. 2 (March): 67-86.

Hunter, Karen A. 1984. Electronic Delivery of Scientific Information. *Drexel Library Quarterly* 20 (Summer): 75-86.

———. 1992. Document Delivery: Issues for Publishers. *Scholarly Publishing Today* 1 (March/April): 5-6. Also published in Rights 6, no. 2 (1992).

Hurd, Douglas P., and Robert E. Molyneux. 1986. An Evaluation of Delivery Times and Costs of a Non-library Document Delivery Service. In *Energies for Transition: ACRL National Conference Proceedings, 4th, Baltimore, 1986*, edited by Danuta A. Nitecki, 182-185. Chicago: ACRL/ALA.

Huston, Geoff. 1991. "AARNET: Australian Academic and Research Network and Kwaihiko Network Resource Guide." [Electronic document.] [Cited 18 February.] Available from wais@archie.au. [547 lines.]

Al Ibrahim, Baha. 1993. Interlibrary Loans in the Arabian Gulf: Issues and Requisites. *Interlending and Document Supply* 21, no. 2:21-25.

ILL-L: Interlibrary Loan Discussion Group [electronic bulletin board]. 1991-1994. Selected comments and general opinions regarding copyright issues. [Selected from citations appearing during 1991-1994.] Available from listserv ILL-L@UVMVM.BITNET.

_____. 1993. Selected comments regarding an appropriate name for an interlibrary loan department to reflect its changing role in document delivery, and selected definitions of "document delivery." [Selected from citations appearing in May 1993.] Available from listserv ILL-L@UVMVM.BITNET.

Information Systems Consultants. 1983. *Document Delivery in the United States: A Report to the Council on Library Resources*, by Richard M. Boss and Judy McQueen. ERIC Document, ED 244626. Washington, DC: Council on Library Resources.

Inoti, V. I. 1992. The Trend of Interlending Activity in Kenya: An Overview. *Journal of Interlibrary Loan and Information Supply* 2, no. 3:39-49.

International Federation of Reproduction Rights Organisations 1991. *IFFRO, 1991*. Salem, MA.

Intner, Sheila S. 1991. Education for the Dual Role Responsibilities of an Access Services Librarian. *The Reference Librarian 34:* 107-130.

ISM Library Information Services. 1993. *AVISO: Comprehensive Interlibrary Loan Management*. Etobicoke, Ontario, Canada.

Jackson, Mary E. 1990. Trends in Resource Sharing. *Wilson Library Bulletin* (April): 54-55; pt. 2 (June): 99-100.

_____. 1991. 48 Hours or 30 Seconds: Document Delivery Choices in Pennsylvania. In *Delivery of Information and Materials Between Libraries: The State of the Art: Proceedings of the June, 1990 ASCLA Multi-Lincs Preconference*, edited by Keith Michael Fiels and Ronald P. Naylor, 9-17. Chicago: ASCLA, ALA.

_____. 1993a. Document Delivery Over the Internet. *Online* 17, no. 2 (March): 14-18, 21.

_____. 1993b. ShaRes: Two Decades of Resource Sharing. *Colloquium News of the RLG Membership* 1, no. 3 (May): 2.

Jacsó, Péter.   1992.   *CD-ROM Software, Dataware, and Hardware: Evaluation, Selection, and Installation.* Database searching series, no. 4. Englewood, CO: Libraries Unlimited.

Jakubekova, Ludmila, and Lydia Sedlackova. 1991. The Functions of the Centre for International Lending of the CSFR. In *East-West Information Transfer: Papers from the Meeting on Interlending and Document Supply between Eastern and Western Europe, held at Gosen, February 1991*, edited by Graham P. Cornish and Monika Segbert, 32-34. N.p.: IFLA Office for International Lending.

Jakubs, Deborah. 1993. The Costs of Cooperation. In Serial Acquisitions and the Third World: The Latin American Perspective: Part 2, edited by Dan C. Hazen. *Serials Review* (Spring): 78-80.

James, David Willis. 1993. Article Document Delivery and Its Implications for Serials Librarians. *Library Acquisitions: Practice and Theory* 17, no. 1 (Spring): 94-95.

Jones, Claire, and William Dowsland. 1990. *Online Sources of European Information: Their Development and Use.* Alderhot, Hants, Eng.; Brookfield, VT: Avebury Gower Publishing Company.

Jones, Paul. 1991. Demand Publishing Inc. Sidebar 4 in Fax-on-Demand, by Christine E. Lachman. *Library Hi Tech*, issue 36, 9, no. 4:14-16.

Josephine, Helen. 1989. *Fee-based Services in ARL Libraries.* SPEC Kit 157. Washington, DC: Association of Research Libraries, Office of Management Studies.

Katz, Ruth. 1987. Trends in the Development of State Networks. *Advances in Library Automation and Networking* 1:169-187.

Kelly, Anette. 1994. EUROPAGATE: Project Summary. *IATUL News* 3, no. 2:6,8.

Kennedy, Sue. 1989. The Role of Commercial Document Delivery Services in Interlibrary Loan. In *Research Access through New Technology*, edited by Mary E. Jackson, 66-81. AMS Studies in Library and Information Science, 1. New York: AMS Press.

Kent, Anthony K., Karen Merry, and David Russon. 1987. *The Use of Serials in Document Delivery Systems.* Paris: International Council for Scientific and Technical Information.

Kesselman, Martin. 1993. Beyond Bitnet: Telnetting to the United Kingdom. *College and Research Libraries News* 54, no. 3 (March): 134-136.

Kessler, Jack. 1992. *Directory to Fulltext Online Resources.* Supplement to Computers in Libraries, 55. Westport, CT: Meckler Corporation.

Keys, Marshall. 1992. On the Future of the OCLC Regional Networks. *Library Administration and Management* 6, no. 1 (Winter): 10-14.

Khalil, Mounir. 1993. Document Delivery: A Better Option. *Library Journal* 118: 2 (1 February 1993): 43-47.

Kingma, Bruce R., and Philip B. Eppard. 1992. Journal Price Escalation and the Market for Information: The Librarians' Solution. *College and Research Libraries* 53, no. 6 (November): 523-535.

Kinney, Jane. 1989. The Role of CD-ROM in Information Isolated Areas. In *Information Knowledge Evolution: Proceedings of the forty-fourth FID Congress held in Helsinki, Finland, 28 August-1 September, 1988,* edited by Sinikka Koskiala and Ritva Launo, 279-290. FID (Federation Internationale de Documentation et d'Information) Publication, 675. Amsterdam; New York: North-Holland/Elsevier Science Publishers.

Kirkegaard, Preben. 1985. Relations Between the State and Libraries in Denmark. In *International Librarianship Today and Tomorrow . . . .*, compiled by Joseph W. and Mary S. Price, 95-106. New York: K. G. Saur.

Kohl, David F. 1993. OhioLINK: Plugging into Progress. *Library Journal* 118, no. 16 (1 October): 42-46.

Kósa, Géza A., and Joan Tucker. 1993. *The ARIEL Electronic Document Transmission System in the Australian Environment.* Geelong, Australia: Deakin University.

Kountz, John. 1992. Tomorrow's Libraries: More Than a Modular Telephone Jack, Less than a Complete Revolution—Perspectives of a Provocateur. *Library Hi Tech* issue 40, 9, no. 4: 39-50.

Krol, Ed. 1994. *The Whole Internet User's Guide and Catalog.* Sebastopol, CA: O'Reilly & Associates.

Krumenaker, Larry. 1993. How to Build a Library Without Walls. *Internet World* 4, no. 5/6 (June/August): 9-12.

Kurzweil, Raymond. 1993. The Virtual Library. *Library Journal* 118, no. 5 (15 March): 54-55.

LaCroix, Michael J. 1987. *MINITEX and ILLINET: Two Library Networks.* Illinois University, Urbana, Graduate School of Library and Information Science. Occasional Papers no. 178.

LASER (London and South Eastern Library Region). 1992. *VISCOUNT: Fact Sheet, 2: Background.* London: LASER.

_____. n.d. *VISCOUNT: Inter-library Communications Network.* London.

Latman, Alan. 1986. *Latman's the Copyright Law . . . Rev. ed. of The Copyright Law,* by Alan Latman, 5th ed. c. 1979. Washington, DC: Bureau of National Affairs.

Lavery, Michael. 1988. The Information Age in a Feudal Society: BASIC Literacy Versus Basic Literacy. In *Unequal Access to Information Resources: Problems and Needs of the World's Information Poor: Proceedings of the Congress For Librarians February 17, 1986 . . . .,* edited by Jovian P. Lang, 41-46. Ann Arbor, MI: Pierian Press.

Leach, Ronald G., and Judith E. Tribble. 1993. Electronic Document Delivery: New Options for Libraries. *The Journal of Academic Librarianship* 18, no. 6 (January): 359-364.

Leath, Janis. 1992. "Document Delivery: An Examination of Commercial Suppliers as an Alternative to Traditional Interlibrary Loan at the University of Wyoming Libraries." ERIC Document ED352036. Laramie, WY: University of Wyoming. Microfiche.

Leeves, Juliet. 1991. *A Guide to Interlibrary Loan Management Systems.* Sheffield, Eng.: Library and Information Cooperation Council.

Lenzini, Rebecca T., and Ward Shaw. 1991. Creating a New Definition of Library Cooperation: Past, Present, and Future Models. *Library Administration and Management* 5, no. 1 (Winter): 37-40.

_____. 1992. UnCover and UnCover 2: An Article Citation Database and Service Featuring Document Delivery. *Interlending and Document Supply* 20, no. 1 (January): 12-15.

Lewis, David W. 1988. Inventing the Electronic University. *College and Research Libraries* 49, no. 4 (July): 291-304.

Lieb, C. H. 1983. Document Supply in the United States. *STM Copyright Bulletin* 19. Amsterdam, NE: International Group of Scientific, Technical and Medical Publishers.   Reprinted in *Modern Copyright Fundamentals*, edited by Ben H. Weil and Barbara Friedman Polansky, 126-138. Medford, NJ: Learned Information.

Line, Maurice B. 1979. Universal Availability of Publications in Developing Countries.   In *Resource Sharing of Libraries in Developing Countries: Proceedings of the 1977 IFLA/UNESCO Pre-session Seminar for Librarians from Developing countries, Antwerp . . . .*, edited by H. D. L. Vervliet, 162-169. Munich; New York: K. G. Saur.

————. 1986. Access to Resources: The International Dimension. *Library Resources and Technical Services* (January/March): 4-12.

————. 1989a. Beyond Networks—National and International Resources. A paper presented at The IATUL/IFLA Seminar, Victoria, Australia, 25 August 1988. *IATUL Quarterly* 3, no. 2:107-112.

————. 1989b. Interlending and Document Supply in a Changing World. In *Interlending and Document Supply: Proceedings of the First International Conference held in London, November 1988*; edited by Graham P. Cornish and Alison Gallico, 1-4. Southampton, UK: IFLA Office for International Lending.

Lippert, Margret. 1992. The Managerial Perspective on Network Planning: A Comparison of CD-ROM Nets and Mounted Tapes. In *Online/CD-ROM '92 Conference Proceedings, Chicago . . . .*, 132-136. Wilton, CT: Eight Bit Books.

Long, Maurice. 1993. Interview with co-author Sheila Walters, 22 March, *British Medical Journal* offices, London, UK.

————. 1994. Faxed letter to co-author Sheila Walters, 28 February.

Love, Erika. 1990. Introduction to *Riding the Electronic Wave—Document Delivery: Proceedings of the Library of Congress Network Advisory Committee Meeting, November 29-December 1, 1989*, 5-7.   Network Planning Paper, 20.   Washington, DC: Network Development and MARC Standards Office, Library of Congress.

Lunau, Carrol D., and Donna J. Dinberg. 1991. Canada and the United States: Resource Sharing Connections. In *Library Computing in Canada: Bilingualism, Multiculturalism, and Transborder Connections*, edited by Nancy Melin Nelson and Eric Flower, 55-66.   Westport, CT: Meckler Corporation.

Lynch, Clifford A. 1993. The Transformation of Scholarly Communication and the Role of the Library in the Age of Networked Information. *Serials Librarian* 23, no. 3/4:5-20.

Mabomba, Roderick S. 1989. Access to Documents by and from Third World Countries. In *Interlending and Document Supply: Proceedings of the First International Conference held in London, November 1988*; edited by Graham P. Cornish and Alison Gallico, 56-61. Southampton, UK: IFLA Office for International Lending.

MacEwan, Bonnie, and Carol E. Chamberlain. 1993. An Interview with Sanford G. Thatcher, Director, the Penn State Press. *Library Acquisitions: Practice and Theory* 17, no. 2: 203-225.

Malinconico, S. Michael. 1992. Information's Brave New World. *Library Journal* 117, no. 8 (1 May): 36-40.

Martin, Marilyn J. 1992. Academic Libraries and Computing Centers: Opportunities for Leadership. *Library Administration and Management* (Spring): 77-81.

Martin, Scott M. 1992. Photocopying and the Doctrine of Fair Use: the Duplication of Error. *Journal of the Copyright Society of the U.S.A.* 39, no. 4 (Summer): 345-395.

Matta, Khalil F., and Naji E. Boutros. 1989. Barriers to Electronic Mail Systems in Developing Countries. *The Information Society* 6:59-68.

McClure, Charles R., Mary McKenna, William E. Moen, and Joe Ryan. 1993. Toward a Virtual Library: Internet and the National Research and Education Network. *The Bowker Annual of Library and Book Trade Information* 38:25-45.

McClure, Charles R., Joe Ryan, and William E. Moen. 1993. The Role of Public Libraries in the Use of Internet/NREN Information Services. *Library and Information Science Research* 15, no. 1 (Winter): 7-34.

McFarland, Robert T. n.d. [1992]. A Comparison of Science Related Document Delivery Services. Photocopy of a paper later published in *Science and Technology Libraries* 13, no. 1 (Fall 1992): 115-eoa.

McKenzie, Richard B. 1993. . . . Just the Fax, Ma'am. *The American Enterprise* 4, no. 2 (March/April), 18-22.

McSean, Tony. 1993. Interview by co-author Sheila Walters 22 March, British Medical Society Library, London, UK

Mehnert, Robert. 1992. National Library of Medicine. *Bowker Annual of Library and Book Trade Information*, 37:152-155.

Ménil, Céline. 1993. FOUDRE: A French Project for Electronic Document Delivery. *Serials* 6, no. 2 (July): 32-34.

Metz, Paul, and Paul M. Gherman. 1991. Serials Pricing and the Role of the Electronic Journal. *College and Research Libraries* 52, no. 4 (July): 315-327.

Miller, Connie, and Patricia Tegler. 1988. An Analysis of Interlibrary Loan and Commercial Document Supply Performance. *Library Quarterly* 58, no. 4 (October 1988): 352-366.

Mitchell, Maurice, and LaVerna M. Saunders. 1991. The Virtual Library: An Agenda for the 1990's. *Computers in Libraries* (April): 8-11.

Morris, Leslie. 1991. Borrowing and Loaning Across a Common Border: The Case for More Interdependence. In *Library Computing in Canada: Bilingualism, Multiculturalism, and Transborder Connections*, edited by Nancy Melin Nelson and Eric Flower, 81-89. Westport, CT: Meckler Corporation.

Morris, Leslie R. 1995. *Interlibrary Loan Policies Directory*. 5th ed. New York: Neal-Schuman Publishers. In press.

Mosko, Nertila. 1991. The Inter-lending Situation in Albania. In *East-West Information Transfer: Papers from the Meeting on Interlending and Document Supply between Eastern and Western Europe, held at Gosen, February 1991*, edited by Graham P. Cornish and Monika Segbert, 13. N.p.: IFLA Office for International Lending.

Naito, H. 1989. Japan's National Science Information System. *Outlook on Research Libraries* 11, no. 11 (November): 1-3.

Ndegwa, J. 1979. Cooperative Storage and Interlending in East Africa. In *Resource Sharing of Libraries in Developing Countries: Proceedings of the 1977 IFLA/UNESCO Pre-session Seminar for Librarians from Developing Countries, Antwerp . . . ., 1977*, edited by H. D. L. Vervliet, 170-178. Munich; New York: K. G. Saur.

Nelson, Milo. 1993. Taking Worldwide Document Delivery Seriously: The British Library is Banking (and Building) on the Future. *Document Delivery World* (April/May): 27-29.

Nevins, Kate, and Darryl Lang. 1993. Interlibrary Loan—A Cooperative Effort Among OCLC Users. *Wilson Library Bulletin* 67, no. 6 (February): 37-40, 110, 112.

Newsome, Karen Liston. 1990. Changing Strategies: Interlibrary Loan in the 1990's. *Illinois Libraries* 72, no. 8 (November): 636-639.

Nicklin, Julie L. 1991. Libraries Drop Thousands of Journals as Budgets Shrink and Prices Rise. *The Chronicle of Higher Education* (11 December): A29.

Nissley, Meta, and Nancy Melin Nelson, eds. 1990. *CD-ROM Licensing and Copyright Issues for Libraries.* Westport, CT: Meckler Corporation.

NREN and the National Information Infrastructure—Competing Visions: A Panel Discussion. 1993. *EDUCOM Review* 28, no. 5 (September/October): 50-53.

OCLC. 1994. "EPA Headquarters Library Logs 52 Millionth ILL Request." In OCLC-News [electronic bulletin board]. Message id 9407112037. AC00471@dgl.ceo.oclc.org 11 July, Subject: "52 Millionth ILL." Available from listserv @ OCLC.ORG.

OCLC Launches E-mail Current Awareness Service. 1993. *OCLC Reference News* 19 (November/December): 1.

OCLC Users in Asia and the Pacific Region. 1994. *OCLC Newsletter,* no. 208 (March/April): 16-27.

O'Flaherty, John J. 1994. EURILIA - European Initiative in Library and Information in Aerospace. *IATUL News* 3, no. 2:9.

Ogburn, Joyce L. 1990. Electronic Resources and Copyright Issues: Consequences For Libraries. *Library Acquisitions: Practice and Theory* 14, no. 3: 257-264.

Okerson, Ann. 1992. Faculty Respond to Serials Prices. *ARL: A Bimonthly Newsletter of Research Library Issues and Actions* 160 (2 January): 1-2.

Okerson, Ann, and Kendon Stubbs. 1992. ARL Annual Statistics 1990-1991: Remembrances of Things Past, Present . . . and Future? *Publishers Weekly* 239 (27 July): 22-23.

Olausson, Carin. 1990. Automation of Academic Libraries and Data Communications Development in Sweden. In *Bibliographic Access in Europe: First International Conference: The Proceedings of a Conference .*

. . *held at the University of Bath 14-17 September 1989*, edited by Lorcan Dempsey, 128-135. Aldershot, UK: Gower Publishing Company.

Olsen, Jan Kennedy. 1990. Cornell University, Mann Library. In *Campus Strategies for Libraries and Electronic Information*, edited by Caroline Arms, 218-242. [Bedford, MA]: Digital Press.

Orenstein, Ruth M., ed. 1994. *BiblioData Fulltext Sources Online for Periodicals, Newspapers, Newsletters, Newswires and TV/Radio Transcripts* 6, no. 1. (January). Needham Heights, MA: BiblioData.

OSI Pilot Demonstration Project. 1990. *IFLA Journal* 16, no. 4:490-491.

Palmour, Vernon, Edward C. Bryant, Nancy W. Caldwell, and Lucy M. Gray, comps. 1972. *A Study of the Characteristics, Costs, and Magnitude of Interlibrary Loans In Academic Libraries*, prepared for the Association of Research Libraries by Westat Research. Westport, CT: Greenwood Publishing Company.

Parisi, Lynn S., and Virginia L. Jones. 1988. *Directory of Online Databases and CD-ROM Resources for High Schools*. Santa Barbara, CA: ABC-Clio.

Paskaleva, Marlena. 1991. The Current Situation of Bulgarian Libraries. In *East-West Information Transfer: Papers from the Meeting on Interlending and Document Supply between Eastern and Western Europe, held at Gosen, February 1991*, edited by Graham P. Cornish and Monika Segbert, 19-21. N.p.: IFLA Office for International Lending.

Pastine, Maureen. 1992. 1992 and Beyond: In Conclusion. *The Reference Librarian* 35:119-129.

Patterson, L. Ray, and Stanley W. Lindberg. 1991. *The Nature of Copyright: A Law of Users' Rights*. Athens, GA: University of Georgia Press.

Pfander, Jeanne, mod. 1991. Shared Resources/Networking (Pt. 2). In *Proceedings/Memorias, Foro Binacional de Bibliotecas/Transborder Library Forum, Rio Rico Resort, Nogales, Arizona, February 1-2, 1991*, n.p. N.p.: Arizona State Library Association; Asociación Sonorense de Bibliotecarios A.C.; and the Asociación Mexican de Bibliotecarios, Sección Jalisco.

Phillips, Janet C. 1993a. Quarterly Cost of Shipping. Memorandum to IDS (Interlibrary Delivery Service of Pennsylvania) Board, 13 May.

_____. 1993b. Letter to co-author, Eleanor Mitchell, 18 May.

Pickup, J. A. 1978. Commercially Funded Services—An Appraisal from the Viewpoint of the Smaller User. *ASLIB Proceedings* 30 (January): 25-33.

Plaister, Jean M.   1991a.   The Interlibrary Loan Network of the United Kingdom. In *East-West Information Transfer: Papers from the Meeting on Interlending and Document Supply between Eastern and Western Europe, held at Gosen, February 1991*, edited by Graham P. Cornish and Monika Segbert, 157-170. N.p.: IFLA Office for International Lending.

_____.   1991b.   Project ION (OSI Pilot/Demonstration Project between Library Networks in Europe for Interlending Services): A Summary . . . Based on Reports Submitted to the European Commission for the Feasibility Study and Phase 1 of the Project. *Libri* 41, no. 4 (October/December): 289-305.

Plassard, Marie-France and Maurice B. Line.   1988.   *The Impact of New Technology on Document Availability and Access.* Rev. ed. Wetherby, Eng.: IFLA International Programme for UAP, British Library Document Supply Centre.

Pleyer, Viktoria.   1991.   Inter-lending with Eastern Europe: The Bavarian State Library's Experience. In *East-West Information Transfer: Papers from the Meeting on Interlending and Document Supply between Eastern and Western Europe, held at Gosen, February 1991*, edited by Graham P. Cornish and Monika Segbert, 67-70. N.p.: IFLA Office for International Lending.

Polterock, Joshua. 1993. Information Search Tools on the Internet. *Gather/Scatter 9*, no. 4 (July-October): 11-12.

Pritchard, Sarah M.   1992.   New Directions for ARL Statistics. *ARL: A Bimonthly Newsletter of Research Library Issues and Actions* 161:1-3.

Quint, Barbara. 1992. Where's Your Parachute? *Wilson Library Bulletin* 66, no. 8 (April): 85-86.

_____. 1993. Win or Lose. *Wilson Library Bulletin* 67, no. 6 (February): 72-74.

Raish, Martin. 1993. *Network Knowledge for the Neophyte Version 3.0.* Binghamton, NY: Binghamton University Libraries.

Raske, Richard E. 1991. Commercial Article Delivery Services. In *Delivery of Information and Materials Between Libraries: The State of the Art: Proceedings of the June, 1990 ASCLA Multi-Lincs Preconference*, edited by Keith Michael Fiels and Ronald P. Naylor, 61-69. Chicago: ASCLA, ALA.

Reed, Mary Hutchings. 1987. *The Copyright Primer for Libraries and Educators.* Chicago, IL: American Library Association and National Education Association.

Repp, Joan M. 1990. From OCLC to OhioLINK: The Ohio Experience. In *NIT '90: Third International Conference on New Information Technology for Library and Information Professionals, Educational Media Specialists and Technologists, . . . Guadalajara, Mexico,* edited by Ching-chih Chen, 263-270. West Newton, MA: MicroUse Information.

Research Libraries Group. 1994. "RLG Hangs Out 'New and Improved' Sign at ALA Booth 760." In RLIN-L, a forum devoted to RLIN issues [electronic bulletin board]. [Cited 1 February.] Available from listserv @ RUTVM1.BITNET.

Rich, R. Bruce. 1992. Living with the Copyright Law: Difficult Yes, Impossible, No. *Bookmark* 50, no. 2 (Winter): 105-108.

Richards, David. 1990. The Research Libraries Group. In *Campus Strategies for Libraries and Electronic Information,* edited by Caroline R. Arms, 57-75. [Bedford, MA]: Digital Press.

Richter, Vit. 1991. Access to Information in Times of Social Change. In *East-West Information Transfer: Papers from the Meeting on Interlending and Document Supply between Eastern and Western Europe, held at Gosen, February 1991,* edited by Graham P. Cornish and Monika Segbert, 25-27. N.p.: IFLA Office for International Lending.

Richwine, Peggy. 1993. "Article Clearinghouse Use." In ILL-L: Interlibrary Loan Discussion Group [electronic bulletin board]. [Cited 27 May:13:03 - 0500.] Available from listserv ILL-L@UVMVM.BITNET.

Rinzler, Carol E. 1983. What's Fair About "Fair Use?" *Publishers Weekly* 223, no. 14 (1 April): 26-28.

Risher, Carol A. 1993. "RE: Electronic Reserve Readings." In CNI-COPY-RIGHT and Intellectual Property Forum [electronic bulletin board]. [Cited 25 March.] Available from listserv CNI-COPYRIGHT@CNI.ORG. Message-ID: 63930325044236/0001750401DC4EM@mci.mail.com. [30 lines.]

Robison, David F. W. 1993. The Changing States of *Current Cites*: The Evolution of an Electronic Journal. *Computers in Libraries* 13, no. 6 (June): 21-26.

Roche, Marilyn M. 1993. *ARL/RLG Interlibrary Loan Cost Study: A Joint Effort by the Association of Research Libraries and the Research Libraries Group.* Washington, DC: Association of Research Libraries.

Rogers, Michael. 1992a. OCLC Reveals Future Networking Strategy. *Library Journal* 117, no. 17 (15 October): 24.

_____. 1992b. Publishers of Science Journals Win Copyright Fair Use Ruling. *Library Journal* 117, no. 14 (1 September): 110.

_____. 1993. Nets Around the Nation. *Library Journal* 118, no. 16 (1 October): 46.

Rosenberg, Jim. 1990. Fax Papers Offer Delivery on Demand. *Editor and Publisher* (4 August): 29.

Rothman, John. 1992. Libraries, Users, and Copyright: Proprietary Rights and Wrongs. *The Bookmark* 50 (Winter): 102-104.

Rouse, William B., and Sandra H. Rouse. 1980. *Management of Library Networks.* New York: John Wiley and Sons.

Rugge, Sue, and Alfred Glossbrenner. 1992. *The Information Broker's Handbook.* Blue Ridge Summit, PA: Windcrest/McGraw Hill.

Runner Ruda, Donna J. 1990. *A Report on the Study of Interlibrary Loans Costs at the Library of the Curtin University of Technology.* Western Library Studies, 15. Perth, WA, Aust.: The Library, Curtin University of Technology.

Rydings, A. 1979. Co-operative Acquisition for Libraries of Developing Countries: Panacea or Placebo? In *Resource Sharing of Libraries in Developing Countries: Proceedings of the 1977 IFLA/UNESCO Pre-session Seminar for Librarians from Developing Countries, Antwerp . . . .*, edited by H. D. L. Vervliet, 72-82. Munich; New York: K. G. Saur.

St. George, Art, and Ron Larsen. 1992, ©1991. "Internet Accessible Library Catalogs and Databases," [electronic document] edited by Carlos Robles and Heather Hughes [cited January 6 1992]. N.p.: University of Maryland, University of New Mexico. Available from LISTSERV@UNMVM.BITNET.

Saldinger, Jeffrey. 1984. Full Service Document Delivery: Our Likely Future. *Wilson Library Bulletin* 58, no. 9 (May): 639-642.

Saunders, Laverna M. 1993. Exploring Library Resources on the Internet. *Internet World* 4, no. 9 (November/December): 44-49.

Schiller, Nancy. 1992. Toward a Realization of the Virtual Library. *ARL: A Bimonthly Newsletter of Research Library Issues and Actions* 163 (19 July): 3-4.

Schuyler, Michael. 1992. *Dial In: An Annual Guide to Online Public Access Catalogs.* Westport, CT: Meckler, 1992. [Title changed in 1993 to *OPAC Directory: An Annual Guide to Online Public Access Catalogs and Databases.*]

Schwuchow, Werner. 1989. The Development of the International Market for Online Information Services. In *Information Knowledge Evolution: Proceedings of the Forty-fourth FID Congress Held in Helsinki, Finland, 28 August-1 September, 1988,* edited by Sinikka Koskiala and Ritva Launo, 357-367. FID (Federation Internationale de Documentation et d'Information) Publication, 675. Amsterdam; New York: North-Holland/Elsevier Science Publishers.

Shaughnessy, Thomas W. 1991. From Ownership to Access: A Dilemma for Library Managers. *Journal of Library Administration* 14, no. 1:1-7.

Shillinglaw, Noel. 1992. Document Supply and Distance Education Library Service. *Interlending and Document Supply* 20, no. 4 (October): 143-151.

Sloan, Bernard G. 1991. ILLINET Online: Resource Sharing in Illinois. *Advances in Library Resource Sharing* 2:100-107.

_____. 1992. Resource Sharing in Times of Retrenchment. *Library Administration and Management* (Winter): 26-28.

_____. 1993a. Fifteen Years of Resource Sharing." In ILL-L: Interlibrary Loan Discussion Group [electronic bulletin board]. [Cited 20 January.] Available from listserv ILL-L@UVMUM.BITNET.

_____. 1993b. Resource Sharing in an Open Network Environment: An Update on the Linked Systems for Resource Sharing Project. *Illinois Libraries* 75, no. 2 (1 March): 62-65.

Smale, Carol. 1994a. National Guidelines for Document Delivery. [Ottawa, Ont: National Library of Canada]. Draft [?] 6 June. Photocopy.

_____. 1994b. E-mail notes to co-author Sheila Walters, 27-28 July.

Smith, Carolyn. 1991. Factors Used to Analyze the Merits of Various Methods of Surface Delivery. In *Delivery of Information and Materials Between Libraries: The State of the Art: Proceedings of the June, 1990 ASCLA*

Multi-Lincs Preconference, edited by Keith Michael Fiels and Ronald P. Naylor, 19-33. Chicago: ASCLA, ALA.

Smith, Eldred.   1991.   Resolving the Acquisitions Dilemma: Into the Electronic Environment.   *College and Research Libraries* 52, no. 3 (May): 231-240.

Smith, Gerry.   N.d.   Document Delivery in Business Information. N.p.: Headland Press?   Photocopy.

Smith, Peter.   1993a.   Front End and Backbone of European Lending. *Library Association Record* 95, no. 2 (February): 94, 97.

_____.   1993b.   Interview by co-author Sheila Walters, 21 March, Newport, Wales.

Soini, Antti.   1990.   LINNEA—Library Information Network for Finnish Academic Libraries.   In *Bibliographic Access in Europe: First International Conference: The Proceedings of a Conference . . . held at the University of Bath 14-17 September 1989*, edited by Lorcan Dempsey, 113-120. Aldershot, UK: Gower Publishing Company.

Soo Lee.   1993.   Memo to Jonathan D. Lauer, ACLCP chair, 14 May.

Starratt, Jay, Carroll Varner, and Pat Cline.   1992.   The Impact of ILLINET Online's Development on Resource Sharing.   In *Academic Libraries: Achieving Excellence in Higher Education: Proceedings of the Sixth National Conference of the Association of College and Research Libraries, Salt Lake City, Utah, April 12-14*, 228-236.   Chicago, IL: ACRL.

Stearns, Susan M.   1991.   FAXON Demand: Creating a Market for Fax-on-Demand.   Sidebar 9 in Fax-on-Demand, by Christine E. Lachman.   *Library Hi Tech* issue 36, 9, no. 4:21-22.

Steele, Colin.   1989.   The Australian Libraries Summit and Document Delivery Past, Present, and Future. *Interlending and Document Supply* 17, no. 3 (1 July): 71-76.

Stern, Barrie T., and Henk C. J. Compier.   1990.   ADONIS—Document Delivery in the CD-ROM Age. *Interlending and Document Supply* 18, no. 3 (July): 79-87.

Stoller, Michael E.   1992.   Electronic Journals in the Humanities: A Survey and Critique.   *Library Trends* 40, no. 4 (Spring): 647-666.

Stone, Peter. 1991. The Development of Library and Information Services on Academic and Research Networks. In *Library Automation and Networking: New Tools for a New Identity = L'automatisation et les réseaux de bibliothèques* . . . . *European Conference of Medical Libraries (1990: Brussels, Belgium)*, 129-142. Munich; London; Paris: K. G. Saur.

Strangelove, Michael, Diane Kovacs, and The Directory Team, Kent State University Libraries. 1993. *Directory of Electronic Journals, Newsletters and Academic Discussion Lists*, edited by Ann Okerson. 3d ed. Washington, DC: Association of Research Libraries.

Strauch, Katina, and Heather Miller. 1993. Academic Libraries: Paring Down and Revving Up. *Library Journal* 118, no. 3 (15 February): 136-139.

Stromquist, John. 1993. Providing Access to the UMI Data Base to Consortium Libraries through Telefacsimile and CD-ROM. A presentation at the ALA Pre-conference "Keying in on Document Delivery: The Management Issues," 25 June, New Orleans, LA.

Stubbs, Kendon. 1993. Introduction to *ARL Statistics 1991-1992*, edited by Nicola Daval and Patricia Brennan, 7-9. Washington, DC: Association of Research Libraries.

———. 1994. *Introduction to ARL Statistics 1992-93*, edited by Nicola Daval and Patricia Brennan, 5-7. Washington, DC: Association of Research Libraries.

Stubbs, Kendon, and Nicola Daval. 1993. ARL Statistics Reflect Impact of Rising Prices. *ARL: A Bimonthly Newsletter of Research Library Issues and Actions* 167 (March): 6.

Summers, F. William. 1989. A Vision of Librarianship: The Impact of Emerging Technologies in the Library World. *Slj: School Library Journal* 35, no. 14 (October): 25-30.

Szabo, Sandor. 1991. The Present Situation of Information Access in Hungary. In *East-West Information Transfer: Papers from the Meeting on Interlending and Document Supply between Eastern and Western Europe, held at Gosen, February 1991*, edited by Graham P. Cornish and Monika Segbert, 107-109. N.p.: IFLA Office for International Lending.

Tackett, Raymond. 1992. Copyright Law Needs to Include `Fair Use' for Course Materials. *Chronicle of Higher Education* (12 February): 33-34.

Talab, R. S. 1986. *Commonsense Copyright: A Guide to the New Technologies*. Jefferson, NC: McFarland and Co.

Taubert, Sigfred, and Peter Weidhaas, eds. 1984. *Africa. The Book Trade of the World*, vol. 2. Munich; New York: K. G. Saur Verlag.

Taylor, C.   1989.   ACLIS National Council on Interlibrary Lending Assessment Study. Unpublished report.

Tee, Lim Huck. 1979. The Southeast Asian University Library Network (SAULNET): A Proposal and a Model for Resource Sharing in ASEAN Countries. In *Resource Sharing of Libraries in Developing Countries: Proceedings of the 1977 IFLA/UNESCO Pre-session Seminar for Librarians from Developing Countries, Antwerp* . . . , edited by H. D. L. Vervliet, 217-233. Munich; New York: K. G. Saur.

Tenopir, Carol. 1992a. Eight Tips for Cost Effective Searching. *Library Journal* 117 (1 October): 65-66.

_____.   1992b.   Online Searching with Internet.   *Library Journal* (December): 102-103.

_____. 1993. Electronic Access to Periodicals. *Library Journal* 118, no. 4 (1 March): 54-55.

Tepper, Laurie C. 1992. Copyright Law and Library Photocopying: An Historical Survey. *Law Library Journal* 84, no. 2 (Spring): 341-363.

Thorburn, Colleen.   1992.   Cataloging Remote Electronic Journals and Databases. *Serials Librarian* 23, no. 1/2: 11-23.

Tomer, Christinger. 1992. Instructional Computing. *Academic and Library Computing* 9, no. 10 (November/December): 8-12.

Turner, Fay. 1990. Interlibrary Loan Protocol: An International Standard for Electronic ILL Messaging. pt. 1. *Journal of Interlibrary Loan and Information Supply* 1, no. 1:110-117.

Tyckoson, David. 1991. Access vs. Ownership: Changing Roles For Librarians. *The Reference Librarian* 34:37-45.

Ulmschneider, John E.   1992.   Library Based Image Transmission Over Internet. *Quarterly Bulletin of the International Association of Agricultural Information Specialists* 37, no. 1/2:77-83.

Ulrich's International Periodicals Directory, 1992/1993. 1992. 31st ed. New Providence, NJ: R. R. Bowker, 3:5031-5042, 5045-5107.

U.S. Court of Appeals. (2d Cir. 1993). 1993a. "American Geophysical Union; Elsevier Science Publishing Co., Inc.; Pergamon Press, LTD; Springer-Verlag, GMBH & Co., K.G.; John Wiley & Sons, Inc.; and Wiley Heyden, LTD., on Behalf of Themselves and Others Similarly Situated v. Texaco Inc. Amicus Curiae Brief of American Library Association." In CNI-Copyright and Intellectual Forum [electronic bulletin board]. March. Available from listserv CNI-COPYRIGHT@CNI.ORG.

_____. 1993b. ". . . Amicus Curiae Brief of Association of Research Libraries." In CNI-Copyright and Intellectual Forum [electronic bulletin board]. May. Available from listserv CNI-COPYRIGHT@CNI.ORG.

_____. 1993c. "Reply Brief of Appellant Texaco." In CNI-Copyright and Intellectual Forum [electronic bulletin board]. May. Available from listserv CNI-COPYRIGHT@CNI.ORG.

_____. House. 1976. *Report, September 3*. 94th Cong., 2d sess. H. Rept. 1476:74-79.

_____. Library of Congress. Copyright Office. 1977. *General Guide to the Copyright Act of 1976*. Washington, DC: Government Printing Office.

_____. 1978. *Reproduction of Copyrighted Works by Educators and Librarians*. Circular R21. Washington, DC: Government Printing Office.

_____. 1983. *Library Reproduction of Copyrighted Works (17 U.S.C.108): Report of the Register of Copyrights*. Washington, DC: Government Printing Office.

_____ Postal Service. 1993. *International Mail Manual*. P1.10/5:993. Washington, DC: U.S. Government Printing Office.

_____. Senate. 1975. *Report on Section 108, Reproduction by Libraries and Archives, November 20, 1975*. 94th Cong., 1st sess. S. Rept. 473.

University of Nebraska—Lincoln Libraries. 1988. Document Delivery Committee. Document Delivery Report. Lincoln, NE. Photocopy.

Using Contents Alert. 1993. *OCLC Reference News* 19 (November/December): 3-5.

van Marle, Gerard A. J. S. 1993. The PICA RAPDOC Project. *IATUL Proceedings*, n.s. 2:54-70.

Vokác, Libena. 1991. The Issue of Copyright in Interlending. In *East-West Information Transfer: Papers from the Meeting on Interlending and*

Document Supply between Eastern and Western Europe, held at Gosen, February 1991, edited by Graham P. Cornish and Monika Segbert, 198-207. N.p.: IFLA Office for International Lending.

Waldhart, Thomas J. 1984. The Growth of Interlibrary Loan Among ARL University Libraries. *The Journal of Academic Librarianship* 10, no. 4 (September): 204-208.

_____. 1985. Performance Evaluation of Interlibrary Loan in the United States: A Review of Research. *Library and Information Services Review* 7, no. 4 (October/December): 313-331.

Walters, Edward M. 1987. The Issues and Needs of a Local Library Consortium. *Journal of Library Administration* 8, no. 3/4:15-29.

Walters, Sheila. 1989. Library Express: The Establishment of an Across-Campus Document Delivery Service. In "Crossing Borders: New Territories in the '90s." Contributed papers presented at the Arizona State Library Association Conference (Tucson, Arizona, November 1989), edited by Carol Hammond, 53. [ERIC document, ED322912.]

_____. 1993a. Commercial Document Delivery Services. A presentation at the Colorado Interlibrary Loan Workshop 29 April 29, Grand Junction, CO. Typescript.

_____. 1993b. Commercial Document Suppliers: The Arizona State University Experience. A presentation at the United Kingdom Serials Group's Annual Conference 23 March, Southampton, UK Typescript.

Walters, Sheila, and Eleanor Mitchell. 1992. "Fewer Subscriptions = Increased Library Services; How ASU and ASU West Met the Challenge." Workshops presented at the North American Serials Interest Group Annual Conference, Chicago, 19 and 21 June.

Wessling, Julie. 1992. Document Delivery: A Primary Service for the Nineties. *Advances in Librarianship* 16:1-31.

_____. 1993. Electronic ILL: The User Interface. *Document Delivery World* 9, no. 3 (April/May):24-26.

_____. 1994. E-mail notes to co-author Sheila Walters, June-July, related to activities at Colorado State University Libraries to integrate "ILL/DDS and Online Systems."

West, Sharon M. 1992. Information Delivery Strategies and the Rural Student. *College and Research Libraries* 53, no. 6 (November): 551-561.

"What is ERNET?"   n.d.   [electronic document]. Available from vikram.doe.ernet.in. 94 lines.

White, Andrew. 1993. See Dyer, Peter Swinnerton, 1992.

White, Andrew, and Noreen White. 1993. Conversations with co-author Sheila Walters, 15-18 March, Southampton, UK

White, Brenda. 1986. *Interlending in the United Kingdom 1985: A Survey of Interlibrary Document Transactions.* Library and Information Research Report, 44. Cambridge, Eng.: Cambridge University Press for The British Library Board.

White, Herbert S. 1992. Collection Development is Just One of the Service Options. *Journal of Academic Librarianship* 18, no. 1 (March): 11-12.

Wiemars, Gene, and John Hankins. 1994. CICNet Project Builds Electronic Journal Collection. *ARL* 173 (March): 8-9.

Wiggins, Gary Dorman. [1985] 1990. "Factors Which Influence the Choice of Document Delivery Mechanisms for Serials by Selected Scientific and Technical Special Librarians". Ph.D. diss. Indiana University, 1985. Ann Arbor, MI: University Microfilms.

Willemsen, Arie W. 1989. Making the Contents of Research Libraries Available to Remote Users. In *Interlending and Document Supply: Proceedings of the First International Conference held in London, November 1988*; edited by Graham P. Cornish and Alison Gallico, 52-55. Southampton, UK: IFLA Office for International Lending.

Williams, Esther W. 1990. Document Delivery in the Pacific. In *Papers Presented at the International Federation of Library Associations (IFLA) General Conference (56th, Stockholm, Sweden, August 18-24, 1990), Division of Special Libraries, Section of Social Science Libraries*, booklet 2, 124-136. ERIC Document ED329283. Microfiche.

_____. 1991. ILDS In the South Pacific: Challenges in the 90s. In *Interlending and Document Supply: Proceedings of the Second International Conference held in London, November 1990*; edited by Alison Gallico, 111-125. Southampton, UK: IFLA Office for International Lending.

Williams, Martha. 1992. Document Delivery Vendors: Benefits and Choices. *Serials Librarian* 23, no. 3/4: 217-224.

Williamson, Mary. 1987. Commercial Document Delivery Services and Interlibrary Loan: A Comparison. Madison, WI: Wisconsin Interlibrary Services. Photocopy.

Williamson, Vicki. 1994. "ACAE Update." E-mail note to co-author Sheila Walters, 3 August.

Wilson, Mark. 1988. How to Set up a Telefacsimile Network—The Pennsylvania Libraries' Experience. *Online* 12, no. 3 (May): 15-16, 19-20.

Wilson, Martin. 1992. Copyright Clearance Center Pilots Electronic Access. *Information Today* (February): 18.

Winick, Les. 1991. New USPS Address Rules. *Linn's Stamp News* 64, no. 3255 (25 March): 26.

Wright, Christopher. 1989. Future Last Resort: Interlibrary Loans and Locations at the Library of Congress. In *Research Access Through New Technology*, edited by Mary E. Jackson, 113-135. AMS Studies in Library and Information Science, 1. New York: AMS Press.

_____. 1993. Announcement made at ALA Annual Conference at the RASD Interlibrary Loan Discussion Group's meeting 26 June, New Orleans, LA.

Wright, Stephen. 1991. Library Automation in Papua New Guinea. *Libri 41*, no. 1:37-50.

York, Vicky, and Audrey Jean Haight. 1992. Government Information: CD-ROM Roundup. *CD-ROM Librarian 7, 10* (November): 14-19.

Zell, Hans M. 1984. Introduction to *Africa. The Book Trade of the World*, 4, edited by Sigfred Taubert and Peter Weidhaas, 15-56. Munich; New York: K. G. Saur Verlag.

Zijlstra, Jaco. 1993. "TULIP." In ILL-L: Interlibrary Loan Discussion Group [electronic bulletin board]. [Cited 27 May.] Available from listserv ILL-L@UVMVM.BITNET.

Appendix

RFP Requirements

Interlibrary Loan And Document Delivery (ILL/DDS) Integrated Online System Features: Suggested Requirements for Inclusion in a Request for Proposals (RFP) from Vendors

☐ The system must be able to provide an integrated interlibrary loan/document delivery (ILL/DDS) module within a reasonable time. The vendor should indicate how soon it can meet any requirement not currently available.

General System Features Relevant to ILL/DDS Functions:

☐ The system must provide the capability to connect seamlessly between other libraries sharing the OPAC.
- The OPAC should provide a union list of holdings of all libraries sharing the online system.
- The union list should encompass items included in local or regionally-created bibliographic databases and gateway services. The vendor must specify how this will be accomplished.
- The system may provide a union list of holdings of all other libraries using the vendor's system.

☐ The vendor must provide a gateway to other library catalogs, bibliographic and full-text databases via the Internet.
- A search interface in compliance with the latest revision of NISO Z39.50, Information Retrieval Service Definition, is required so that the same search logic can be utilized on all accessible databases.
- The vendor should be an active member of a Z39.50 implementor's group or interoperability test group.
- Authorized users must have access via local network and/or gateways to bibliographic utilities (e.g., OCLC and RLIN) and commercial databases (e.g., UnCover, UMI's ProQuest databases, or *Monthly Catalog of Government Publications*)

☐ Vendor must offer facsimile, Ariel, and full-text electronic delivery capabilities.

☐ The system should be able to provide as many documents as possible in an online, real-time mode at workstations in the library, or via high-speed dial-in or network links. Such electronic document delivery services should interface through the OPAC.

- The system must support remote log-in, file transfer and e-mail transfer capabilities in conformance with the Internet suite of protocols from any TCP/IP port within the local network.
- The vendor should assure compliance with emerging EDI (Electronic Document Interchange) standards (ANSI X.12) for ordering, invoicing, shipping, claiming, etc. used by publishers, book jobbers, and library acquisitions departments for paperless acquisitions.
- The system must provide the capability to mark and capture citations from any database on the system and check against library holdings and/or other union lists or commercial databases.
- The screen capture programming should use "grab-it" functions modified to identify fields for matching with standard machine-readable cataloging fields.

☐ The system should provide a "view/print only" status record of all library transactions for individual patrons, including books checked out, status of ILL/docdel requests, order requests, etc. Editing/updating functions should be restricted to authorized library staff only.

- The system should automatically notify patrons whenever the status of a library transaction requires action (e.g., material available for pick up; item being recalled; overdue notice).

Required Features Specific to Interlibrary Loan/Document Delivery Functions

☐ ILL/document delivery requests are prepared, transmitted received and processed interactively with status of transaction available online at all times.

☐ Status is automatically updated throughout processing stages, with only minor rekeying required at each stage; automatic downloading/uploading between OCLC, RLIN, or other specified ILL subsystem is required.

☐ From any PAC terminal (dumb or PC), Internet session or dial-in session, authorized users must be able to generate an e-mail request for any captured citation for the purpose of placing a hold/recall, search, interlibrary loan document delivery, or order request without rekeying data.

- Requests must be transmitted and received in a standard format (e.g., ALA, IFLA, OCLC, or locally designed) using terminology defined in ANSI/NISO Z39.63-1989 (or latest revision) for Interlibrary Loan Data Elements.
- Full-screen editing of records should be allowed when generating requests. For example, if a record indicates a four volume set, a patron should be able to request only vol. 2, or vol. 2, chap. 3, or pp. 7-15.

- Standard data fields should be filled in automatically, and restricted fields verified. Frequently occurring data may be entered in code form for subsequent translation.

☐ The system should allow the user to initiate in PAC an online interactive document ordering blank template from which a document can be requested even if the item is not found in any OPAC database.

- The system should alert the end-user if a key bibliographic data field is omitted, providing an option to correct the request or advise of consequences (i.e. incomplete citations have a high unfilled rate or require more time to fill). The system should reject requests that are too incomplete to be processed, directing the end-user for assistance in verifying citation.

☐ A copyright warning should appear on all ILL/document delivery requests; programming should require end-user action to indicate compliance.

☐ The automatic order functions must include links to the patron database and incorporate appropriate screening and security parameters.

☐ Patron-generated requests should be transparently referred to the appropriate processing unit based on holdings and/or shelf/circulation status, and any protocols established by the library; if not transparent, the patron should be offered a menu of delivery options.

☐ The messaging system must be compatible with a designated ILL subsystem and capable of automatic downloading/uploading with that system. The interface should allow any ILL request entered by a patron in PAC (either manually or by electronic screen capture) to be electronically uploaded to the designated ILL Subsystem and appear in a review file of the home ILL department, with a complete bibliographic citation, complete patron information, and identified locations in the lender's string.

- Incoming requests may be forwarded to other libraries without re-keying data.

- A gateway to telecommunication messaging services available on the Internet should be available.

- The system must comply with X.400 Message Handling Services standard.

☐ Local, regional, and national ILL protocols are supported, with automatic, load-levelled routing to specified suppliers, with ability to override if desired, so that outgoing requests can be automatically routed online to any library within the local system or any member library of a designated online ILL subsystem, or to other libraries with e-mail connections.

☐ If a request cannot be sent online, the system should generate a mail request form in ALA format with supplier's address automatically supplied in standard USPS format suitable for mailing in a window envelope. The local library's return address should be automatically recorded on outgoing request.

- Other formats may be specified when necessary on outgoing requests.

☐ Requests may be entered intermittently and held for batch transmission during off-peak hours.

☐ Messages may be transmitted or received in background mode, allowing other activities to continue.

☐ The system should allow automatic forwarding of electronically received materials to patrons, when possible.

☐ Incoming ILL and document delivery requests are linked to OPAC and Circulation database. Local call numbers are assigned automatically to incoming requests with all possible locations and shelf status noted.

☐ Requests which can not be filled (e.g., item is noncirculating; library does not own requested title/volume) will be automatically forwarded to the next lender in the string or returned to the requesting library if the library is the last or only supplier listed in the lending string.

- This requires an interface with a designated ILL subsystem that will allow the local ILL lending unit to download requests received through the ILL network and pass them through the local PAC database with a matching routing (based on OCLC or other standard numbers) so that call numbers are automatically linked with incoming ILL requests.
- Incoming ILL/DDS requests may be linked with specified commercial databases licensed for local use, so that incoming requests (both from end-users within the local system AND from other libraries outside the system) which meet imbedded parameters (price, library-owned, etc.) could be profiled for automatic matching and forwarding to those systems, thus preserving the library collections for primary users and reducing the volume of requests handled by ILL/DDS.

An ILL/DDS database must be maintained with subfiles of ILL/DDS patrons, suppliers, and transactions.

☐ Subfiles can be updated online.

☐ Directory files must comply with X.500 Directory Services standard.

☐ Data can be printed in a variety of ways and on a variety of forms, such as pre-printed ALA ILL request forms, mailing labels, overdue notices, stack retrieval forms, pick-up notifications, cancellation letters, overdue notices, recalls, billing notifications.

☐ The online record of ILL/DDS patrons should be linked to the circulation database of registered library users, containing complete mail, telephone, fax, and e-mail data; library status (blocked, temporary clearance, etc.); patron ID number, patron type, department or library affiliation.

- Automatic block of ILL/DDS request if patron status does not qualify requestor for ILL/DDS service (i.e. courtesy card holder, outstanding fines, etc.), with the ability to override block by authorized operator should be available.

☐ Online directory of suppliers (lending libraries, commercial suppliers and intra-system suppliers) and borrowers (other libraries and intra-system branches).

- Linked to library codes of designated online bibliographic utilities, national library codes, Ariel IP addresses, and/or locally assigned supplier codes.
- Contains complete mail, phone, fax, Ariel and e-mail information, with key personnel contacts.
- Must be able to provide secondary addresses (i.e. billing address, instead of ILL Department).
- Contains data relevant to charge or reciprocal status, fees, order restrictions, circulation period, and renewal period.
- Subject specialties of suppliers may be a searchable field.
- Automatic block if order request does not meet supplier's parameters. Ex.: If supplier's lending fee is $25.00, a request with a max cost of $15.00 would be blocked.

☐ The system must signal a copyright violation alert when:
  —the CONTU "rule of five" guideline is exceeded;
  —requests from a single requestor exceed CONTU guidelines; or,
  —when a decision "not to own" has been made for titles regularly acquired in excess of CONTU guidelines.

- An override feature should allow for ordering documents that exceed copyright guidelines, but that mark the record for copyright reporting and payment of any required royalties.

☐ The system should automatically produce pre-pasted labels to use on book wrappers.

☐ If possible, the system should automatically produce a pre-pasted (or with removable tape) exit "wrapper" which contains the appropriate due dates and supplier information needed to return loans.

The system must track all ILL/DDS transactions, including nonsystem-produced requests, with automatic updating of records from requests coming in and going out.

Subfiles must be producible for borrowing, lending, special handling, intra-campus document delivery, and/or intercampus document delivery activities.

☐ The system shall support a file of borrowing transaction records which can be searched and/or/not sorted by: patron name; patron type;

department or campus affiliation; type of request (loan, copy, etc.); supplier (including supplier type); payment type (free, fee or subsidized); bibliographic citation fields, various order numbers (OCLC IL number, RLIN IL number, system assigned request numbers); subject classification; request date; filled or cancelled date; in-state/out-of-state; method of transmitting request (OCLC; fax, mail, e-mail, etc.); method of receipt (in library, mail, courier, fax, Ariel or e-mail; full-text online, etc.).

- The system must allow for multiple re-sends until fulfillment or cancellation.
- The system should allow for reactivating cancelled requests if re-submitted.
- The system should allow for multiple suppliers for the same request if request was not filled completely.

☐ The system shall support files of ILL/DDS lending records to accommodate requests for the loan of items from the local system to other institutions; between branch campuses; and on-campus document delivery.

- Lending files shall include the following fields which can be searched and/or/not sorted by: transaction numbers (OCLC IL number, RLIN IL number, or system assigned request number); requesting library/-type/state; date of request; date request was received; date request was filled/cancelled; item code; format code; billing code; full bibliographic citation; and nonavailability code.
- The system must allow an ILL/DDS ID to be used to create circulation records for materials checked out to other libraries or branch campuses and it must have capability to link to multiple billing addresses.
- ILL lending requests which do not indicate copyright compliance will be automatically returned with reason indicated.

☐ Quantitative search results should be available on screen or printed separately without viewing/printing actual records. (Ex.: The system should be able to search the database and display how many patrons from a specific department used ILL without having to print or view the complete list, unless desired.)

☐ Status of ILL/docdel requests is continually tracked, with follow-up reminders generated at specified intervals for a variety of functions (i.e., no response to request; not received after supplier updated to shipped; overdue; returned material not updated to received; no action on active request.

☐ Automatic update functions may be overridden (i.e., if item was received on Friday afternoon, but updating does not occur until Monday morning, actual date of receipt can be input rather than current date being automatically input).

☐ Constant data can be changed for batch processing.

☐ It should be optional whether to include or exclude nonwork days; i.e., overall borrowing turnaround includes every day, although actual processing time might exclude nonwork days; lending and document delivery turnaround is based on work days only; automatically assigned due dates or overdue notices should exclude nonworkdays.

☐ Turnaround time should be accessible for each phase of the process, as well as total turnaround from date of request through date of fulfillment.

☐ Older requests must be able to be moved to an archival file based on library defined parameters.

☐ Requests or portions thereof may be printed or queued for printing at any time. Printing may be suspended, resumed, or cancelled.

☐ ILL/DDS data management functions are required of the system.

- Statistics are kept automatically. Examples include number of requests transmitted, supplier, cost, average turnaround time (overall and per supplier).
- Invoices should be produced whenever item supplied has a fee involved.
- Data from all searchable fields in the patron/transaction databases/files must be capable of manipulation for creating specialized reports, surveys, or time/cost studies, as well as producing standard reports regularly.
- Vendor must supply, as part of the proposal, samples of all ILL/DDS reports that the system can produce.
- Partial or complete transaction history for individual patrons should be generated as needed.

Reports to be generated monthly with annual cumulation or for designated time frame, include:

☐ Collection development data sorted by any combination of the following fields: LC classification, main entry, patron type or library affiliation, document type, supplier, access costs, etc.
- The system will produce a purchase alert report for any title requested more than a designated number of times within a designated time frame.
- Accounts receivable/payable, with spreadsheet analysis.

☐ Statistical reports for each ILL/DDS unit or subunit, with total cumulation for department.

☐ Copyright compliance log, containing the following searchable data elements: journal title; ISSN; year of publication; pages copied; requestor's name, department, and campus affiliation; supplier; transaction number(s); title usage statistics; fair use or not; CCG/CCL compliance; supplier's cost, royalty cost, and subscription cost; and subject classification (LC).

- The copyright log should indicate when copy requests are for items owned, copyright-cleared, or when copyright royalties are pre-paid.
- The system should produce a copyright log that can be edited, or printed in whole or part, for submission to the Copyright Clearance Center, publishers, or Collection Development.
- Reports sorted by requesting branch and/or owning branch should be produced, but all records should be linked to holdings of entire library system.
- Copyright logs should be retained for five years.

Resources

Alberta University Library. 1992. University of Alberta—A Checklist for Automated Systems. In *System Migration in ARL Libraries*, prepared by Ling-yuh W. Pattie and Michael Lach, for the ARL Systems and Procedures Exchange Center, 25-26. SPEC Kit, no. 185. Washington, DC: Association of Research Libraries, Office of Management Services.

Arizona State University. University Libraries. Public Access Catalog Task Group. 1994. Interlibrary Loan and Document Delivery (ILL/DDS) Online System Features. [Working paper prepared by S. Walters and E. Mitchell.] Draft revised 6 July. [Temp, AZ]. Typescript.

Data Research Associates. N.d. *Sample Request For Proposals for an Automated Library System for the (Library Name)*. Version 1.0. St. Louis, MO. Photocopy.

Emory University Libraries. 1992. Emory University Libraries Request for Proposal (RFP), Section III.L. Interlibrary Loan, draft 2 (1 July): 152-53. [Atlanta, GA: Robert W. Woodruff Library, Emory University]. Telefacsimile.

Jackson, Mary E. 1994. "The North American Interlibrary Loan and Document Delivery (NAILDD Project: A Status Report March, 1994.)" In *Summary of ARL's NAILDD Project* [electronic document]. [Cited 3 May.] Available from /news/archives/bit.listserv.pacs-1/blp.9404. 109 lines.

Payne, Nancy E. 1994. Various correspondence and e-mail with co-author Sheila Walters during June, related to "Using UTCAT to Place ILS Requests" [Austin, TX; Inter-Library Service, University of Texas.]

Wessling, Julie, 1994. E-mail notes to co-author Sheila Walters, June-July, related to activities at Colorado State University Libraries to integrate "ILL/DDS and Online Systems."

Whittier-Ellingson, Margaret. 1994. E-mail notes to co-author Sheila Walters, June-July, related to "ILL/DDS and Online Systems" at Emory University Libraries.

Index